T0418819

MULTISCALE MODELING APPROACHES FOR COMPOSITES

MULTISCALE MODELING APPROACHES FOR COMPOSITES

George Chatzigeorgiou
CNRS, LEM3–UMR 7239, Arts et Métiers Institute of Technology, Université de Lorraine, Metz, France

Fodil Meraghni
Arts et Métiers Institute of Technology, LEM3–UMR 7239, Metz, France

Nicolas Charalambakis
Aristotle University of Thessaloniki, Department of Civil Engineering, Thessaloniki, Greece

Elsevier
Radarweg 29, PO Box 211, 1000 AE Amsterdam, Netherlands
The Boulevard, Langford Lane, Kidlington, Oxford OX5 1GB, United Kingdom
50 Hampshire Street, 5th Floor, Cambridge, MA 02139, United States

Library of Congress Cataloging-in-Publication Data
A catalog record for this book is available from the Library of Congress

British Library Cataloguing-in-Publication Data
A catalogue record for this book is available from the British Library

ISBN: 978-0-12-823143-2

For information on all Elsevier publications
visit our website at https://www.elsevier.com/books-and-journals

Publisher: Matthew Deans
Acquisitions Editor: Dennis McGonagle
Editorial Project Manager: Rachel Pomery
Production Project Manager: Manju Thirumalaivasan
Designer: Mark Rogers

Typeset by VTeX

Contents

I

Tensors and continuum mechanics concepts

II

Micromechanics for composite media

7. Classical laminate theory

III

Special topics in homogenization

8. Composite sphere/cylinder assemblage

9. Green's tensor

10. Hashin–Shtrikman bounds

11. Mathematical homogenization theory

12. Nonlinear composites

A. Fiber orientation in composites

For additional information on the topics covered in the book, visit the companion site: https://www.elsevier.com/books-and-journals/book-companion/9780128231432

About the authors

G. Chatzigeorgiou is Research Scientist of CNRS (CR–HDR). He is hosted in the Arts et Métiers Institute of Technology in Metz and is a member of the LEM3–UMR CNRS 7239 laboratory. His research is devoted to constitutive modeling of multifunctional materials and homogenization theories of composites. He coauthored more than 50 papers in peer–reviewed journals, one book, and more than 30 papers in proceedings of international conferences.

F. Meraghni is Distinguished Professor at Arts et Métiers Institute of Technology (Campus of Metz). He is the head of Composites and Smart Materials Research Group at LEM3–UMR CNRS 7239 working on multiscale modeling using micromechanical approaches for homogenization of polymer composites and shape memory alloys. As the principal investigator (PI), Prof. Meraghni and his group are involved in several research programs in collaboration with French industry or funded by the French Agency of Research (ANR) and other funding institutions. He is a coauthor of more than 100 peer–reviewed papers published in journals and one book, and he advised/coadvised 30 PhD students. Currently, he is the Director of the Doctoral School of Arts et Métiers.

N. Charalambakis is Professor Emeritus at the Civil Engineering Department of Aristotle University of Thessaloniki (AUTH) and a member of the Center for Research and Development on Advanced Materials AUTH and Texas A&M (CERDAM). He authored or coauthored more than 70 papers in peer–reviewed journals, one book, and more than 40 papers in proceedings of international conferences.

Foreword

It is a particular pleasure to write the foreword of this extraordinary book written by three esteemed colleagues. Multiscale approaches allow building a connection between the continuum properties of a material and its microstructure. This physical way for describing overall material behavior is particularly well adapted for modern materials like composites. The present book fills a wide gap between fundamental treatises about micromechanics, like Toshio Mura's book or the Sia Nemat–Nasser and Muneo Hori one and the abundant literature published in scientific journals. A first attempt to provide a textbook written pedagogically and providing an integrated approach to the various topics of homogenization was due to Jianmin Qu and Mohammed Cherkaoui fifteen years ago. It is worth noting that Mohammed Cherkaoui, Fodil Meraghni, and George Chatzigeorgiou are linked to the micromechanical school established at the University of Metz in the beginning of the 1980s by the late Professor Marcel Berveiller. Under his guidance, this school has developed a vision focused on engineering applications of multiscale modeling. Therefore it is not surprising that textbooks oriented toward education of engineers about homogenization techniques find some common roots at Metz.

A notable feature of the present book is its full dedication to composite materials. This clearly distinguishes it from the one authored by Qu and Cherkaoui, mainly oriented toward polycrystalline materials. The text provides a comprehensive development of both mean–field and full–field scale–transition approaches; main characteristics and advantages of each of these approaches are presented in a very pedagogic way. Restrictions and field of application for each method are made clear. Having these two homogenization schemes presented in the same book is a very positive point.

Every chapter is written using both rigorous and detailed mathematical treatment and an engineering approach, which will help many people to become familiar with the powerful tools offered by micromechanics. Linear elasticity and infinitesimal strain framework are considered. Voigt notation is nicely introduced and used offering a more engineering description of rank four tensors.

The reader will be happy to find numerous examples in each chapter. Detailed solutions are given with every example, including corresponding python scripts for numerical applications. That makes the book very useful for professionals willing to improve their knowledge about scale transition and also for graduate students taking a micromechanical class. Major takeaways in each chapter are also clearly defined and highlighted using color boxes. The objective of the book is to clearly provide enough element to the reader allowing him for making sound decision about the choice of the self–consistent Mori–Tanaka method or periodic homogenization for a given practical application.

The final chapters are devoted to advanced topics in multiscale modeling, including a detailed exposition of the Hashin–Shtrikman bounds and a very interesting and dense chapter about mathematical homogenization theory for composites. The book ends with a short chapter considering homogenization in nonlinear materials. These advanced topics will clearly be of great interest for PhD students and researchers.

To conclude, I believe the book will be of significant use to professional researchers and graduate students in engineering science interested in modeling composites and heterogeneous materials.

Etienne Patoor
Georgia Tech Lorraine, Metz, France
June 2021

Preface

Multiscale methods for composite materials have received growing attention between the mechanics community in the recent years. The reason for this interest lies on the significant increase of composite materials usage in many engineering applications. The modern needs of civil, mechanical, aerospace, and bioengineering industries have led to the development of novel material systems with complex microstructural characteristics. To identify the mechanical behavior of these systems requires advanced theoretical and numerical tools provided by multiscale methods.

The scope of the present book is providing, in a pedagogic manner, the main concepts of the commonly utilized multiscale methods in the study of composite materials. The manuscript aims at serving as a guide for identifying and utilizing the appropriate homogenization approach according to the needs of each specific application. The book is oriented to a large spectrum of readers, including engineers from the industrial sector and Master and Ph.D. students. Various homogenization methodologies are discussed in a simple and didactic way, whereas the latest chapters of the book contain more advanced topics, suitable for students or researchers with special interests. The adopted methodology in writing the chapters is a combination of a solid theoretical background, practical tools for each approach, and an assessment of the methods. Examples are included in every chapter to illustrate the use of the discussed approaches. In addition, Python scripts similar to Matlab codes are provided for several numerical examples.

The book is organized in three complementary parts. Its content is divided into 12 chapters, supplemented by an Appendix.

The first part of the book recalls important concepts and definitions from tensor calculus and continuum mechanics, before passing to the main topic, the multiscale approaches. This part contains two chapters. In Chapter 1, the tensors and their properties are introduced. Moreover, the Voigt notation that allows us to treat second– and fourth–order tensors with the help of matrices is presented. Chapter 2 deals with concepts from continuum mechanics theory such as the strains and stresses, as well as the relation between them in the case of elastic media. A reduction of the elasticity problem from 3 to 2 dimensions is required for the study of laminate composites, and thus it is also discussed there.

The second part is considered the "heart" of the book, since it introduces various micromechanics concepts and presents the most frequently used mean–field and full–field homogenization theories. Chapter 3 discusses the general principles of all multiscale approaches, including important average theorems, the Hill–Mandel principle and the notion of the concentration tensors. The oldest micromechanics methods (the Voigt and Reuss bound estimates) and their application in several engineering structures are the topic of Chapter 4. Chapter 5 presents certain widely used Eshelby–based, mean–field homogenization techniques, such as the Eshelby dilute, the Mori–Tanaka method, and the self–consistent method. Chapter 6 introduces the key elements of the most common full–field multiscale approach, the periodic homogenization theory. Finally, the laminate composite plates and their corresponding modeling theory are the subject of Chapter 7. The latter includes a comprehensive example with step–by–step detailed solution.

The third part of the book serves for introducing in depth several concepts of multiscale modeling that were not discussed in the previous part. Most of these concepts require advanced knowledge of mathematics and mechanics, which the interested reader needs to have for the better understanding of the following chapters. In addition, the part presents certain additional specialized micromechanics techniques.

The topic of Chapter 8 is the so–called composite sphere and composite cylinder assemblage methods, which were developed for the study of particulate and long fiber composites, respectively. The Green's tensor, a powerful tool used in the Eshelby theory, and its applications in micromechanics problems are the subject of Chapter 9. The Hashin–Shtrikman bounds, which were presented rapidly in Chapter 5, are reintroduced and discussed in depth in Chapter 10. The mathematical foundation of full–field multiscale approaches such as the periodic homogenization theory is the topic of Chapter 11. Finally, Chapter 12 contains a brief discussion of the application of multiscale approaches to nonlinear composites; details for certain simple mean–field methods dedicated to studying inelastic and damageable heterogeneous media are included.

The book closes with an appendix, which explains the general methodology for integrating the reinforcement orientation into mean–field homogenization schemes (specifically, the Mori–Tanaka approach).

Chatzigeorgiou, Meraghni, and Charalambakis
May 2021

Acknowledgments

We want to express our sincere gratitude to Professor Etienne Patoor for extensive discussions during the last 15 years and Professor André Eberhardt for assistance, encouragement, and for providing us an image for the cover of the book.

We also thank Professor Dimitris Lagoudas for our fruitful discussions about micromechanics and Professor François Murat for our long discussions about the periodic homogenization. Additionally, we highly appreciate the scientific exchanges and ideas we had over the years with our national and international partners Professor Yves Chemisky (University of Bordeaux), Professor André Chrysochoos (University of Montpellier), Professor Fransisco (Paco) Chinesta (ENSAM Paris), Assistant Professor Joseph Fitoussi (ENSAM Paris), Dr. Gilles Robert (Domo Chemicals), Dr. Renan Léon (Valeo), Assistant Professor Theocharis Baxevanis (University of Houston), Assistant Professor Darren Hartl (Texas A&M University), Associate Professor Gary Don Seidel (VirginiaTech), Professor Paul Steinmann (FAU Erlangen–Nürnberg), Assistant Professor Ali Javili (Bilkent University), Professor Björn Kiefer (TU Bergakademie Freiberg), Professor Marek–Jerzy Pindera (University of Virginia) and Professor Andreas Menzel (TU Dortmund).

It is of course important for us to warmly thank all the members of the SMART team of LEM3 laboratory for their support. Special thanks belong to our colleagues, Assistant Professors Adil Benaarbia, Francis Praud and Boris Piotrowski, as well as our Postdoc Qiang Chen and our Ph.D. student Soheil Satouri, for their valuable comments and remarks.

Chatzigeorgiou, Meraghni, and Charalambakis
May 2021

PART I

Tensors and continuum mechanics concepts

CHAPTER

1

Tensors

OUTLINE

1.1 Tensors in Cartesian coordinates

A tensor is defined as a geometric object that describes linear relations between scalars, vectors, or other tensors, and thus it can be seen as a multilinear map. In mathematics and physics a tensor is a very general object, intrinsically defined from a vector space that is independent of coordinate systems. In depth analysis about tensors, tensor operations, and differential geometry for continuum mechanics applications has been presented elsewhere [1,2]. In this chapter, we focus on those key aspects that help the reader to follow the discussion throughout the book.

Bold letters are used to denote tensors to distinguish them from scalars. This chapter frequently adopts the index notation. When the tensors appear in their indicial notation, they have normal form (no bold font). A

https://doi.org/10.1016/B978-0-12-823143-2.00010-2

tensor $\boldsymbol{A}$ of order n is expressed as

$$A_{i_1 i_2 i_3 \ldots i_n}.$$

In 3–D space, the indices i_1, i_2, etc. take the values 1, 2, or 3. The most common tensors are:

- A tensor of order 0, which denotes a scalar (no bold letter is used in this case).
- A tensor of order 1, which represents a 3×1 vector.
- A tensor of order 2, which is represented by a 3×3 matrix.[1]

In the index notation, the Einstein summation convention for repeated indices is used. For example:

- The product $a_i b_i$ of two vectors $\boldsymbol{a}$ and $\boldsymbol{b}$ implies

$$a_i b_i = a_1 b_1 + a_2 b_2 + a_3 b_3.$$

- The product $A_{ij} B_{jk}$ of two second–order tensors $\boldsymbol{A}$ and $\boldsymbol{B}$ implies

$$A_{ij} B_{jk} = A_{i1} B_{1k} + A_{i2} B_{2k} + A_{i3} B_{3k}.$$

- The product $A_{ijkl} B_{kl}$ of a fourth–order tensor $\boldsymbol{A}$ and a second–order tensor $\boldsymbol{B}$ implies

$$\begin{aligned} A_{ijkl} B_{kl} = {} & A_{ij11} B_{11} + A_{ij22} B_{22} + A_{ij33} B_{33} + A_{ij12} B_{12} + A_{ij21} B_{21} \\ & + A_{ij13} B_{13} + A_{ij31} B_{31} + A_{ij23} B_{23} + A_{ij32} B_{32}. \end{aligned}$$

The repeated indices are "dummy" in the sense that we can change the letter used for them with another letter. Thus $a_i b_i$ and $a_j b_j$ represent the same summation.

The symbol $\boldsymbol{I}$ denotes the second–order identity tensor with components given by the Kronecker delta δ_{ij}:

$$I_{ij} = \delta_{ij}, \quad \delta_{ij} = \delta_{ji} = \begin{cases} 1, & i = j, \\ 0, & i \neq j. \end{cases} \tag{1.1}$$

Clearly, $\delta_{ii} = \delta_{11} + \delta_{22} + \delta_{33} = 3$.

The third–order Cartesian permutation tensor $\boldsymbol{\epsilon}$ is defined as

$$\epsilon_{ijk} = \begin{cases} 1 & \text{if } (i, j, k) \text{ is } (1,2,3),\ (2,3,1),\ \text{or } (3,1,2), \\ -1 & \text{if } (i, j, k) \text{ is } (1,3,2),\ (2,1,3),\ \text{or } (3,2,1), \\ 0 & \text{if } i = j,\ j = k,\ \text{or } k = i, \end{cases} \tag{1.2}$$

[1]For symmetric second–order tensors, we can utilize the Voigt notation and write them as 6×1 vectors; see later in this chapter.

and has the property $\epsilon_{ijk}\epsilon_{pqk} = \delta_{ip}\delta_{jq} - \delta_{iq}\delta_{jp}$.

The symmetric fourth–order identity tensor $\mathcal{I}$ is defined as

$$\mathcal{I}_{ijkl} = \frac{1}{2}[\delta_{ik}\delta_{jl} + \delta_{il}\delta_{jk}]. \tag{1.3}$$

From its structure it is clear that $\mathcal{I}$ exhibits the minor symmetries $\mathcal{I}_{ijkl} = \mathcal{I}_{jikl} = \mathcal{I}_{ijlk}$ and the major symmetry $\mathcal{I}_{ijkl} = \mathcal{I}_{klij}$.

Using the index notation leads very often to large expressions. To reduce the size of the expressions, the following symbols are frequently adopted (especially, in the micromechanics part of this book):

1. A tensor $\boldsymbol{A}$ of order n, a tensor $\boldsymbol{B}$ of order n, and a scalar c produce the nth–order tensors

$$\boldsymbol{C} = \boldsymbol{A} + \boldsymbol{B} \quad \text{with} \quad C_{i_1 i_2 \dots i_n} = A_{i_1 i_2 \dots i_n} + B_{i_1 i_2 \dots i_n},$$
$$\boldsymbol{C} = c\boldsymbol{A} \quad \text{with} \quad C_{i_1 i_2 \dots i_n} = cA_{i_1 i_2 \dots i_n}.$$

2. A tensor $\boldsymbol{A}$ of order $n+1$ and a tensor $\boldsymbol{B}$ of order $m+1$ produce the $(n+m)$th–order tensor

$$\boldsymbol{C} = \boldsymbol{A} \cdot \boldsymbol{B} \quad \text{with} \quad C_{i_1 i_2 \dots i_{n+m}} = A_{i_1 i_2 \dots i_n q} B_{q i_{n+1} i_{n+2} \dots i_{n+m}}.$$

   This operation is called the single contraction product.

3. A tensor $\boldsymbol{A}$ of order $n+2$ and a tensor $\boldsymbol{B}$ of order $m+2$ produce the $(n+m)$th–order tensor

$$\boldsymbol{C} = \boldsymbol{A} : \boldsymbol{B} \quad \text{with} \quad C_{i_1 i_2 \dots i_{n+m}} = A_{i_1 i_2 \dots i_n pq} B_{pq i_{n+1} i_{n+2} \dots i_{n+m}}.$$

   This operation is called the double contraction product.

4. A tensor $\boldsymbol{A}$ of order n and a tensor $\boldsymbol{B}$ of order m produce the $(n+m)$th–order tensor

$$\boldsymbol{C} = \boldsymbol{A} \otimes \boldsymbol{B} \quad \text{with} \quad C_{i_1 i_2 \dots i_{n+m}} = A_{i_1 i_2 \dots i_n} B_{i_{n+1} i_{n+2} \dots i_{n+m}}.$$

   This operation is called the dyadic product.

5. A tensor $\boldsymbol{A}$ of order n and a tensor $\boldsymbol{B}$ of order 1 produce the nth–order tensors

$$\boldsymbol{C} = \boldsymbol{A} \times \boldsymbol{B} \quad \text{with} \quad \boldsymbol{A} \times \boldsymbol{B} = [\boldsymbol{A} \otimes \boldsymbol{B}] : \boldsymbol{\epsilon},$$
$$\boldsymbol{C} = \boldsymbol{B} \times \boldsymbol{A} \quad \text{with} \quad \boldsymbol{B} \times \boldsymbol{A} = \boldsymbol{\epsilon} : [\boldsymbol{B} \otimes \boldsymbol{A}].$$

   These two operations are called the cross products.

1.2 Cartesian systems and tensor rotation

A Cartesian coordinate system is described by three basis vectors $\boldsymbol{e}_1$, $\boldsymbol{e}_2$, and $\boldsymbol{e}_3$. Each one of these vectors has unit length. A Cartesian system is

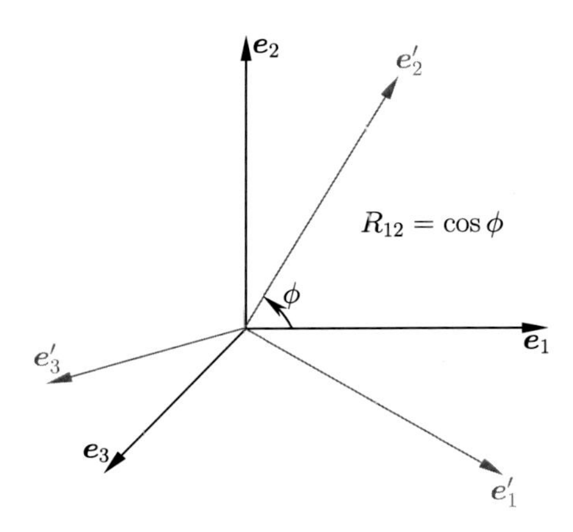

FIGURE 1.1 Transformation of a Cartesian coordinate system through a rotation.

orthogonal, which means that

$$\boldsymbol{e}_i \cdot \boldsymbol{e}_j = \delta_{ij}. \tag{1.4}$$

Let us consider two Cartesian coordinate systems with basis vectors ($\boldsymbol{e}_1, \boldsymbol{e}_2, \boldsymbol{e}_3$) and ($\boldsymbol{e}'_1, \boldsymbol{e}'_2, \boldsymbol{e}'_3$) having the same point of origin (Fig. 1.1). The relation between them can be expressed with the help of a second–order tensor $\boldsymbol{R}$ as

$$\begin{aligned}
\boldsymbol{e}'_1 &= R_{11}\boldsymbol{e}_1 + R_{21}\boldsymbol{e}_2 + R_{31}\boldsymbol{e}_3, \\
\boldsymbol{e}'_2 &= R_{12}\boldsymbol{e}_1 + R_{22}\boldsymbol{e}_2 + R_{32}\boldsymbol{e}_3, \\
\boldsymbol{e}'_3 &= R_{13}\boldsymbol{e}_1 + R_{23}\boldsymbol{e}_2 + R_{33}\boldsymbol{e}_3.
\end{aligned}$$

This tensor $\boldsymbol{R}$ is called a rotator and has the following properties:

$$R_{ij}R_{ik} = R_{ji}R_{ki} = \delta_{jk} \quad \text{and} \quad \det\boldsymbol{R} = 1, \tag{1.5}$$

where det denotes the determinant of $\boldsymbol{R}$,

$$\begin{aligned}
\det\boldsymbol{R} = {} & R_{13}R_{21}R_{31} - R_{13}R_{22}R_{31} - R_{11}R_{23}R_{31} \\
& + R_{12}R_{23}R_{31} - R_{12}R_{21}R_{33} + R_{11}R_{22}R_{33}.
\end{aligned}$$

Due to property (1.5) of the rotators, the vectors $\boldsymbol{e}_i$ can be expressed in terms of $\boldsymbol{e}'_i$ as

$$\begin{aligned}
\boldsymbol{e}_1 &= R_{11}\boldsymbol{e}'_1 + R_{12}\boldsymbol{e}'_2 + R_{13}\boldsymbol{e}'_3, \\
\boldsymbol{e}_2 &= R_{21}\boldsymbol{e}'_1 + R_{22}\boldsymbol{e}'_2 + R_{23}\boldsymbol{e}'_3, \\
\boldsymbol{e}_3 &= R_{31}\boldsymbol{e}'_1 + R_{32}\boldsymbol{e}'_2 + R_{33}\boldsymbol{e}'_3.
\end{aligned}$$

The components of $\boldsymbol{R}$ are the cosines of the angles between the vectors of the two bases,

$$R_{ij} = \cos\left(\boldsymbol{e}_i, \boldsymbol{e}'_j\right) = \boldsymbol{e}_i \cdot \boldsymbol{e}'_j.$$

For the elementary basis vectors

$$\boldsymbol{e}_1 = \begin{bmatrix} 1 \\ 0 \\ 0 \end{bmatrix}, \quad \boldsymbol{e}_2 = \begin{bmatrix} 0 \\ 1 \\ 0 \end{bmatrix}, \quad \boldsymbol{e}_3 = \begin{bmatrix} 0 \\ 0 \\ 1 \end{bmatrix},$$

we obtain that the ith column of the tensor $\boldsymbol{R}$ represents the basis vector $\boldsymbol{e}'_i$.

The rotator tensor represents a transformation from one coordinate system to another through rotation. The introduction of $\boldsymbol{R}$ allows us to provide a proper definition for the tensors:

BOX 1.1 Tensor

A tensor $\boldsymbol{A}$ of order n is a quantity that can be transformed, with the help of a rotator $\boldsymbol{R}$, from one coordinate system with basis $\boldsymbol{e}_i$ to another coordinate system with basis $\boldsymbol{e}'_i$ according to the formulas

$$A'_{i_1 i_2 \dots i_n} = R_{j_1 i_1} R_{j_2 i_2} \dots R_{j_n i_n} A_{j_1 j_2 \dots j_n}, \qquad A_{i_1 i_2 \dots i_n} = R_{i_1 j_1} R_{i_2 j_2} \dots R_{i_n j_n} A'_{j_1 j_2 \dots j_n}. \tag{1.6}$$

1.3 Tensor calculus

Considering Cartesian coordinates and the position vector $\boldsymbol{x}$, differential geometry provides the following definitions for an nth–order tensor $\boldsymbol{A}$ [3,4]:

1. The gradient of $\boldsymbol{A}$ is a tensor of order $n+1$ defined as

$$\frac{\partial A_{i_1 i_2 \dots i_n}}{\partial x_{i_{n+1}}}.$$

2. The divergence of $\boldsymbol{A}$ is a tensor of order $n-1$ defined as

$$\frac{\partial A_{i_1 i_2 \dots i_{n-1} i_n}}{\partial x_{i_n}}.$$

3. The curl of $\boldsymbol{A}$ is a tensor of order n defined as

$$-\frac{\partial A_{i_1 i_2 \dots i_{n-1} j}}{\partial x_k} \epsilon_{jki_n}.$$

Moreover, for a tensor $\boldsymbol{A}$ of order $n > 0$ and a tensor $\boldsymbol{B}$ of order $m > 0$, the derivative of $\boldsymbol{A}$ with respect to $\boldsymbol{B}$ is the tensor of order $n + m$ defined as

$$\frac{\partial A_{i_1 i_2 \ldots i_n}}{\partial B_{i_{n+1} i_{n+2} \ldots i_{n+m}}}.$$

Clearly, the typical chain rule for derivatives holds also in the case of tensors. For example, for two tensors $\boldsymbol{A}$ and $\boldsymbol{B}$ of order 2, the gradient of their product $A_{ij} B_{jk}$ is written as

$$\frac{\partial}{\partial x_m}\left(A_{ij} B_{jk}\right) = \frac{\partial A_{ij}}{\partial x_m} B_{jk} + A_{ij} \frac{\partial B_{jk}}{\partial x_m}.$$

1.4 Examples in tensor operations

Example 1

Prove the following tensorial identities:

1. For a second–order tensor $\boldsymbol{A}$, $A_{ij}\delta_{jk} = A_{ij}\delta_{kj} = A_{ik}$.
2. For a second–order tensor $\boldsymbol{A}$, $A_{ij}\delta_{ij} = A_{ii}$.
3. If $\boldsymbol{A}$ is symmetric second–order tensor ($A_{ij} = A_{ji}$), then $A_{kl}\mathcal{I}_{klij} = \mathcal{I}_{ijkl}A_{kl} = A_{ij}$.
4. If the fourth–order tensor A_{ijkl} exhibits the minor symmetries $A_{ijkl} = A_{jikl} = A_{ijlk}$, then $A_{ijmn}\mathcal{I}_{mnkl} = \mathcal{I}_{ijmn}A_{mnkl} = A_{ijkl}$.
5. For an arbitrary vector $\boldsymbol{a}$, $\dfrac{\partial a_i}{\partial a_j} = \delta_{ij}$.
6. If $\boldsymbol{A}$ is a symmetric second–order tensor ($A_{ij} = A_{ji}$), then $\dfrac{\partial A_{ij}}{\partial A_{kl}} = \mathcal{I}_{ijkl}$.

Solution:

1. For the first identity, we have

$$A_{ij}\delta_{jk} = A_{i1}\delta_{1k} + A_{i2}\delta_{2k} + A_{i3}\delta_{3k},$$
$$A_{ij}\delta_{kj} = A_{i1}\delta_{k1} + A_{i2}\delta_{k2} + A_{i3}\delta_{k3}.$$

There are three cases:

- $k = 1$. Then

$$A_{ij}\delta_{jk} = A_{i1}\delta_{11} + A_{i2}\delta_{21} + A_{i3}\delta_{31} = A_{i1} = A_{ik},$$
$$A_{ij}\delta_{kj} = A_{i1}\delta_{11} + A_{i2}\delta_{12} + A_{i3}\delta_{13} = A_{i1} = A_{ik}.$$

- $k = 2$. Then

$$A_{ij}\delta_{jk} = A_{i1}\delta_{12} + A_{i2}\delta_{22} + A_{i3}\delta_{32} = A_{i2} = A_{ik},$$
$$A_{ij}\delta_{kj} = A_{i1}\delta_{21} + A_{i2}\delta_{22} + A_{i3}\delta_{23} = A_{i2} = A_{ik}.$$

- $k = 3$. Then

$$A_{ij}\delta_{jk} = A_{i1}\delta_{13} + A_{i2}\delta_{23} + A_{i3}\delta_{33} = A_{i3} = A_{ik},$$
$$A_{ij}\delta_{kj} = A_{i1}\delta_{31} + A_{i2}\delta_{32} + A_{i3}\delta_{33} = A_{i3} = A_{ik}.$$

Clearly, this result can be extended for nth–order tensors. We can easily show that at any index i_k of an nth–order tensor, multiplication with the Kronecker delta yields

$$\begin{aligned} A_{i_1 i_2 \dots i_{k-1} j i_{k+1} \dots i_{n-1} i_n}\delta_{i_k j} &= A_{i_1 i_2 \dots i_{k-1} j i_{k+1} \dots i_{n-1} i_n}\delta_{j i_k} \\ &= A_{i_1 i_2 \dots i_{k-1} i_k i_{k+1} \dots i_{n-1} i_n}. \end{aligned}$$

2. Using the symmetry of δ_{ij} and the previous identity, we obtain

$$A_{ij}\delta_{ij} = A_{ij}\delta_{ji} = A_{ii}.$$

3. Using the symmetry of $\boldsymbol{A}$ yields

$$A_{kl}\mathcal{I}_{klij} = \frac{1}{2}A_{kl}[\delta_{ki}\delta_{lj} + \delta_{kj}\delta_{li}] = \frac{1}{2}[A_{ij} + A_{ji}] = A_{ij}.$$

Since $\mathcal{I}_{ijkl} = \mathcal{I}_{klij}$, it becomes evident that the second relation

$$\mathcal{I}_{ijkl}A_{kl} = A_{ij}$$

also holds.

4. Using the extension of the first identity to fourth–order tensors yields

$$\begin{aligned} A_{ijmn}\mathcal{I}_{mnkl} &= \frac{1}{2}A_{ijmn}[\delta_{mk}\delta_{nl} + \delta_{ml}\delta_{nk}] = \frac{1}{2}[A_{ijml}\delta_{mk} + A_{ijmk}\delta_{ml}] \\ &= \frac{1}{2}[A_{ijkl} + A_{ijlk}] = A_{ijkl} \end{aligned}$$

and

$$\begin{aligned} \mathcal{I}_{ijmn}A_{mnkl} &= \frac{1}{2}A_{mnkl}[\delta_{im}\delta_{jn} + \delta_{in}\delta_{jm}] = \frac{1}{2}[A_{inkl}\delta_{jn} + A_{jnkl}\delta_{in}] \\ &= \frac{1}{2}[A_{ijkl} + A_{jikl}] = A_{ijkl}. \end{aligned}$$

Note that $\mathcal{I}$ itself also possesses the minor symmetries, and thus the same identity holds for it too.

5. The partial derivative $\dfrac{\partial a_i}{\partial a_j}$ takes the value 1 when $i = j$ and is zero when $i \neq j$. This is similar to the definition of Kronecker delta.

6. The partial derivative $\frac{\partial A_{ij}}{\partial A_{kl}}$ for an arbitrary tensor $\boldsymbol{A}$ is equal to 1 when $i = k$ and $j = l$ and zero otherwise. So,

$$\frac{\partial A_{ij}}{\partial A_{kl}} = \delta_{ik}\delta_{jl}.$$

The latter result though does not account for symmetries. To include the symmetry of $\boldsymbol{A}$, we can write

$$\begin{aligned}\frac{\partial A_{ij}}{\partial A_{kl}} &= \frac{1}{4}\frac{\partial}{\partial A_{kl}}(A_{ij} + A_{ji}) + \frac{1}{4}\frac{\partial}{\partial A_{lk}}(A_{ij} + A_{ji}) \\ &= \frac{1}{2}[\delta_{ik}\delta_{jl} + \delta_{jk}\delta_{il}] = \mathcal{I}_{ijkl}.\end{aligned}$$

The tensor $\mathcal{I}$ is symmetric and thus represents the partial derivative of symmetric second–order tensor with itself.

Example 2: Hill notation

The fourth–order tensors $\boldsymbol{A}$ and $\boldsymbol{B}$, expressed in terms of the scalars A^b, A^s, B^b, and B^s in the form

$$\begin{aligned}A_{ijkl} &= 3A^b\mathcal{I}^{\text{hyd}}_{ijkl} + 2A^s\mathcal{I}^{\text{dev}}_{ijkl}, \\ B_{ijkl} &= 3B^b\mathcal{I}^{\text{hyd}}_{ijkl} + 2B^s\mathcal{I}^{\text{dev}}_{ijkl},\end{aligned}$$

where

$$\mathcal{I}^{\text{hyd}}_{ijkl} = \frac{1}{3}\delta_{ij}\delta_{kl}, \quad \mathcal{I}^{\text{dev}}_{ijkl} = \mathcal{I}_{ijkl} - \mathcal{I}^{\text{hyd}}_{ijkl},$$

are called isotropic tensors. Prove the following identities:

1. $A_{ijmn}B_{mnkl} = B_{ijmn}A_{mnkl} = 9A^bB^b\mathcal{I}^{\text{hyd}}_{ijkl} + 4A^sB^s\mathcal{I}^{\text{dev}}_{ijkl}$.

2. If $A_{ijmn}B_{mnkl} = B_{ijmn}A_{mnkl} = \mathcal{I}_{ijkl}$, then

$$B_{ijkl} = \frac{1}{3A^b}\mathcal{I}^{\text{hyd}}_{ijkl} + \frac{1}{2A^s}\mathcal{I}^{\text{dev}}_{ijkl}.$$

Remark:

Due to these interesting identities, a special notation for isotropic fourth–order tensors is proposed in [5]. According to this notation, an isotropic tensor $\boldsymbol{A}$ can be represented in the form $\boldsymbol{A} = (3A^b, 2A^s)$. This allows computations of scalar type when dealing with isotropic tensor algebra.

Solution:

Before passing to the main proofs, we need the following identities:

$$\mathcal{I}^{\text{hyd}}_{ijmn}\mathcal{I}^{\text{hyd}}_{mnkl} = \mathcal{I}^{\text{hyd}}_{ijkl},$$
$$\mathcal{I}^{\text{hyd}}_{ijmn}\mathcal{I}^{\text{dev}}_{mnkl} = \mathcal{I}^{\text{dev}}_{ijmn}\mathcal{I}^{\text{hyd}}_{mnkl} = 0_{ijkl},$$
$$\mathcal{I}^{\text{dev}}_{ijmn}\mathcal{I}^{\text{dev}}_{mnkl} = \mathcal{I}^{\text{dev}}_{ijkl},$$

where 0_{ijkl} is the fourth–order null tensor with all zero terms. For the first expression,

$$\begin{aligned}\mathcal{I}^{\text{hyd}}_{ijmn}\mathcal{I}^{\text{hyd}}_{mnkl} &= \frac{1}{3}\delta_{ij}\delta_{mn}\frac{1}{3}\delta_{mn}\delta_{kl} = \frac{1}{9}\delta_{ij}\delta_{kl}\delta_{mm} \\ &= \frac{1}{9}\delta_{ij}\delta_{kl}[\delta_{11}+\delta_{22}+\delta_{33}] = \frac{3}{9}\delta_{ij}\delta_{kl} = \mathcal{I}^{\text{hyd}}_{ijkl}.\end{aligned}$$

For the second and third expressions, note that $\mathcal{I}^{\text{hyd}}$ exhibits the minor symmetries $\mathcal{I}^{\text{hyd}}_{ijkl} = \mathcal{I}^{\text{hyd}}_{jikl} = \mathcal{I}^{\text{hyd}}_{ijlk}$, and thus the fourth identity of Example 1 holds. So

$$\begin{aligned}\mathcal{I}^{\text{hyd}}_{ijmn}\mathcal{I}^{\text{dev}}_{mnkl} &= \mathcal{I}^{\text{hyd}}_{ijmn}[\mathcal{I}_{mnkl} - \mathcal{I}^{\text{hyd}}_{mnkl}] = \mathcal{I}^{\text{hyd}}_{ijkl} - \mathcal{I}^{\text{hyd}}_{ijkl} = 0_{ijkl}, \\ \mathcal{I}^{\text{dev}}_{ijmn}\mathcal{I}^{\text{hyd}}_{mnkl} &= [\mathcal{I}_{ijmn} - \mathcal{I}^{\text{hyd}}_{ijmn}]\mathcal{I}^{\text{hyd}}_{mnkl} = \mathcal{I}^{\text{hyd}}_{ijkl} - \mathcal{I}^{\text{hyd}}_{ijkl} = 0_{ijkl}, \\ \mathcal{I}^{\text{dev}}_{ijmn}\mathcal{I}^{\text{dev}}_{mnkl} &= [\mathcal{I}_{ijmn} - \mathcal{I}^{\text{hyd}}_{ijmn}][\mathcal{I}_{mnkl} - \mathcal{I}^{\text{hyd}}_{mnkl}] \\ &= \mathcal{I}_{ijkl} - \mathcal{I}^{\text{hyd}}_{ijkl} - \mathcal{I}^{\text{hyd}}_{ijkl} + \mathcal{I}^{\text{hyd}}_{ijkl} = \mathcal{I}^{\text{dev}}_{ijkl}.\end{aligned}$$

With these expressions we can now pass to the main proofs.

1. For the first identity,

$$\begin{aligned}A_{ijmn}B_{mnkl} &= [3A^b\mathcal{I}^{\text{hyd}}_{ijmn} + 2A^s\mathcal{I}^{\text{dev}}_{ijmn}][3B^b\mathcal{I}^{\text{hyd}}_{mnkl} + 2B^s\mathcal{I}^{\text{dev}}_{mnkl}] \\ &= 9A^bB^b\mathcal{I}^{\text{hyd}}_{ijmn}\mathcal{I}^{\text{hyd}}_{mnkl} + 4A^sB^s\mathcal{I}^{\text{dev}}_{ijmn}\mathcal{I}^{\text{dev}}_{mnkl} \\ &+ 6A^bB^s\mathcal{I}^{\text{hyd}}_{ijmn}\mathcal{I}^{\text{dev}}_{mnkl} + 6A^sB^b\mathcal{I}^{\text{dev}}_{ijmn}\mathcal{I}^{\text{hyd}}_{mnkl} \\ &= 9A^bB^b\mathcal{I}^{\text{hyd}}_{ijkl} + 4A^sB^s\mathcal{I}^{\text{dev}}_{ijkl}.\end{aligned}$$

By similar computations we also get the identity

$$B_{ijmn}A_{mnkl} = 9A^bB^b\mathcal{I}^{\text{hyd}}_{ijkl} + 4A^sB^s\mathcal{I}^{\text{dev}}_{ijkl}.$$

2. For the second identity, using the first yields

$$9A^bB^b\mathcal{I}^{\text{hyd}}_{ijkl} + 4A^sB^s\mathcal{I}^{\text{dev}}_{ijkl} = \mathcal{I}^{\text{hyd}}_{ijkl} + \mathcal{I}^{\text{dev}}_{ijkl}.$$

From the last expression it becomes evident that

$$9A^bB^b = 1 \quad \text{and} \quad 4A^sB^s = 1,$$

or

$$3B^b = \frac{1}{3A^b} \quad \text{and} \quad 2B^s = \frac{1}{2A^s}.$$

Example 3

A symmetric second–order tensor $a_{ij} = a_{ji}$ can be split into deviatoric and volumetric/hydrostatic parts, which are also symmetric:

$$a_{ij}^{\text{hyd}} = a_{ji}^{\text{hyd}} = \frac{1}{3}a_{kk}\delta_{ij},$$
$$a_{ij}^{\text{dev}} = a_{ji}^{\text{dev}} = a_{ij} - a_{ij}^{\text{hyd}}.$$

For two symmetric second–order tensors $\boldsymbol{a}$ and $\boldsymbol{b}$, prove the following identities:

1. $a_{ij}^{\text{hyd}} b_{ij}^{\text{dev}} = 0.$

2. $a_{ij} b_{ij}^{\text{dev}} = a_{ij}^{\text{dev}} b_{ij}^{\text{dev}}.$

3. $\mathcal{I}_{ijkl}^{\text{hyd}} a_{kl} = a_{kl}\mathcal{I}_{klij}^{\text{hyd}} = a_{ij}^{\text{hyd}}$ and $\mathcal{I}_{ijkl}^{\text{dev}} a_{kl} = a_{kl}\mathcal{I}_{klij}^{\text{dev}} = a_{ij}^{\text{dev}}$, where $\mathcal{I}_{ijkl}^{\text{hyd}}$ and $\mathcal{I}_{ijkl}^{\text{dev}}$ are defined in Example 2.

4. $\dfrac{\partial a^{\text{ndv}}}{\partial a_{ij}} = \dfrac{a_{ij}^{\text{dev}}}{a^{\text{ndv}}}$, where $a^{\text{ndv}} = \sqrt{a_{kl}^{\text{dev}} a_{kl}^{\text{dev}}}.$

Solution:

1. For the first identity,

$$a_{ij}^{\text{hyd}} b_{ij}^{\text{dev}} = \frac{1}{3}a_{kk}\delta_{ij}\left[b_{ij} - \frac{1}{3}b_{mm}\delta_{ij}\right] = \frac{1}{3}a_{kk}b_{ii} - \frac{1}{9}a_{kk}b_{mm}\delta_{ii}$$
$$= \frac{1}{3}a_{kk}b_{mm} - \frac{1}{3}a_{kk}b_{mm} = 0.$$

2. Using the first identity, we easily prove the second:

$$a_{ij}^{\text{dev}} b_{ij}^{\text{dev}} = \left[a_{ij} - a_{ij}^{\text{hyd}}\right] b_{ij}^{\text{dev}} = a_{ij} b_{ij}^{\text{dev}}.$$

3. We have

$$\mathcal{I}_{ijkl}^{\text{hyd}} a_{kl} = \frac{1}{3}\delta_{ij}\delta_{kl}a_{kl} = \frac{1}{3}a_{kk}\delta_{ij} = a_{ij}^{\text{hyd}},$$
$$\mathcal{I}_{ijkl}^{\text{dev}} a_{kl} = \left[\mathcal{I}_{ijkl} - \mathcal{I}_{ijkl}^{\text{hyd}}\right] a_{kl} = a_{ij} - a_{ij}^{\text{hyd}} = a_{ij}^{\text{dev}}.$$

Since $\mathcal{I}^{\text{hyd}}_{ijkl} = \mathcal{I}^{\text{hyd}}_{klij}$ and $\mathcal{I}^{\text{dev}}_{ijkl} = \mathcal{I}^{\text{dev}}_{klij}$, we also get the relations

$$a_{kl}\mathcal{I}^{\text{hyd}}_{klij} = a^{\text{hyd}}_{ij}, \quad a_{kl}\mathcal{I}^{\text{dev}}_{klij} = a^{\text{dev}}_{ij}.$$

4. We split the proof of the fourth identity into two parts:

$$\begin{aligned}\frac{\partial a^{\text{ndv}}}{\partial a^{\text{dev}}_{mn}} &= \frac{1}{2a^{\text{ndv}}}\frac{\partial}{\partial a^{\text{dev}}_{mn}}\left(a^{\text{dev}}_{kl}a^{\text{dev}}_{kl}\right) = \frac{1}{a^{\text{ndv}}}\mathcal{I}_{klmn}a^{\text{dev}}_{kl} \\ &= \frac{a^{\text{dev}}_{kl}}{a^{\text{ndv}}}\frac{1}{2}[\delta_{km}\delta_{ln} + \delta_{kn}\delta_{lm}] = \frac{1}{a^{\text{ndv}}}\frac{1}{2}[a^{\text{dev}}_{mn} + a^{\text{dev}}_{nm}] = \frac{a^{\text{dev}}_{mn}}{a^{\text{ndv}}},\end{aligned}$$

where we used the sixth identity of Example 1, and

$$\begin{aligned}\frac{\partial a^{\text{dev}}_{mn}}{\partial a_{ij}} &= \mathcal{I}_{mnij} - \frac{1}{3}\delta_{mn}\frac{\partial a_{kk}}{\partial a_{ij}} = \mathcal{I}_{mnij} - \frac{1}{3}\delta_{mn}\mathcal{I}_{kkij} \\ &= \frac{1}{2}[\delta_{mi}\delta_{nj} + \delta_{mj}\delta_{ni}] - \frac{1}{6}\delta_{mn}[\delta_{ki}\delta_{kj} + \delta_{kj}\delta_{ki}] \\ &= \frac{1}{2}[\delta_{mi}\delta_{nj} + \delta_{mj}\delta_{ni}] - \frac{1}{3}\delta_{mn}\delta_{ij} = \mathcal{I}^{\text{dev}}_{mnij}.\end{aligned}$$

Since a^{dev}_{ij} is symmetric, the second identity of identities 3 holds, that is,

$$a^{\text{dev}}_{kl}\mathcal{I}^{\text{dev}}_{klij} = a^{\text{dev}}_{ij}.$$

Thus

$$\frac{\partial a^{\text{ndv}}}{\partial a_{ij}} = \frac{\partial a^{\text{ndv}}}{\partial a^{\text{dev}}_{mn}}\frac{\partial a^{\text{dev}}_{mn}}{\partial a_{ij}} = \frac{a^{\text{dev}}_{mn}}{a^{\text{ndv}}}\mathcal{I}^{\text{dev}}_{mnij} = \frac{a^{\text{dev}}_{ij}}{a^{\text{ndv}}}.$$

1.5 Voigt notation: general aspects

Until now, the notion of tensors has been presented. Numerical computations using tensorial or indicial notation can be quite cumbersome, especially when dealing with operations related to fourth–order tensors. In the present chapter, we discuss the well–known Voigt notation, which allows us to transform tensorial operations to matrix operations. The reader is required to have some basic knowledge about vector and matrix operations.

The Voigt notation is a practical way to represent second– and fourth–order symmetric tensors in vector and matrix form, respectively. The Voigt representation substitutes two indices i, j with one index I according to the rule[2]

$$I = \begin{cases} i, & i = j, \\ 1 + [i + j], & i \neq j. \end{cases}$$

Let us write the Voigt representation in detail:

first tensorial index	second tensorial index	resultant Voigt index
i	j	I
1	1	1
2	2	2
3	3	3
1	2	4
2	1	4
1	3	5
3	1	5
2	3	6
3	2	6

With this interchange rule we have the following:

- A symmetric second–order tensor $\boldsymbol{a}$ is characterized by the property $a_{ij} = a_{ji}$. It is usually represented as the 3×3 matrix

$$\boldsymbol{a} = \begin{bmatrix} a_{11} & a_{12} & a_{13} \\ a_{12} & a_{22} & a_{23} \\ a_{13} & a_{23} & a_{33} \end{bmatrix}.$$

Since $\boldsymbol{a}$ has only six independent components, we can also express it as a 6×1 vector. There are two types of vectors considered in the Voigt notation, the "s" type and the "e" type,

$$\boldsymbol{\sigma} = \begin{bmatrix} \sigma_{11} \\ \sigma_{22} \\ \sigma_{33} \\ \sigma_{12} \\ \sigma_{13} \\ \sigma_{23} \end{bmatrix} = \begin{bmatrix} \sigma_1 \\ \sigma_2 \\ \sigma_3 \\ \sigma_4 \\ \sigma_5 \\ \sigma_6 \end{bmatrix} \quad \text{and} \quad \tilde{\boldsymbol{\varepsilon}} = \begin{bmatrix} \varepsilon_{11} \\ \varepsilon_{22} \\ \varepsilon_{33} \\ 2\varepsilon_{12} \\ 2\varepsilon_{13} \\ 2\varepsilon_{23} \end{bmatrix} = \begin{bmatrix} \varepsilon_1 \\ \varepsilon_2 \\ \varepsilon_3 \\ 2\varepsilon_4 \\ 2\varepsilon_5 \\ 2\varepsilon_6 \end{bmatrix},$$

respectively. The factor 2 appearing on the "e" type is very useful when performing various tensor operations, as it will become clear further.

[2]In the classical continuum mechanics studies the indices $i = 2, j = 3$ or $i = 3, j = 2$ correspond to $I = 4$, and the indices $i = 1, j = 2$ or $i = 2, j = 1$ correspond to $I = 6$. Here we adopt a slight modification in the Voigt convention.

- A fourth–order tensor that has minor symmetries (i.e., $A_{ijkl} = A_{jikl} = A_{ijlk}$) represents a linear relation between symmetric second–order tensors; it has only 36 independent components and can be represented as a 6×6 matrix. The two Voigt representations of second–order tensors necessitate the identification of four Voigt representations of fourth–order tensors. The standard 6×6 matrix form is

$$\boldsymbol{A} = \begin{bmatrix} A_{1111} & A_{1122} & A_{1133} & A_{1112} & A_{1113} & A_{1123} \\ A_{2211} & A_{2222} & A_{2233} & A_{2212} & A_{2213} & A_{2223} \\ A_{3311} & A_{3322} & A_{3333} & A_{3312} & A_{3313} & A_{3323} \\ A_{1211} & A_{1222} & A_{1233} & A_{1212} & A_{1213} & A_{1223} \\ A_{1311} & A_{1322} & A_{1333} & A_{1312} & A_{1313} & A_{1323} \\ A_{2311} & A_{2322} & A_{2333} & A_{2312} & A_{2313} & A_{2323} \end{bmatrix},$$

written for simplicity as

$$\boldsymbol{A} = \begin{bmatrix} A_{11} & A_{12} & A_{13} & A_{14} & A_{15} & A_{16} \\ A_{21} & A_{22} & A_{23} & A_{24} & A_{25} & A_{26} \\ A_{31} & A_{32} & A_{33} & A_{34} & A_{35} & A_{36} \\ A_{41} & A_{42} & A_{43} & A_{44} & A_{45} & A_{46} \\ A_{51} & A_{52} & A_{53} & A_{54} & A_{55} & A_{56} \\ A_{61} & A_{62} & A_{63} & A_{64} & A_{65} & A_{66} \end{bmatrix},$$

where the four indices i, j, k, l are substituted with two indices I, J following the notation:

first tensorial index	second tensorial index	resultant Voigt index	first tensorial index	second tensorial index	resultant Voigt index
i	j	I	k	l	J
1	1	1	1	1	1
2	2	2	2	2	2
3	3	3	3	3	3
1	2	4	1	2	4
2	1	4	2	1	4
1	3	5	1	3	5
3	1	5	3	1	5
2	3	6	2	3	6
3	2	6	3	2	6

The three additional matrix forms that can be identified are [6]

$$\widetilde{A} = \begin{bmatrix} A_{11} & A_{12} & A_{13} & A_{14} & A_{15} & A_{16} \\ A_{21} & A_{22} & A_{23} & A_{24} & A_{25} & A_{26} \\ A_{31} & A_{32} & A_{33} & A_{34} & A_{35} & A_{36} \\ 2A_{41} & 2A_{42} & 2A_{43} & 2A_{44} & 2A_{45} & 2A_{46} \\ 2A_{51} & 2A_{52} & 2A_{53} & 2A_{54} & 2A_{55} & 2A_{56} \\ 2A_{61} & 2A_{62} & 2A_{63} & 2A_{64} & 2A_{65} & 2A_{66} \end{bmatrix},$$

$$\breve{A} = \begin{bmatrix} A_{11} & A_{12} & A_{13} & 2A_{14} & 2A_{15} & 2A_{16} \\ A_{21} & A_{22} & A_{23} & 2A_{24} & 2A_{25} & 2A_{26} \\ A_{31} & A_{32} & A_{33} & 2A_{34} & 2A_{35} & 2A_{36} \\ A_{41} & A_{42} & A_{43} & 2A_{44} & 2A_{45} & 2A_{46} \\ A_{51} & A_{52} & A_{53} & 2A_{54} & 2A_{55} & 2A_{56} \\ A_{61} & A_{62} & A_{63} & 2A_{64} & 2A_{65} & 2A_{66} \end{bmatrix},$$

$$\widehat{A} = \begin{bmatrix} A_{11} & A_{12} & A_{13} & 2A_{14} & 2A_{15} & 2A_{16} \\ A_{21} & A_{22} & A_{23} & 2A_{24} & 2A_{25} & 2A_{26} \\ A_{31} & A_{32} & A_{33} & 2A_{34} & 2A_{35} & 2A_{36} \\ 2A_{41} & 2A_{42} & 2A_{43} & 4A_{44} & 4A_{45} & 4A_{46} \\ 2A_{51} & 2A_{52} & 2A_{53} & 4A_{54} & 4A_{55} & 4A_{56} \\ 2A_{61} & 2A_{62} & 2A_{63} & 4A_{64} & 4A_{65} & 4A_{66} \end{bmatrix}.$$

1.6 Operations using the Voigt notation

Using the representations described above for second– and fourth–order tensors, we can simplify various tensorial operations.

Considering the two types of second–order tensors, we can write the scalar tensorial product

$$W = \sigma_{ij}\varepsilon_{ij}$$

with the help of matrix operations as

$$W = \boldsymbol{\sigma}^T \cdot \widetilde{\boldsymbol{\varepsilon}}, \quad \text{or} \quad W = \widetilde{\boldsymbol{\varepsilon}}^T \cdot \boldsymbol{\sigma},$$

where the symbol $(\cdot)$ in the Voigt notation denotes the classical matrix multiplication, and the superscript T is the usual transpose operator. With regard to fourth–order tensors, we easily show that the tensorial products

$$\sigma_{ij} = L_{ijkl}\varepsilon_{kl}, \quad \varepsilon_{ij} = M_{ijkl}\sigma_{kl}, \quad \varepsilon^r_{ij} = T_{ijkl}\varepsilon^0_{kl}, \quad \sigma^r_{ij} = H_{ijkl}\sigma^0_{kl} \tag{1.7}$$

can be written using matrix multiplication as

$$\boldsymbol{\sigma} = \boldsymbol{L} \cdot \widetilde{\boldsymbol{\varepsilon}}, \qquad \widetilde{\boldsymbol{\varepsilon}} = \widehat{\boldsymbol{M}} \cdot \boldsymbol{\sigma}, \quad \widetilde{\boldsymbol{\varepsilon}}^r = \widetilde{\boldsymbol{T}} \cdot \widetilde{\boldsymbol{\varepsilon}}^0, \quad \boldsymbol{\sigma}^r = \breve{\boldsymbol{H}} \cdot \boldsymbol{\sigma}^0. \tag{1.8}$$

In certain occasions, tensorial products appear in which the second–order tensor is the first and the fourth–order tensor is the second, that is,

$$\sigma_{kl} = \varepsilon_{ij} P_{ijkl}, \quad \varepsilon_{kl} = \sigma_{ij} Q_{ijkl}, \quad \varepsilon^r_{kl} = \varepsilon^0_{ij} F_{ijkl}, \quad \sigma^r_{kl} = \sigma^0_{ij} G_{ijkl}.$$

Using matrix multiplication, we write these expressions as

$$\boldsymbol{\sigma}^T = \widetilde{\boldsymbol{\varepsilon}}^T \cdot \boldsymbol{P}, \quad \widetilde{\boldsymbol{\varepsilon}}^T = \boldsymbol{\sigma}^T \cdot \widehat{\boldsymbol{Q}}, \quad [\widetilde{\boldsymbol{\varepsilon}}^r]^T = [\widetilde{\boldsymbol{\varepsilon}}^0]^T \cdot \check{\boldsymbol{F}}, \quad [\boldsymbol{\sigma}^r]^T = [\boldsymbol{\sigma}^0]^T \cdot \widetilde{\boldsymbol{G}}.$$

Motivated by relations (1.8), we can name the different forms of a fourth–order tensor $\boldsymbol{A}$ as follows:

- the form $\boldsymbol{A}$ as "s–e" type,
- the form $\widehat{\boldsymbol{A}}$ as "e–s" type,
- the form $\widetilde{\boldsymbol{A}}$ as "e–e" type, and
- the form $\check{\boldsymbol{A}}$ as "s–s" type.

Additionally to the above expressions, we can write the dyadic products

$$A_{ijkl} = \sigma_{ij}\varepsilon_{kl}, \quad B_{ijkl} = \varepsilon_{ij}\sigma_{kl}, \quad C_{ijkl} = \varepsilon_{ij}\varepsilon_{kl}, \quad D_{ijkl} = \sigma_{ij}\sigma_{kl}$$

using matrix multiplications as

$$\check{\boldsymbol{A}} = \boldsymbol{\sigma} \cdot \widetilde{\boldsymbol{\varepsilon}}^T, \quad \widetilde{\boldsymbol{B}} = \widetilde{\boldsymbol{\varepsilon}} \cdot \boldsymbol{\sigma}^T, \quad \widehat{\boldsymbol{C}} = \widetilde{\boldsymbol{\varepsilon}} \cdot \widetilde{\boldsymbol{\varepsilon}}^T, \quad \boldsymbol{D} = \boldsymbol{\sigma} \cdot \boldsymbol{\sigma}^T.$$

An obvious property is that the four matrix representations for the fourth–order tensors respect the usual matrix summation and multiplication by a scalar c:

$$C_{ijkl} = A_{ijkl} + B_{ijkl} \implies \overset{\#}{\boldsymbol{C}} = \overset{\#}{\boldsymbol{A}} + \overset{\#}{\boldsymbol{B}},$$
$$B_{ijkl} = cA_{ijkl} \implies \overset{\#}{\boldsymbol{B}} = c\overset{\#}{\boldsymbol{A}},$$

where $\overset{\#}{\{\bullet\}}$ is any of the four types.

Multiplication identities of fourth–order tensors

For fourth–order tensors that respect minor symmetries, the double contraction product of A_{ijkl} and B_{ijkl} provides the tensor C_{ijkl} for which we have

$$C_{ijkl} = A_{ijmn} B_{mnkl}.$$

In matrix notation this is written as

$$\boldsymbol{C} = \boldsymbol{A} \cdot \widetilde{\boldsymbol{B}}.$$

Using this relation, we can also show that

$$\boldsymbol{A}\cdot\widetilde{\boldsymbol{B}}=\check{\boldsymbol{A}}\cdot\boldsymbol{B}=\boldsymbol{C},\qquad \widetilde{\boldsymbol{A}}\cdot\widetilde{\boldsymbol{B}}=\widehat{\boldsymbol{A}}\cdot\boldsymbol{B}=\widetilde{\boldsymbol{C}},$$
$$\check{\boldsymbol{A}}\cdot\check{\boldsymbol{B}}=\boldsymbol{A}\cdot\widehat{\boldsymbol{B}}=\check{\boldsymbol{C}},\qquad \widetilde{\boldsymbol{A}}\cdot\widehat{\boldsymbol{B}}=\widehat{\boldsymbol{A}}\cdot\check{\boldsymbol{B}}=\widehat{\boldsymbol{C}}.$$

The above multiplications are expressed in symbolic (type) form as

$$\text{“s–e”}\cdot\text{“e–e”} \text{ or } \text{“s–s”}\cdot\text{“s–e”}\rightarrow\text{“s–e”},$$
$$\text{“e–e”}\cdot\text{“e–e”} \text{ or } \text{“e–s”}\cdot\text{“s–e”}\rightarrow\text{“e–e”},$$
$$\text{“s–s”}\cdot\text{“s–s”} \text{ or } \text{“s–e”}\cdot\text{“e–s”}\rightarrow\text{“s–s”},$$
$$\text{“e–e”}\cdot\text{“e–s”} \text{ or } \text{“e–s”}\cdot\text{“s–s”}\rightarrow\text{“e–s”}.$$

Transpose identities of fourth–order tensors

For fourth–order tensors that respect minor symmetries, the transpose of $\boldsymbol{B}$ is defined as the tensor $\boldsymbol{A}$ with

$$A_{ijkl}=B^T_{ijkl}=B_{klij}.$$

We can easily show that, in the Voigt notation,

$$\boldsymbol{B}^T=\boldsymbol{A},\qquad \widehat{\boldsymbol{B}}^T=\widehat{\boldsymbol{A}},\qquad \widetilde{\boldsymbol{B}}^T=\check{\boldsymbol{A}},\qquad \check{\boldsymbol{B}}^T=\widetilde{\boldsymbol{A}}.$$

The latter operations are expressed in symbolic (type) form as

$$\text{“s–e”}^T\rightarrow\text{“s–e”},\quad \text{“e–s”}^T\rightarrow\text{“e–s”},\quad \text{“e–e”}^T\rightarrow\text{“s–s”},\quad \text{“s–s”}^T\rightarrow\text{“e–e”}.$$

Inversion identities of fourth–order tensors

For fourth–order tensors that respect minor symmetries, the inverse of L_{ijkl} is the tensor $M_{ijkl}=L^{-1}_{ijkl}$, for which

$$L_{ijmn}M_{mnkl}=\mathcal{I}_{ijkl}=\frac{1}{2}[\delta_{ik}\delta_{jl}+\delta_{il}\delta_{jk}].$$

In matrix notation, this expression can be written as

$$\boldsymbol{L}\cdot\widetilde{\boldsymbol{M}}=\boldsymbol{\mathcal{I}}=\begin{bmatrix}1&0&0&0&0&0\\0&1&0&0&0&0\\0&0&1&0&0&0\\0&0&0&0.5&0&0\\0&0&0&0&0.5&0\\0&0&0&0&0&0.5\end{bmatrix}.$$

The above relation can be written in the following ways:

$$\boldsymbol{L}\cdot\widehat{\boldsymbol{M}}=\check{\boldsymbol{\mathcal{I}}},\quad \widehat{\boldsymbol{L}}\cdot\boldsymbol{M}=\widetilde{\boldsymbol{\mathcal{I}}},\quad \widetilde{\boldsymbol{L}}\cdot\widetilde{\boldsymbol{M}}=\widetilde{\boldsymbol{\mathcal{I}}},\quad \check{\boldsymbol{L}}\cdot\check{\boldsymbol{M}}=\check{\boldsymbol{\mathcal{I}}},$$

$$\widetilde{\boldsymbol{I}} = \check{\boldsymbol{I}} = \begin{bmatrix} 1 & 0 & 0 & 0 & 0 & 0 \\ 0 & 1 & 0 & 0 & 0 & 0 \\ 0 & 0 & 1 & 0 & 0 & 0 \\ 0 & 0 & 0 & 1 & 0 & 0 \\ 0 & 0 & 0 & 0 & 1 & 0 \\ 0 & 0 & 0 & 0 & 0 & 1 \end{bmatrix}.$$

Thus, in the Voigt notation, we have the following properties:

$$\boldsymbol{L}^{-1} = \widehat{\boldsymbol{M}}, \quad \widehat{\boldsymbol{L}}^{-1} = \boldsymbol{M}, \quad \widetilde{\boldsymbol{L}}^{-1} = \widetilde{\boldsymbol{M}}, \quad \check{\boldsymbol{L}}^{-1} = \check{\boldsymbol{M}}.$$

The latter operations are expressed in symbolic (type) form as

$$\text{"s–e"}^{-1} \rightarrow \text{"e–s"}, \quad \text{"e–s"}^{-1} \rightarrow \text{"s–e"},$$
$$\text{"e–e"}^{-1} \rightarrow \text{"e–e"}, \quad \text{"s–s"}^{-1} \rightarrow \text{"s–s"}.$$

A summary of the various properties and the connection between indicial and Voigt notations is presented in Table 1.1.

1.7 Tensor rotation in Voigt notation

General case

As already discussed in the chapter, a second–order tensor $\boldsymbol{A}$ in the global coordinate system with basis $\boldsymbol{e}_i$ can be rotated to obtain the tensor $\boldsymbol{A}'$ in a local coordinate system with basis $\boldsymbol{e}'_i$ through the relation

$$A'_{ij} = R_{mi} R_{nj} A_{mn} \,. \tag{1.9}$$

A usual rotator tensor $\boldsymbol{R}$ is a second–order orthogonal tensor ($R_{ij} R_{ik} = R_{ji} R_{ki} = I_{jk}$) which can be written as a 3×3 matrix

$$\boldsymbol{R} = \begin{bmatrix} R_{11} & R_{12} & R_{13} \\ R_{21} & R_{22} & R_{23} \\ R_{31} & R_{32} & R_{33} \end{bmatrix}.$$

If a rotation ϑ is performed around the axis i with i =1, 2, or 3, then the rotator tensor is written as [1]

$$\boldsymbol{R}^1 = \begin{bmatrix} 1 & 0 & 0 \\ 0 & \cos\vartheta & -\sin\vartheta \\ 0 & \sin\vartheta & \cos\vartheta \end{bmatrix}, \quad \boldsymbol{R}^2 = \begin{bmatrix} \cos\vartheta & 0 & \sin\vartheta \\ 0 & 1 & 0 \\ -\sin\vartheta & 0 & \cos\vartheta \end{bmatrix},$$

$$\boldsymbol{R}^3 = \begin{bmatrix} \cos\vartheta & -\sin\vartheta & 0 \\ \sin\vartheta & \cos\vartheta & 0 \\ 0 & 0 & 1 \end{bmatrix},$$

TABLE 1.1 Tensor operations in indicial, tensorial, and Voigt notations. The symbol # above a fourth–order tensor denotes any of the four Voigt types.

Indicial	Tensorial	Voigt
$W = \sigma_{ij}\varepsilon_{ij} = \varepsilon_{ij}\sigma_{ij}$	$W = \boldsymbol{\sigma} : \boldsymbol{\varepsilon} = \boldsymbol{\varepsilon} : \boldsymbol{\sigma}$	$W = \boldsymbol{\sigma}^T \cdot \widetilde{\boldsymbol{\varepsilon}} = \widetilde{\boldsymbol{\varepsilon}}^T \cdot \boldsymbol{\sigma}$
$\sigma_{ij} = L_{ijkl}\varepsilon_{kl}$	$\boldsymbol{\sigma} = \boldsymbol{L} : \boldsymbol{\varepsilon}$	$\boldsymbol{\sigma} = \boldsymbol{L} \cdot \widetilde{\boldsymbol{\varepsilon}}$
$\varepsilon_{ij} = M_{ijkl}\sigma_{kl}$	$\boldsymbol{\varepsilon} = \boldsymbol{M} : \boldsymbol{\sigma}$	$\widetilde{\boldsymbol{\varepsilon}} = \widehat{\boldsymbol{M}} \cdot \boldsymbol{\sigma}$
$\varepsilon^r_{ij} = T_{ijkl}\varepsilon^0_{kl}$	$\boldsymbol{\varepsilon}^r = \boldsymbol{T} : \boldsymbol{\varepsilon}^0$	$\widetilde{\boldsymbol{\varepsilon}}^r = \widetilde{\boldsymbol{T}} \cdot \widetilde{\boldsymbol{\varepsilon}}^0$
$\sigma^r_{ij} = H_{ijkl}\sigma^0_{kl}$	$\boldsymbol{\sigma}^r = \boldsymbol{H} : \boldsymbol{\sigma}^0$	$\boldsymbol{\sigma}^r = \breve{\boldsymbol{H}} \cdot \boldsymbol{\sigma}^0$
$\sigma_{ij} = \varepsilon_{ij}L_{ijkl}$	$\boldsymbol{\sigma} = \boldsymbol{\varepsilon} : \boldsymbol{L}$	$\boldsymbol{\sigma}^T = \widetilde{\boldsymbol{\varepsilon}}^T \cdot \boldsymbol{L}$
$\varepsilon_{ij} = \sigma_{ij}M_{ijkl}$	$\boldsymbol{\varepsilon} = \boldsymbol{\sigma} : \boldsymbol{M}$	$\widetilde{\boldsymbol{\varepsilon}}^T = \boldsymbol{\sigma}^T \cdot \widehat{\boldsymbol{M}}$
$\sigma^r_{ij} = \sigma^0_{ij}T_{ijkl}$	$\boldsymbol{\sigma}^r = \boldsymbol{\sigma}^0 : \boldsymbol{T}$	$[\boldsymbol{\sigma}^r]^T = [\boldsymbol{\sigma}^0]^T \cdot \widetilde{\boldsymbol{T}}$
$\varepsilon^r_{ij} = \varepsilon^0_{ij}H_{ijkl}$	$\boldsymbol{\varepsilon}^r = \boldsymbol{\varepsilon}^0 : \boldsymbol{H}$	$[\widetilde{\boldsymbol{\varepsilon}}^r]^T = [\widetilde{\boldsymbol{\varepsilon}}^0]^T \cdot \breve{\boldsymbol{H}}$
$L_{ijkl} = \sigma_{ij}\sigma_{kl}$	$\boldsymbol{L} = \boldsymbol{\sigma} \otimes \boldsymbol{\sigma}$	$\boldsymbol{L} = \boldsymbol{\sigma} \cdot \boldsymbol{\sigma}^T$
$M_{ijkl} = \varepsilon_{ij}\varepsilon_{kl}$	$\boldsymbol{M} = \boldsymbol{\varepsilon} \otimes \boldsymbol{\varepsilon}$	$\widehat{\boldsymbol{M}} = \widetilde{\boldsymbol{\varepsilon}} \cdot \widetilde{\boldsymbol{\varepsilon}}^T$
$T_{ijkl} = \varepsilon_{ij}\sigma_{kl}$	$\boldsymbol{T} = \boldsymbol{\varepsilon} \otimes \boldsymbol{\sigma}$	$\widetilde{\boldsymbol{T}} = \widetilde{\boldsymbol{\varepsilon}} \cdot \boldsymbol{\sigma}^T$
$H_{ijkl} = \sigma_{ij}\varepsilon_{kl}$	$\boldsymbol{H} = \boldsymbol{\sigma} \otimes \boldsymbol{\varepsilon}$	$\breve{\boldsymbol{H}} = \boldsymbol{\sigma} \cdot \widetilde{\boldsymbol{\varepsilon}}^T$
$C_{ijkl} = A_{ijkl} + B_{ijkl}$	$\boldsymbol{C} = \boldsymbol{A} + \boldsymbol{B}$	$\overset{\#}{\boldsymbol{C}} = \overset{\#}{\boldsymbol{A}} + \overset{\#}{\boldsymbol{B}}$
$B_{ijkl} = cA_{ijkl}$	$\boldsymbol{B} = c\boldsymbol{A}$	$\overset{\#}{\boldsymbol{B}} = c\overset{\#}{\boldsymbol{A}}$
$L_{ijkl} = A_{ijmn}B_{mnkl}$	$\boldsymbol{L} = \boldsymbol{A} : \boldsymbol{B}$	$\boldsymbol{L} = \boldsymbol{A} \cdot \widetilde{\boldsymbol{B}} = \breve{\boldsymbol{A}} \cdot \boldsymbol{B}$
$M_{ijkl} = A_{ijmn}B_{mnkl}$	$\boldsymbol{M} = \boldsymbol{A} : \boldsymbol{B}$	$\widehat{\boldsymbol{M}} = \widetilde{\boldsymbol{A}} \cdot \hat{\boldsymbol{B}} = \hat{\boldsymbol{A}} \cdot \breve{\boldsymbol{B}}$
$T_{ijkl} = A_{ijmn}B_{mnkl}$	$\boldsymbol{T} = \boldsymbol{A} : \boldsymbol{B}$	$\widetilde{\boldsymbol{T}} = \widetilde{\boldsymbol{A}} \cdot \widetilde{\boldsymbol{B}} = \hat{\boldsymbol{A}} \cdot \boldsymbol{B}$
$H_{ijkl} = A_{ijmn}B_{mnkl}$	$\boldsymbol{H} = \boldsymbol{A} : \boldsymbol{B}$	$\breve{\boldsymbol{H}} = \breve{\boldsymbol{A}} \cdot \breve{\boldsymbol{B}} = \boldsymbol{A} \cdot \hat{\boldsymbol{B}}$
$L^T_{ijkl} = A_{ijkl}$	$\boldsymbol{L}^T = \boldsymbol{A}$	$\boldsymbol{L}^T = \boldsymbol{A}$
$M^T_{ijkl} = A_{ijkl}$	$\boldsymbol{M}^T = \boldsymbol{A}$	$\widehat{\boldsymbol{M}}^T = \hat{\boldsymbol{A}}$
$T^T_{ijkl} = A_{ijkl}$	$\boldsymbol{T}^T = \boldsymbol{A}$	$\widetilde{\boldsymbol{T}}^T = \breve{\boldsymbol{A}}$
$H^T_{ijkl} = A_{ijkl}$	$\boldsymbol{H}^T = \boldsymbol{A}$	$\breve{\boldsymbol{H}}^T = \widetilde{\boldsymbol{A}}$
$L^{-1}_{ijkl} = A_{ijkl}$	$\boldsymbol{L}^{-1} = \boldsymbol{A}$	$\boldsymbol{L}^{-1} = \hat{\boldsymbol{A}}$
$M^{-1}_{ijkl} = A_{ijkl}$	$\boldsymbol{M}^{-1} = \boldsymbol{A}$	$\widehat{\boldsymbol{M}}^{-1} = \boldsymbol{A}$
$T^{-1}_{ijkl} = A_{ijkl}$	$\boldsymbol{T}^{-1} = \boldsymbol{A}$	$\widetilde{\boldsymbol{T}}^{-1} = \widetilde{\boldsymbol{A}}$
$H^{-1}_{ijkl} = A_{ijkl}$	$\boldsymbol{H}^{-1} = \boldsymbol{A}$	$\breve{\boldsymbol{H}}^{-1} = \breve{\boldsymbol{A}}$

respectively. When $\boldsymbol{A}$ is expressed also as a 3×3 matrix, (1.9) can be rewritten in matrix form as

$$\boldsymbol{A}' = \boldsymbol{R}^T \cdot \boldsymbol{A} \cdot \boldsymbol{R}\,.$$

Although this matrix representation is convenient, extending it to fourth–order tensors is impossible as long as $\boldsymbol{R}$ appears in the 3×3 form.

For a symmetric second–order tensor $\boldsymbol{A}$, a rotation yields another symmetric second–order tensor $\boldsymbol{A}'$, and we can identify a Voigt fourth–order rotator tensor with minor symmetries that links the two second–order tensors through the relation

$$A'_{ij} = Q_{ijmn} A_{mn}\,, \qquad Q_{ijmn} = R_{mi} R_{nj}.$$

In the above expression, Q_{ijmn} denotes the rotation from global to local coordinates and is written in matrix form as

$$\boldsymbol{Q} = \begin{bmatrix} R_{11}^2 & R_{21}^2 & R_{31}^2 & R_{11}R_{21} & R_{11}R_{31} & R_{21}R_{31} \\ R_{12}^2 & R_{22}^2 & R_{32}^2 & R_{12}R_{22} & R_{12}R_{32} & R_{22}R_{32} \\ R_{13}^2 & R_{23}^2 & R_{33}^2 & R_{13}R_{23} & R_{13}R_{33} & R_{23}R_{33} \\ R_{11}R_{12} & R_{21}R_{22} & R_{31}R_{32} & \dfrac{R_{12}R_{21}+R_{11}R_{22}}{2} & \dfrac{R_{12}R_{31}+R_{11}R_{32}}{2} & \dfrac{R_{22}R_{31}+R_{21}R_{32}}{2} \\ R_{11}R_{13} & R_{21}R_{23} & R_{31}R_{33} & \dfrac{R_{13}R_{21}+R_{11}R_{23}}{2} & \dfrac{R_{13}R_{31}+R_{11}R_{33}}{2} & \dfrac{R_{23}R_{31}+R_{21}R_{33}}{2} \\ R_{12}R_{13} & R_{22}R_{23} & R_{32}R_{33} & \dfrac{R_{13}R_{22}+R_{12}R_{23}}{2} & \dfrac{R_{13}R_{32}+R_{12}R_{33}}{2} & \dfrac{R_{23}R_{32}+R_{22}R_{33}}{2} \end{bmatrix}.$$

Due to the orthogonality of $\boldsymbol{R}$, we can easily verify that

$$A_{ij} = Q_{mnij} A'_{mn}.$$

Using the properties of the fourth–order tensors identified previously, we can express a rotation of an "s" type second–order tensor in the Voigt notation as

$$\boldsymbol{\sigma}' = \check{\boldsymbol{Q}} \cdot \boldsymbol{\sigma}\,, \quad \boldsymbol{\sigma} = \widetilde{\boldsymbol{Q}}^T \cdot \boldsymbol{\sigma}'.$$

Similarly, we can express a rotation of an "e" type second–order tensor in the Voigt notation as

$$\widetilde{\boldsymbol{\varepsilon}}' = \widetilde{\boldsymbol{Q}} \cdot \widetilde{\boldsymbol{\varepsilon}}\,, \quad \widetilde{\boldsymbol{\varepsilon}} = \check{\boldsymbol{Q}}^T \cdot \widetilde{\boldsymbol{\varepsilon}}'.$$

With the help of $\boldsymbol{Q}$ and its various Voigt forms, four types of rotations for fourth–order tensors appear:

- For a fourth–order tensor $\boldsymbol{L}$ of "s–e" type ($\boldsymbol{\sigma} = \boldsymbol{L} \cdot \widetilde{\boldsymbol{\varepsilon}}$), we have

$$\boldsymbol{\sigma}' = \check{\boldsymbol{Q}} \cdot \boldsymbol{\sigma} = \check{\boldsymbol{Q}} \cdot \boldsymbol{L} \cdot \widetilde{\boldsymbol{\varepsilon}} = \check{\boldsymbol{Q}} \cdot \boldsymbol{L} \cdot \check{\boldsymbol{Q}}^T \cdot \widetilde{\boldsymbol{\varepsilon}}',$$
$$\boldsymbol{\sigma} = \widetilde{\boldsymbol{Q}}^T \cdot \boldsymbol{\sigma}' = \widetilde{\boldsymbol{Q}}^T \cdot \boldsymbol{L}' \cdot \widetilde{\boldsymbol{\varepsilon}}' = \widetilde{\boldsymbol{Q}}^T \cdot \boldsymbol{L}' \cdot \widetilde{\boldsymbol{Q}} \cdot \widetilde{\boldsymbol{\varepsilon}}.$$

These expressions lead to the conclusion that

$$\boldsymbol{L}' = \check{\boldsymbol{Q}} \cdot \boldsymbol{L} \cdot \check{\boldsymbol{Q}}^T \quad \text{and} \quad \boldsymbol{L} = \widetilde{\boldsymbol{Q}}^T \cdot \boldsymbol{L}' \cdot \widetilde{\boldsymbol{Q}}.$$

- For a fourth–order tensor $\widehat{\boldsymbol{M}}$ of "e–s" type ($\widetilde{\boldsymbol{\varepsilon}} = \widehat{\boldsymbol{M}} \cdot \boldsymbol{\sigma}$), we have

$$\widetilde{\boldsymbol{\varepsilon}}' = \widetilde{\boldsymbol{Q}} \cdot \widetilde{\boldsymbol{\varepsilon}} = \widetilde{\boldsymbol{Q}} \cdot \widehat{\boldsymbol{M}} \cdot \boldsymbol{\sigma} = \widetilde{\boldsymbol{Q}} \cdot \widehat{\boldsymbol{M}} \cdot \widetilde{\boldsymbol{Q}}^T \cdot \boldsymbol{\sigma}',$$
$$\widetilde{\boldsymbol{\varepsilon}} = \check{\boldsymbol{Q}}^T \cdot \widetilde{\boldsymbol{\varepsilon}}' = \check{\boldsymbol{Q}}^T \cdot \widehat{\boldsymbol{M}}' \cdot \boldsymbol{\sigma}' = \check{\boldsymbol{Q}}^T \cdot \widehat{\boldsymbol{M}}' \cdot \check{\boldsymbol{Q}} \cdot \boldsymbol{\sigma}.$$

These expressions lead to the conclusion that

$$\widehat{\boldsymbol{M}}' = \widetilde{\boldsymbol{Q}} \cdot \widehat{\boldsymbol{M}} \cdot \widetilde{\boldsymbol{Q}}^T \quad \text{and} \quad \widehat{\boldsymbol{M}} = \check{\boldsymbol{Q}}^T \cdot \widehat{\boldsymbol{M}}' \cdot \check{\boldsymbol{Q}}.$$

- For a fourth–order tensor $\widetilde{\boldsymbol{T}}$ of "e–e" type ($\widetilde{\boldsymbol{\varepsilon}}^r = \widetilde{\boldsymbol{T}} \cdot \widetilde{\boldsymbol{\varepsilon}}^0$), we have

$$\widetilde{\boldsymbol{\varepsilon}}^{r'} = \widetilde{\boldsymbol{Q}} \cdot \widetilde{\boldsymbol{\varepsilon}}^r = \widetilde{\boldsymbol{Q}} \cdot \widetilde{\boldsymbol{T}} \cdot \widetilde{\boldsymbol{\varepsilon}}^0 = \widetilde{\boldsymbol{Q}} \cdot \widetilde{\boldsymbol{T}} \cdot \check{\boldsymbol{Q}}^T \cdot \widetilde{\boldsymbol{\varepsilon}}^{0'},$$
$$\widetilde{\boldsymbol{\varepsilon}}^r = \check{\boldsymbol{Q}}^T \cdot \widetilde{\boldsymbol{\varepsilon}}^{r'} = \check{\boldsymbol{Q}}^T \cdot \widetilde{\boldsymbol{T}}' \cdot \widetilde{\boldsymbol{\varepsilon}}^{0'} = \check{\boldsymbol{Q}}^T \cdot \widetilde{\boldsymbol{T}}' \cdot \widetilde{\boldsymbol{Q}} \cdot \widetilde{\boldsymbol{\varepsilon}}^0.$$

These expressions lead to the conclusion that

$$\widetilde{\boldsymbol{T}}' = \widetilde{\boldsymbol{Q}} \cdot \widetilde{\boldsymbol{T}} \cdot \check{\boldsymbol{Q}}^T \quad \text{and} \quad \widetilde{\boldsymbol{T}} = \check{\boldsymbol{Q}}^T \cdot \widetilde{\boldsymbol{T}}' \cdot \widetilde{\boldsymbol{Q}}.$$

- For a fourth–order tensor $\check{\boldsymbol{H}}$ of "s–s" type ($\boldsymbol{\sigma}^r = \check{\boldsymbol{H}} \cdot \boldsymbol{\sigma}^0$), we have

$$\boldsymbol{\sigma}^{r'} = \check{\boldsymbol{Q}} \cdot \boldsymbol{\sigma}^r = \check{\boldsymbol{Q}} \cdot \check{\boldsymbol{H}} \cdot \boldsymbol{\sigma}^0 = \check{\boldsymbol{Q}} \cdot \check{\boldsymbol{H}} \cdot \widetilde{\boldsymbol{Q}}^T \cdot \boldsymbol{\sigma}^{0'},$$
$$\boldsymbol{\sigma}^r = \widetilde{\boldsymbol{Q}}^T \cdot \boldsymbol{\sigma}^{r'} = \widetilde{\boldsymbol{Q}}^T \cdot \check{\boldsymbol{H}}' \cdot \boldsymbol{\sigma}^{0'} = \widetilde{\boldsymbol{Q}}^T \cdot \check{\boldsymbol{H}}' \cdot \check{\boldsymbol{Q}} \cdot \boldsymbol{\sigma}^0.$$

These expressions lead to the conclusion that

$$\check{\boldsymbol{H}}' = \check{\boldsymbol{Q}} \cdot \check{\boldsymbol{H}} \cdot \widetilde{\boldsymbol{Q}}^T \quad \text{and} \quad \check{\boldsymbol{H}} = \widetilde{\boldsymbol{Q}}^T \cdot \check{\boldsymbol{H}}' \cdot \check{\boldsymbol{Q}}.$$

A summary of proper tensor rotations using the Voigt notation is given in Table 1.2.

TABLE 1.2 Rotation of second– and fourth–order tensors using the Voigt notation. $\boldsymbol{Q}$ is the Voigt 6×6 matrix rotator.

Tensor order	From global to local	From local to global
2	$\boldsymbol{\sigma}' = \check{\boldsymbol{Q}} \cdot \boldsymbol{\sigma}$	$\boldsymbol{\sigma} = \widetilde{\boldsymbol{Q}}^T \cdot \boldsymbol{\sigma}'$
2	$\widetilde{\boldsymbol{\varepsilon}}' = \widetilde{\boldsymbol{Q}} \cdot \widetilde{\boldsymbol{\varepsilon}}$	$\widetilde{\boldsymbol{\varepsilon}} = \check{\boldsymbol{Q}}^T \cdot \widetilde{\boldsymbol{\varepsilon}}'$
4	$\boldsymbol{L}' = \check{\boldsymbol{Q}} \cdot \boldsymbol{L} \cdot \check{\boldsymbol{Q}}^T$	$\boldsymbol{L} = \widetilde{\boldsymbol{Q}}^T \cdot \boldsymbol{L}' \cdot \widetilde{\boldsymbol{Q}}$
4	$\widehat{\boldsymbol{M}}' = \widetilde{\boldsymbol{Q}} \cdot \widehat{\boldsymbol{M}} \cdot \widetilde{\boldsymbol{Q}}^T$	$\widehat{\boldsymbol{M}} = \check{\boldsymbol{Q}}^T \cdot \widehat{\boldsymbol{M}}' \cdot \check{\boldsymbol{Q}}$
4	$\widetilde{\boldsymbol{T}}' = \widetilde{\boldsymbol{Q}} \cdot \widetilde{\boldsymbol{T}} \cdot \check{\boldsymbol{Q}}^T$	$\widetilde{\boldsymbol{T}} = \check{\boldsymbol{Q}}^T \cdot \widetilde{\boldsymbol{T}}' \cdot \widetilde{\boldsymbol{Q}}$
4	$\check{\boldsymbol{H}}' = \check{\boldsymbol{Q}} \cdot \check{\boldsymbol{H}} \cdot \widetilde{\boldsymbol{Q}}^T$	$\check{\boldsymbol{H}} = \widetilde{\boldsymbol{Q}}^T \cdot \check{\boldsymbol{H}}' \cdot \check{\boldsymbol{Q}}$

Reduction to 2–D

In plates and laminate structures the thickness is very small compared to the other two dimensions. This allows us, under certain conditions, to approximately represent all mechanical fields and equations using 2–D type of tensors. Specifically, setting the axis 3 as the thickness direction, the stress ("s" type) and strain ("e" type) tensors are expressed as

$$\boldsymbol{\sigma} = \begin{bmatrix} \sigma_{11} \\ \sigma_{22} \\ \sigma_{12} \end{bmatrix}, \qquad \widetilde{\boldsymbol{\varepsilon}} = \begin{bmatrix} \varepsilon_{11} \\ \varepsilon_{22} \\ 2\varepsilon_{12} \end{bmatrix}.$$

A fourth–order tensor $\boldsymbol{A}$ of "s–e" type should also appear in the reduced form

$$\boldsymbol{A} = \begin{bmatrix} A_{1111} & A_{1122} & A_{1112} \\ A_{2211} & A_{2222} & A_{2212} \\ A_{1211} & A_{1222} & A_{1212} \end{bmatrix}.$$

The other three types of $\boldsymbol{A}$ are produced in similar manner with the 3–D case:

$$\widetilde{\boldsymbol{A}} = \begin{bmatrix} A_{1111} & A_{1122} & A_{1112} \\ A_{2211} & A_{2222} & A_{2212} \\ 2A_{1211} & 2A_{1222} & 2A_{1212} \end{bmatrix},$$

$$\check{\boldsymbol{A}} = \begin{bmatrix} A_{1111} & A_{1122} & 2A_{1112} \\ A_{2211} & A_{2222} & 2A_{2212} \\ A_{1211} & A_{1222} & 2A_{1212} \end{bmatrix},$$

$$\widehat{\boldsymbol{A}} = \begin{bmatrix} A_{1111} & A_{1122} & 2A_{1112} \\ A_{2211} & A_{2222} & 2A_{2212} \\ 2A_{1211} & 2A_{1222} & 4A_{1212} \end{bmatrix}.$$

All the operations described in Table 1.1 hold for the 2–D tensors. Rotating these tensors in the plane of the structure requires a single angle and the use of the corresponding rotator, which in the classical notation is written as

$$\boldsymbol{R} = \begin{bmatrix} \cos\vartheta & -\sin\vartheta & 0 \\ \sin\vartheta & \cos\vartheta & 0 \\ 0 & 0 & 1 \end{bmatrix}.$$

The $\boldsymbol{R}$ in the Voigt notation for 2–D problems is written as the 3×3 matrix

$$\boldsymbol{Q} = \begin{bmatrix} \cos^2\vartheta & \sin^2\vartheta & \cos\vartheta\sin\vartheta \\ \sin^2\vartheta & \cos^2\vartheta & -\cos\vartheta\sin\vartheta \\ -\cos\vartheta\sin\vartheta & \cos\vartheta\sin\vartheta & \dfrac{\cos^2\vartheta - \sin^2\vartheta}{2} \end{bmatrix}.$$

Considering this reduced rotation matrix, all the relations presented in Table 1.2 hold for 2–D tensors.

1.8 Examples in Voigt notation operations

Example 1

Assume that $\boldsymbol{A}$ is a fourth–order tensor that connects two "e" type second–order tensors ($\varepsilon^1_{ij} = A_{ijkl}\varepsilon^2_{kl}$) and $\boldsymbol{L}$ and $\boldsymbol{L}^*$ are fourth–order tensors that connect "s" and "e" type second–order tensors ($\sigma^1_{ij} = L_{ijkl}\varepsilon^1_{kl}$ and $\sigma^2_{ij} = L^*_{ijkl}\varepsilon^2_{kl}$).

1. Identify the product

$$B_{ijkl} = L^*_{ijmn}A_{mnpq}L^{-1}_{pqkl} \tag{1.10}$$

using the Voigt notation.

2. Let the tensors $\boldsymbol{A}$, $\boldsymbol{L}$, and $\boldsymbol{L}^*$ in their classical matrix form be equal to

$$\boldsymbol{A} = \begin{bmatrix} 1.6656 & -0.0009 & 0.2148 & 0.0083 & 0.0106 & 0.0202 \\ -0.0009 & 1.6656 & 0.2148 & 0.0083 & 0.0106 & 0.0202 \\ 0 & 0 & 1 & 0 & 0 & 0 \\ 0.0075 & 0.0075 & 0 & 0.4166 & 0 & 0 \\ 0.0066 & 0.0066 & 0 & 0 & 0.4166 & 0 \\ 0.0098 & 0.0098 & 0 & 0 & 0 & 0.4166 \end{bmatrix},$$

$$\boldsymbol{L} = \begin{bmatrix} 344.08 & 329.46 & 285.62 & 0 & 0 & 0 \\ 329.46 & 344.08 & 285.62 & 0 & 0 & 0 \\ 285.62 & 285.62 & 42600.79 & 0 & 0 & 0 \\ 0 & 0 & 0 & 7.31 & 0 & 0 \\ 0 & 0 & 0 & 0 & 7.38 & 0 \\ 0 & 0 & 0 & 0 & 0 & 7.38 \end{bmatrix},$$

$$\boldsymbol{L}^* = \begin{bmatrix} 204.205 & 197.881 & 197.881 & 0 & 0 & 0 \\ 197.881 & 204.205 & 197.881 & 0 & 0 & 0 \\ 197.881 & 197.881 & 204.205 & 0 & 0 & 0 \\ 0 & 0 & 0 & 3.162 & 0 & 0 \\ 0 & 0 & 0 & 0 & 3.162 & 0 \\ 0 & 0 & 0 & 0 & 0 & 3.162 \end{bmatrix}.$$

Compute the components B_{1112}, B_{2322}, and B_{1313}.

Solution:

Taking into account the nature of the given fourth–order tensors, we can write the product of (1.10) in the Voigt notation as

$$\boldsymbol{L}^* \cdot \widetilde{\boldsymbol{A}} \cdot \boldsymbol{L}^{-1}.$$

According to the properties presented in Table 1.1, we have:

- $\boldsymbol{L}^{-1}$ renders to be of "e–s" type ($\widehat{\boldsymbol{M}}$).
- The product $\widetilde{\boldsymbol{A}} \cdot \widehat{\boldsymbol{M}}$ yields the "e–s" type $\widehat{\boldsymbol{C}}$.
- The product $\boldsymbol{L}^* \cdot \widehat{\boldsymbol{C}}$ yields the "s–s" type $\breve{\boldsymbol{B}}$.

As a conclusion, the Voigt notation transforms the tensorial relation (1.10) into the matrix–type relation

$$\breve{\boldsymbol{B}} = \boldsymbol{L}^* \cdot \widetilde{\boldsymbol{A}} \cdot \boldsymbol{L}^{-1}. \tag{1.11}$$

Concerning the second question, initially, we have to transform the tensor $\boldsymbol{A}$ into its "e–e" type form

$$\widetilde{\boldsymbol{A}} = \begin{bmatrix} 1.6656 & -0.0009 & 0.2148 & 0.0083 & 0.0106 & 0.0202 \\ -0.0009 & 1.6656 & 0.2148 & 0.0083 & 0.0106 & 0.0202 \\ 0 & 0 & 1 & 0 & 0 & 0 \\ 0.015 & 0.015 & 0 & 0.8332 & 0 & 0 \\ 0.0132 & 0.0132 & 0 & 0 & 0.8332 & 0 \\ 0.0196 & 0.0196 & 0 & 0 & 0 & 0.8332 \end{bmatrix}.$$

The Voigt notation product (1.11) yields

$$\breve{\boldsymbol{B}} = \begin{bmatrix} 0.857 & 0.137 & 9.56E-6 & 0.457 & 0.578 & 1.101 \\ 0.137 & 0.857 & 9.56E-6 & 0.457 & 0.578 & 1.101 \\ 0.489 & 0.489 & 2.32E-4 & 0.449 & 0.568 & 1.083 \\ 7.08E-5 & 7.08E-5 & -9.5E-7 & 0.36 & 0 & 0 \\ 6.23E-5 & 6.23E-5 & -8.36E-7 & 0 & 0.357 & 0 \\ 9.25E-5 & 9.25E-5 & -1.24E-6 & 0 & 0 & 0.357 \end{bmatrix}.$$

Thus the classical matrix form of $\boldsymbol{B}$ is

$$\boldsymbol{B}=\begin{bmatrix} 0.857 & 0.137 & 9.56E-6 & 0.229 & 0.289 & 0.551 \\ 0.137 & 0.857 & 9.56E-6 & 0.229 & 0.289 & 0.551 \\ 0.489 & 0.489 & 2.32E-4 & 0.225 & 0.284 & 0.542 \\ 7.08E-5 & 7.08E-5 & -9.5E-7 & 0.18 & 0 & 0 \\ 6.23E-5 & 6.23E-5 & -8.36E-7 & 0 & 0.179 & 0 \\ 9.25E-5 & 9.25E-5 & -1.24E-6 & 0 & 0 & 0.179 \end{bmatrix}.$$

In the matrix form of $\boldsymbol{B}$ the required components correspond to the elements B_{14}, B_{52}, and B_{55}. So

$$B_{1112}=0.229, \quad B_{2322}=9.25E-5, \quad B_{1313}=0.179.$$

Example 2

Consider two "s–e" type tensors $\boldsymbol{L}$ and $\boldsymbol{L}^*$ and an "e–e" type tensor $\widetilde{\boldsymbol{S}}$. Using the "e–e" type of the symmetric identity tensor $\widetilde{\boldsymbol{\mathcal{I}}}$ and operations in the Voigt notation, show that the fourth–order tensor

$$T_{ijkl}=\Big[\mathcal{I}_{ijkl}+S_{ijmn}L^{-1}_{mnpq}[L_{pqkl}-L^*_{pqkl}]\Big]^{-1} \tag{1.12}$$

is of "e–e" type.

Solution:

To solve this problem, we will use the properties presented in Table 1.1.

- Since $\boldsymbol{L}$ is of "s–e" type, its inverse $M_{mnpq}=L^{-1}_{mnpq}$, computed through classical matrix inversion, is of "e–s" type ($\widehat{\boldsymbol{M}}$).
- Since $\boldsymbol{L}$ and $\boldsymbol{L}^*$ are of "s–e" type, the difference $L^d_{pqkl}=L_{pqkl}-L^*_{pqkl}$, computed through classical matrix subtraction, is of "s–e" type too ($\boldsymbol{L}^d$).
- The product $\widehat{\boldsymbol{M}}\cdot\boldsymbol{L}^d$ yields the "e–e" type $\widetilde{\boldsymbol{C}}$.
- The product $\widetilde{\boldsymbol{S}}\cdot\widetilde{\boldsymbol{C}}$ yields the "e–e" type $\widetilde{\boldsymbol{D}}$.
- The sum $\widetilde{\boldsymbol{\mathcal{I}}}+\widetilde{\boldsymbol{D}}$ yields the "e–e" type $\widetilde{\boldsymbol{E}}$.
- The inverse $\widetilde{\boldsymbol{E}}^{-1}$ yields the "e–e" type $\widetilde{\boldsymbol{T}}$.

As a conclusion, the Voigt notation transforms the tensorial relation (1.12) into the matrix–type relation

$$\widetilde{\boldsymbol{T}}=\Big[\widetilde{\boldsymbol{\mathcal{I}}}+\widetilde{\boldsymbol{S}}\cdot\boldsymbol{L}^{-1}\cdot[\boldsymbol{L}-\boldsymbol{L}^*]\Big]^{-1}.$$

Example 3

In elastoplastic materials it is very common to introduce the notion of "tangent modulus", which is an algorithmic path–dependent fourth–order tensor that defines a relation between the increments of total stress and total strain. A typical formula for calculating the tangent modulus $\boldsymbol{D}$ of a J_2 elastoplastic material with linear isotropic hardening is the following:

$$D_{ijkl} = \frac{1}{a} L_{ijmn} \Lambda_{mn} L_{klpq} \Lambda_{pq} , \qquad a = \Lambda_{rs} L_{rstu} \Lambda_{tu} + H, \tag{1.13}$$

where $\boldsymbol{L}$ is the fourth–order elasticity tensor, $\boldsymbol{\Lambda}$ is the second–order plastic flow direction tensor, and H is the hardening related scalar.

Using the Voigt notation, express (1.13) in matrix form.

Solution:

Both $\boldsymbol{D}$ and $\boldsymbol{L}$ are "s–e" type fourth–order tensors. Starting by the triple product $\Lambda_{rs} L_{rstu} \Lambda_{tu}$, we should notice that the tensor $\boldsymbol{\Lambda}$ should appear in "e" type form for having a proper matrix multiplication. Indeed, with the help of Table 1.1, we observe that:

- Using the "e" type form of $\widetilde{\boldsymbol{\Lambda}}$, the tensor $Z_{rs} = L_{rstu} \Lambda_{tu}$ is written as $\boldsymbol{Z} = \boldsymbol{L} \cdot \widetilde{\boldsymbol{\Lambda}}$ and is of "s" type.
- The double contraction between $\boldsymbol{\Lambda}$ and $\boldsymbol{Z}$ can be expressed as $\widetilde{\boldsymbol{\Lambda}}^T \cdot \boldsymbol{Z}$.

With regard to the tensorial product $L_{ijmn} \Lambda_{mn} L_{klpq} \Lambda_{pq}$ of (1.13), we can substitute it with the dyadic product $Z_{ij} Z_{kl}$. Since $\boldsymbol{Z}$ is of "s" type, the operation $\boldsymbol{Z} \cdot \boldsymbol{Z}^T$ gives an "s–e" type fourth–order tensor. Thus we conclude that $\boldsymbol{D}$ can be computed from the relation

$$\boldsymbol{D} = \frac{1}{a} \boldsymbol{Z} \cdot \boldsymbol{Z}^T , \qquad a = \left[\widetilde{\boldsymbol{\Lambda}}^T \cdot \boldsymbol{Z} + H \right] , \qquad \boldsymbol{Z} = \boldsymbol{L} \cdot \widetilde{\boldsymbol{\Lambda}}.$$

References

[1] W.M. Lai, D. Rubin, E. Krempl, Introduction to continuum mechanics, 4th ed., Elsevier, Burlington, 2010.

[2] P.G. Ciarlet, An Introduction to Differential Geometry with Applications to Elasticity, Journal of Elasticity 78-79 (1–3) (2005) 1–215.

[3] P. Steinmann, On boundary potential energies in deformational and configurational mechanics, Journal of the Mechanics and Physics of Solids 56 (3) (2008) 772–800.

[4] P. Steinmann, Geometrical Foundations of Continuum Mechanics: An Application to First– and Second–Order Elasticity and Elasto–Plasticity, Springer-Verlag, Berlin, 2015.

[5] R. Hill, Continuum micro–mechanics of elastoplastic polycrystals, Journal of the Mechanics and Physics of Solids 13 (1965) 89–101.

[6] G. Chatzigeorgiou, N. Charalambakis, Y. Chemisky, F. Meraghni, Thermomechanical Behavior of Dissipative Composite Materials, ISTE Press – Elsevier, London, 2018.

CHAPTER

2

Continuum mechanics

OUTLINE

2.1 Strain

To understand the notion of strains, let us consider a continuum body, which initially (at time t_0) occupies the space $\mathcal{B}_0$. Under mechanical loading, the body moves and deforms, and at time t, it occupies the space $\mathcal{B}$. The movement of any material point Q with respect to a fixed point O (arbitrary origin of coordinates) can be described in two ways: with a) a position vector $\boldsymbol{X}$ of the initial configuration and b) the position vector $\boldsymbol{x}$ of the final configuration (Fig. 2.1). Whereas $\boldsymbol{X}$ is fixed for every material point, $\boldsymbol{x}$ changes with time. Both vectors refer to Cartesian coordinate systems and are connected through the (generally, nonlinear) relations

$$x_i := x_i(\boldsymbol{X}, t) \tag{2.1}$$

Multiscale Modeling Approaches for Composites
https://doi.org/10.1016/B978-0-12-823143-2.00011-4

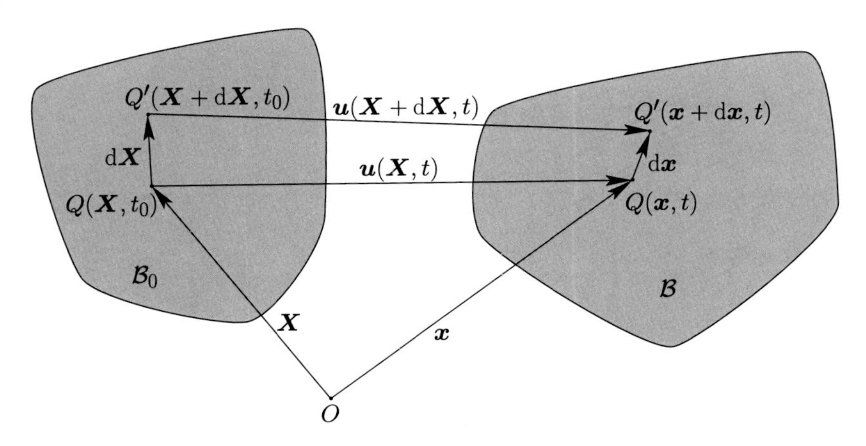

FIGURE 2.1 Continuum body movement and deformation due to mechanical loading.

for $i = 1, 2, 3$ when referring to 3–D spaces. The position change of the point Q can be defined with the help of the displacement vector $\boldsymbol{u}$, whose spatial components are given by the expression

$$u_i(\boldsymbol{X}, t) = x_i(\boldsymbol{X}, t) - X_i. \tag{2.2}$$

Furthermore, let Q' be a neighbor point of Q with infinitesimal distance $\mathrm{d}\boldsymbol{X}$ at time t_0. At time t, the mechanical loading causes Q' to move to a distance $\mathrm{d}\boldsymbol{x}$ from $\boldsymbol{X}$. The change of the distance vector $\mathrm{d}\boldsymbol{X}$ to $\mathrm{d}\boldsymbol{x}$ is expressed in a differential form by the indicial relation

$$\mathrm{d}x_i = \frac{\partial x_i}{\partial X_j}\mathrm{d}X_j = \left[\delta_{ij} + \frac{\partial u_i}{\partial X_j}\right]\mathrm{d}X_j = F_{ij}\mathrm{d}X_j, \tag{2.3}$$

where the second–order tensor $\boldsymbol{F}$ is called the deformation gradient. Note that under specific loading conditions, the movement and deformation of the body are unique, which means that there is a one–to–one mapping between $\boldsymbol{X}$ and $\boldsymbol{x}$. The difference of the lengths of the vectors $\mathrm{d}\boldsymbol{X}$ and $\mathrm{d}\boldsymbol{x}$ is given by

$$\begin{aligned} \mathrm{d}x_i\mathrm{d}x_i - \mathrm{d}X_i\mathrm{d}X_i &= [F_{ij}\mathrm{d}X_j][F_{ik}\mathrm{d}X_k] - \mathrm{d}X_i\mathrm{d}X_i \\ &= \mathrm{d}X_j[F_{ij}F_{ik} - \delta_{jk}]\mathrm{d}X_k = 2\mathrm{d}X_jE_{jk}\mathrm{d}X_k, \end{aligned} \tag{2.4}$$

where $\boldsymbol{E}$ is the second–order Green–Lagrange strain tensor. This tensor has the following properties:

1. It is zero when rigid body motion occurs. A rigid body motion can be either a translation ($F_{ij} = \delta_{ij}$) or a rotation ($F_{ij}F_{ik} = \delta_{jk}$). Both choices of $\boldsymbol{F}$ yield $E_{ij} = 0$.

2. It can be determined exclusively by the displacement vector $\boldsymbol{u}$. Indeed, from Eqs. (2.2) and (2.3) we can obtain

$$
\begin{aligned}
E_{ij} &= \frac{1}{2}\left[F_{ki}F_{kj} - \delta_{ij}\right] \\
&= \frac{1}{2}\left[\left[\delta_{ki} + \frac{\partial u_k}{\partial X_i}\right]\left[\delta_{kj} + \frac{\partial u_k}{\partial X_j}\right] - \delta_{ij}\right] \\
&= \frac{1}{2}\left[\frac{\partial u_i}{\partial X_j} + \frac{\partial u_j}{\partial X_i} + \frac{\partial u_k}{\partial X_i}\frac{\partial u_k}{\partial X_j}\right].
\end{aligned}
\tag{2.5}
$$

In the framework of small deformations and small rotations, the quadratic term

$$\frac{\partial u_k}{\partial X_i}\frac{\partial u_k}{\partial X_j}$$

is very small compared to the first–order terms and can be ignored. In addition, the position changes can be considered insignificant, allowing us to avoid distinction between $\boldsymbol{x}$ and $\boldsymbol{X}$. Thus we can utilize only one position vector, for instance, $\boldsymbol{x}$. With this in mind, the following definition follows:

BOX 2.1 Strain tensor

The infinitesimal strain tensor $\boldsymbol{\varepsilon}$ is of second order, and its components are defined by the relations

$$\varepsilon_{ij} = \frac{1}{2}\left[\frac{\partial u_i}{\partial x_j} + \frac{\partial u_j}{\partial x_i}\right]. \tag{2.6}$$

For the rest of this book, the term "strain" will refer to the infinitesimal strain tensor. From its definition it is clear that $\boldsymbol{\varepsilon}$ is symmetric. Thus, as already discussed in the previous chapter, it can be represented as a 3×3 matrix

$$\boldsymbol{\varepsilon} = \begin{bmatrix} \varepsilon_{11} & \varepsilon_{12} & \varepsilon_{13} \\ \varepsilon_{12} & \varepsilon_{22} & \varepsilon_{23} \\ \varepsilon_{13} & \varepsilon_{23} & \varepsilon_{33} \end{bmatrix}$$

or as a 6×1 vector in the Voigt notation

$$\widetilde{\boldsymbol{\varepsilon}} = \begin{bmatrix} \varepsilon_{11} \\ \varepsilon_{22} \\ \varepsilon_{33} \\ 2\varepsilon_{12} \\ 2\varepsilon_{13} \\ 2\varepsilon_{23} \end{bmatrix}.$$

Geometrically, the six components of the strain denote the following:

- ε_{11} is the change in length divided by the initial length of an infinitesimal element originally in the x_1 direction,
- ε_{22} is the change in length divided by the initial length of an infinitesimal element originally in the x_2 direction,
- ε_{33} is the change in length divided by the initial length of an infinitesimal element originally in the x_3 direction,
- $2\varepsilon_{12}$ is the change in angle between two infinitesimal elements originally in the x_1 and x_2 directions,
- $2\varepsilon_{13}$ is the change in angle between two infinitesimal elements originally in the x_1 and x_3 directions,
- $2\varepsilon_{23}$ is the change in angle between two infinitesimal elements originally in the x_2 and x_3 directions.

Using definition (2.6) of the strain tensor, the following compatibility conditions hold:

$$\frac{\partial^2 \varepsilon_{ij}}{\partial x_k \partial x_l} + \frac{\partial^2 \varepsilon_{kl}}{\partial x_i \partial x_j} = \frac{\partial^2 \varepsilon_{ik}}{\partial x_j \partial x_l} + \frac{\partial^2 \varepsilon_{jl}}{\partial x_i \partial x_k}, \quad i, j, k, l = 1, 2, 3. \tag{2.7}$$

2.2 Stress

Consider the body of Fig. 2.2. A plane S having unit normal vector $\boldsymbol{n}$ crosses the body and passes through an arbitrary internal point Q. This plane separates the medium into two portions, denoted as I and II in

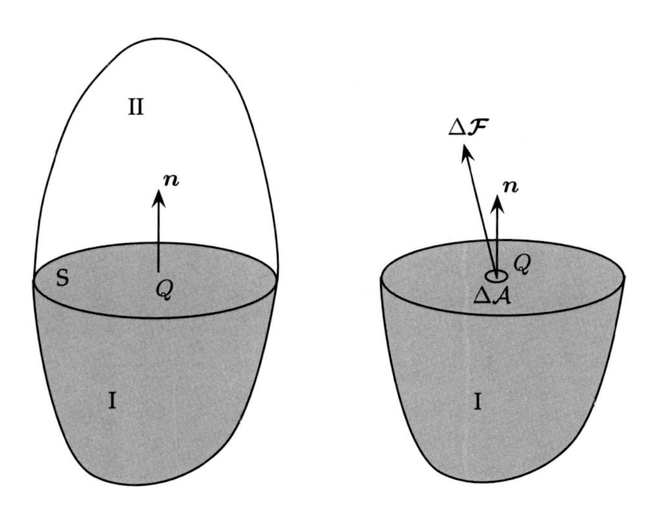

FIGURE 2.2 Continuum body under mechanical loading: a surface cut and the reaction force passing through a point Q of the surface S.

Fig. 2.2. Considering the portion I, on the plane S, there exists a resultant force $\Delta\mathcal{F}$ acting on a small area $\Delta\mathcal{A}$ containing Q. The traction vector $\boldsymbol{t}$ acting from II to I at the point Q on the plane S is defined as the limit of the ratio $\Delta\mathcal{F}/\Delta\mathcal{A}$ as $\Delta\mathcal{A}$ becomes infinitesimally small,

$$\boldsymbol{t} = \lim_{\Delta\mathcal{A}\to 0} \frac{\Delta\mathcal{F}}{\Delta\mathcal{A}}. \tag{2.8}$$

The stress principle of Cauchy states the following [1]: The traction vector at any given place and time has a common value on all parts of material having a common tangent plane at Q and lying on the same side of it. In other words, if $\boldsymbol{n}$ is the unit outward normal (i.e., a vector of unit length pointing outward, away from the material) to the tangent plane, then

$$\boldsymbol{t} := \boldsymbol{t}(\boldsymbol{x}, t, \boldsymbol{n}),$$

where the scalar t denotes time.

BOX 2.2 Stress tensor

The dependence of the traction vector on the outward normal vector $\boldsymbol{n}$ can be expressed in indicial notation with the help of a linear transformation $\boldsymbol{\sigma}$ as

$$t_i = \sigma_{ij} n_j. \tag{2.9}$$

The second order tensor $\boldsymbol{\sigma}$ is called Cauchy's stress tensor.

For the rest of the book, the term "stress" will refer to Cauchy's stress tensor.

For a continuum body, the conservation principles should hold. Specifically, under the absence of body (or volume) moments, the conservation of angular momentum leads to the conclusion that the stress is a symmetric second–order tensor [1,2]. Thus it can be represented either as a 3×3 matrix

$$\boldsymbol{\sigma} = \begin{bmatrix} \sigma_{11} & \sigma_{12} & \sigma_{13} \\ \sigma_{12} & \sigma_{22} & \sigma_{23} \\ \sigma_{13} & \sigma_{23} & \sigma_{33} \end{bmatrix}$$

or as a 6×1 vector in the Voigt notation

$$\boldsymbol{\sigma} = \begin{bmatrix} \sigma_{11} \\ \sigma_{22} \\ \sigma_{33} \\ \sigma_{12} \\ \sigma_{13} \\ \sigma_{23} \end{bmatrix}.$$

In addition, the conservation of linear momentum, ignoring inertia effects and body forces, yields the equilibrium equations [1,2], which in indicial notation are written as

$$\frac{\partial \sigma_{ij}}{\partial x_j} = 0. \tag{2.10}$$

2.3 Elasticity

2.3.1 General aspects

Linear elasticity is the simplest case of material behavior for engineers. It represents fully reversible material response, which is characterized by linear relation between kinetics (stress) and kinematics (strain) variables. The mechanical state of an elastic material is uniquely described by the boundary conditions at which it is subjected, without prior knowledge of the loading path: for a known state of strain (or stress), the material point experiences a specific state of stress (or strain).

When dealing with large deformation processes, the elastic response may present certain nonlinearity. For the scope of the current chapter, the discussion is limited to small deformation and small rotation processes, in which the elastic materials are always considered to exhibit elastic behavior.

The elastic materials in the 3–D space are described from a constitutive relation between the stress and strain second–order tensors. This relation is defined with the help of a unique scalar potential W, denoting the elastic energy. In fact, the stress is given as the derivative of the energy potential with respect to strain following the expression

$$\sigma_{ij} = \frac{\partial W}{\partial \varepsilon_{ij}}. \tag{2.11}$$

An appropriate form of this potential considering small deformations can be found as follows [3]: In a purely elastic material the elastic energy W can be written in a Taylor series expanded form

$$W = L^0 + L^1_{ij}\varepsilon_{ij} + \frac{1}{2}\varepsilon_{ij}L_{ijkl}\varepsilon_{kl} + \cdots,$$

where L^0 is a constant, L^1_{ij} is a second–order tensor, and L_{ijkl} is a fourth–order tensor. Owing to the assumption of small deformations, higher–order terms on strain can be neglected. The constant L^0 can be arbitrarily selected equal to 0, whereas $\boldsymbol{L}^1$ can also be zero, as long as there are no pre–stress conditions inside the material. Thus the only remaining term is quadratic. This, combined with (2.11), leads to the following:

BOX 2.3 Elasticity tensor

In a linear elastic medium, the strain tensor $\boldsymbol{\varepsilon}$ and the stress tensor $\boldsymbol{\sigma}$ are connected with the help of the fourth–order elasticity tensor $\boldsymbol{L}$ according to the formula

$$\boldsymbol{\sigma} = \boldsymbol{L} : \boldsymbol{\varepsilon} \quad \text{or} \quad \sigma_{ij} = L_{ijkl}\varepsilon_{kl}. \tag{2.12}$$

The latter expression, which is a generalization of the Hooke law in the 3–D space, and the existence of a unique W provide valuable information about the structure of the elasticity tensor:

- Due to the symmetry of the stress tensor ($\sigma_{ij} = \sigma_{ji}$), $\boldsymbol{L}$ should present minor symmetry in the first two indices, that is, $L_{ijkl} = L_{jikl}$.
- Due to the symmetry of the strain tensor ($\varepsilon_{ij} = \varepsilon_{ji}$), Eq. (2.12) can be written as

$$\sigma_{ij} = \frac{1}{2}\left[L_{ijkl} + L_{ijlk}\right]\varepsilon_{kl}.$$

  Since there is no real experiment that allows the distinction of L_{ijkl} and L_{ijlk}, a reasonable assumption is that the $\boldsymbol{L}$ tensor exhibits also minor symmetry in the last two indices, that is, $L_{ijkl} = L_{ijlk}$.
- The uniqueness of the elastic energy potential

$$W = \frac{1}{2}\varepsilon_{ij}L_{ijkl}\varepsilon_{kl}$$

  clearly indicates major symmetry in $\boldsymbol{L}$, allowing interchange between the first and the last two indices, that is, $L_{ijkl} = L_{klij}$.

In conclusion, a real elastic material under small deformation processes is described by a fourth–order elasticity tensor that exhibits minor and major symmetries,

$$L_{ijkl} = L_{jikl} = L_{ijlk} = L_{klij}.$$

Due to the latter result, the elasticity tensor can have only 21 independent material coefficients: the 81 components of a general fourth–order tensor are reduced to 36 from the minor symmetries and to 21 when the major symmetry is also accounted. The 21 independent coefficients, whose units are similar to those of the stress field (i.e., force per area), can be further reduced if special symmetries inside the material microstructure exist. We examine such cases in the next subsection.

2.3.2 Special symmetries

As already discussed, the fourth–order elasticity tensor of real materials possesses minor and major symmetries. This allows us to write such a tensor in the Voigt form of a 6×6 symmetric matrix

$$\boldsymbol{L} = \begin{bmatrix} L_{11} & L_{12} & L_{13} & L_{14} & L_{15} & L_{16} \\ L_{12} & L_{22} & L_{23} & L_{24} & L_{25} & L_{26} \\ L_{13} & L_{23} & L_{33} & L_{34} & L_{35} & L_{36} \\ L_{14} & L_{24} & L_{34} & L_{44} & L_{45} & L_{46} \\ L_{15} & L_{25} & L_{35} & L_{45} & L_{55} & L_{56} \\ L_{16} & L_{26} & L_{36} & L_{46} & L_{56} & L_{66} \end{bmatrix}. \tag{2.13}$$

The form (2.13) is the most general case, representing a fully anisotropic material with 21 independent material parameters. The materials appearing in nature and utilized in the engineering applications very frequently present symmetries that reduce the number of independent parameters [1].

2.3.2.1 *Monoclinic materials*

Let us assume that the geometrical characteristics of a material microstructure cause the appearance of a plane of symmetry. Without loss of generality, we can consider that such plane of symmetry has normal axis parallel to the x_3 direction. In such case, it becomes clear that the elasticity tensor of a material point does not change under full rotation from the x_3 direction to the $-x_3$ direction. In other words, a change of the basis

$$\text{from} \quad \begin{bmatrix} \boldsymbol{e}_1 \\ \boldsymbol{e}_2 \\ \boldsymbol{e}_3 \end{bmatrix} \quad \text{to} \quad \begin{bmatrix} \boldsymbol{e}_1 \\ \boldsymbol{e}_2 \\ -\boldsymbol{e}_3 \end{bmatrix},$$

corresponding to the rotation tensor $\boldsymbol{R}^3$ (or the rotator tensor $\check{\boldsymbol{Q}}^3$ of "s–s" type in the Voigt notation) with components

$$\boldsymbol{R}^3 = \begin{bmatrix} 1 & 0 & 0 \\ 0 & 1 & 0 \\ 0 & 0 & -1 \end{bmatrix} \quad \text{or} \quad \check{\boldsymbol{Q}}^3 = \begin{bmatrix} 1 & 0 & 0 & 0 & 0 & 0 \\ 0 & 1 & 0 & 0 & 0 & 0 \\ 0 & 0 & 1 & 0 & 0 & 0 \\ 0 & 0 & 0 & 1 & 0 & 0 \\ 0 & 0 & 0 & 0 & -1 & 0 \\ 0 & 0 & 0 & 0 & 0 & -1 \end{bmatrix}$$

leaves the elasticity tensor unchanged, that is,

$$\boldsymbol{L}' = \check{\boldsymbol{Q}}^3 \cdot \boldsymbol{L} \cdot \left[\check{\boldsymbol{Q}}^3\right]^T = \boldsymbol{L}.$$

The rotation leads to the form

$$\boldsymbol{L}' = \begin{bmatrix} L_{11} & L_{12} & L_{13} & L_{14} & -L_{15} & -L_{16} \\ L_{12} & L_{22} & L_{23} & L_{24} & -L_{25} & -L_{26} \\ L_{13} & L_{23} & L_{33} & L_{34} & -L_{35} & -L_{36} \\ L_{14} & L_{24} & L_{34} & L_{44} & -L_{45} & -L_{46} \\ -L_{15} & -L_{25} & -L_{35} & -L_{45} & L_{55} & L_{56} \\ -L_{16} & -L_{26} & -L_{36} & -L_{46} & L_{56} & L_{66} \end{bmatrix}.$$

Since $\boldsymbol{L}' = \boldsymbol{L}$ and $\boldsymbol{L}$ is given by the general expression (2.13), the component L_{15} should vanish:

$$-L_{15} = L_{15} \quad \text{or} \quad L_{15} = 0.$$

Following the same procedure for the rest of the components, we conclude that:

BOX 2.4 Monoclinic material

A monoclinic material has 13 independent nonzero properties. When the plane of material symmetry has normal axis parallel to the x_3 direction, the elasticity tensor is written in the Voigt notation as

$$\boldsymbol{L}_{\text{mcl}} = \begin{bmatrix} L_{11} & L_{12} & L_{13} & L_{14} & 0 & 0 \\ L_{12} & L_{22} & L_{23} & L_{24} & 0 & 0 \\ L_{13} & L_{23} & L_{33} & L_{34} & 0 & 0 \\ L_{14} & L_{24} & L_{34} & L_{44} & 0 & 0 \\ 0 & 0 & 0 & 0 & L_{55} & L_{56} \\ 0 & 0 & 0 & 0 & L_{56} & L_{66} \end{bmatrix}. \qquad (2.14)$$

2.3.2.2 *Orthotropic materials*

Let us assume that the geometrical characteristics of a material microstructure cause the appearance of three planes of symmetry. Without loss of generality, we can consider that the planes of symmetry have normal axes parallel to the x_1, x_2, x_3 directions. This orthotropic material is already monoclinic, with two additional symmetric planes. In this case the elasticity tensor of a material point does not change under full rotation from the x_1 direction to the $-x_1$ direction. In other words, a change of the basis

$$\text{from} \quad \begin{bmatrix} \boldsymbol{e}_1 \\ \boldsymbol{e}_2 \\ \boldsymbol{e}_3 \end{bmatrix} \quad \text{to} \quad \begin{bmatrix} -\boldsymbol{e}_1 \\ \boldsymbol{e}_2 \\ \boldsymbol{e}_3 \end{bmatrix},$$

corresponding to the rotation tensor $\boldsymbol{R}^1$ (or the rotator tensor $\check{\boldsymbol{Q}}^1$ of "s–s" type in the Voigt notation) with components

$$\boldsymbol{R}^1 = \begin{bmatrix} -1 & 0 & 0 \\ 0 & 1 & 0 \\ 0 & 0 & 1 \end{bmatrix} \quad \text{or} \quad \check{\boldsymbol{Q}}^1 = \begin{bmatrix} 1 & 0 & 0 & 0 & 0 & 0 \\ 0 & 1 & 0 & 0 & 0 & 0 \\ 0 & 0 & 1 & 0 & 0 & 0 \\ 0 & 0 & 0 & -1 & 0 & 0 \\ 0 & 0 & 0 & 0 & -1 & 0 \\ 0 & 0 & 0 & 0 & 0 & 1 \end{bmatrix},$$

leaves the elasticity tensor unchanged, that is,

$$\boldsymbol{L}'_{\text{mcl}} = \check{\boldsymbol{Q}}^1 \cdot \boldsymbol{L}_{\text{mcl}} \cdot \left[\check{\boldsymbol{Q}}^1\right]^T = \boldsymbol{L}_{\text{mcl}}.$$

The rotation leads to the form

$$\boldsymbol{L}'_{\text{mcl}} = \begin{bmatrix} L_{11} & L_{12} & L_{13} & -L_{14} & 0 & 0 \\ L_{12} & L_{22} & L_{23} & -L_{24} & 0 & 0 \\ L_{13} & L_{23} & L_{33} & -L_{34} & 0 & 0 \\ -L_{14} & -L_{24} & -L_{34} & L_{44} & 0 & 0 \\ 0 & 0 & 0 & 0 & L_{55} & -L_{56} \\ 0 & 0 & 0 & 0 & -L_{56} & L_{66} \end{bmatrix}.$$

A full rotation from the x_2 direction to the $-x_2$ direction yields the same result. Since $\boldsymbol{L}'_{\text{mcl}} = \boldsymbol{L}_{\text{mcl}}$ and $\boldsymbol{L}_{\text{mcl}}$ is given by expression (2.14), the component L_{14} should vanish:

$$-L_{14} = L_{14} \quad \text{or} \quad L_{14} = 0.$$

Following the same procedure for the rest of the components, we conclude that:

BOX 2.5 Orthotropic material

An orthotropic material has 9 independent nonzero properties, and its elasticity tensor is written in the Voigt notation as

$$\boldsymbol{L}_{\text{orth}} = \begin{bmatrix} L_{11} & L_{12} & L_{13} & 0 & 0 & 0 \\ L_{12} & L_{22} & L_{23} & 0 & 0 & 0 \\ L_{13} & L_{23} & L_{33} & 0 & 0 & 0 \\ 0 & 0 & 0 & L_{44} & 0 & 0 \\ 0 & 0 & 0 & 0 & L_{55} & 0 \\ 0 & 0 & 0 & 0 & 0 & L_{66} \end{bmatrix}. \tag{2.15}$$

The compliance tensor (inverse of elasticity tensor) of an orthotropic material can be expressed in terms of the three Young moduli E_1, E_2, E_3, the three shear moduli μ_{12}, μ_{13}, μ_{23}, and the three Poisson ratios ν_{12}, ν_{31}, ν_{23} in the Voigt form:

$$\widehat{\boldsymbol{M}}_{\text{orth}} = \boldsymbol{L}_{\text{orth}}^{-1} = \begin{bmatrix} \dfrac{1}{E_1} & -\dfrac{\nu_{12}}{E_1} & -\dfrac{\nu_{31}}{E_3} & 0 & 0 & 0 \\ -\dfrac{\nu_{12}}{E_1} & \dfrac{1}{E_2} & -\dfrac{\nu_{23}}{E_2} & 0 & 0 & 0 \\ -\dfrac{\nu_{31}}{E_3} & -\dfrac{\nu_{23}}{E_2} & \dfrac{1}{E_3} & 0 & 0 & 0 \\ 0 & 0 & 0 & \dfrac{1}{\mu_{12}} & 0 & 0 \\ 0 & 0 & 0 & 0 & \dfrac{1}{\mu_{13}} & 0 \\ 0 & 0 & 0 & 0 & 0 & \dfrac{1}{\mu_{23}} \end{bmatrix}. \tag{2.16}$$

The three alternative Poisson ratios ν_{21}, ν_{13}, ν_{32} can be also utilized. These material parameters satisfy the relations

$$\frac{\nu_{21}}{E_2} = \frac{\nu_{12}}{E_1}, \quad \frac{\nu_{13}}{E_1} = \frac{\nu_{31}}{E_3}, \quad \frac{\nu_{32}}{E_3} = \frac{\nu_{23}}{E_2}. \tag{2.17}$$

Orthotropic materials appear frequently in nature (the wood is a characteristic example).

2.3.2.3 *Transversely isotropic materials*

Transversely isotropic materials are orthotropic materials with material symmetry at every direction inside a specific plane. Assuming as plane of isotropy the one with normal axis parallel to the x_3 direction, the elasticity tensor of a material point does not change under an arbitrary rotation ϑ from the x_3 direction. In other words, a change of the basis

$$\text{from} \quad \begin{bmatrix} \boldsymbol{e}_1 \\ \boldsymbol{e}_2 \\ \boldsymbol{e}_3 \end{bmatrix} \quad \text{to} \quad \begin{bmatrix} \cos\vartheta\ \boldsymbol{e}_1 + \sin\vartheta\ \boldsymbol{e}_2 \\ -\sin\vartheta\ \boldsymbol{e}_1 + \cos\vartheta\ \boldsymbol{e}_2 \\ \boldsymbol{e}_3 \end{bmatrix},$$

corresponding to the rotation tensor $\boldsymbol{R}^{\vartheta}$ (or the rotator tensor $\breve{\boldsymbol{Q}}^{\vartheta}$ of "s–s" type in the Voigt notation) with components

$$\boldsymbol{R}^{\vartheta} = \begin{bmatrix} c & -s & 0 \\ s & c & 0 \\ 0 & 0 & 1 \end{bmatrix} \quad \text{or} \quad \breve{\boldsymbol{Q}}^{\vartheta} = \begin{bmatrix} c^2 & s^2 & 0 & 2cs & 0 & 0 \\ s^2 & c^2 & 0 & -2cs & 0 & 0 \\ 0 & 0 & 1 & 0 & 0 & 0 \\ -cs & cs & 0 & c^2 - s^2 & 0 & 0 \\ 0 & 0 & 0 & 0 & c & s \\ 0 & 0 & 0 & 0 & -s & c \end{bmatrix},$$

where $c = \cos\vartheta$ and $s = \sin\vartheta$, leaves the elasticity tensor unchanged, that is,

$$\boldsymbol{L}'_{\text{orth}} = \breve{\boldsymbol{Q}}^{\vartheta} \cdot \boldsymbol{L}_{\text{orth}} \cdot \left[\breve{\boldsymbol{Q}}^{\vartheta}\right]^T = \boldsymbol{L}_{\text{orth}}.$$

For $\vartheta = \pi/3$, the rotation leads to the form

$$\boldsymbol{L}'_{\text{orth}} = \begin{bmatrix} L'_{11} & L'_{12} & L'_{13} & L'_{14} & 0 & 0 \\ L'_{12} & L'_{22} & L'_{23} & L'_{24} & 0 & 0 \\ L'_{13} & L'_{23} & L'_{33} & L'_{34} & 0 & 0 \\ L'_{14} & L'_{24} & L'_{34} & L'_{44} & 0 & 0 \\ 0 & 0 & 0 & 0 & L'_{55} & L'_{56} \\ 0 & 0 & 0 & 0 & L'_{56} & L'_{66} \end{bmatrix}$$

with

$$\begin{aligned}
L'_{11} &= \frac{1}{16}[L_{11} + 6L_{12} + 9L_{22} + 12L_{44}], \\
L'_{12} &= \frac{1}{16}[3L_{11} + 10L_{12} + 3L_{22} - 12L_{44}], \\
L'_{14} &= -\frac{\sqrt{3}}{16}[L_{11} + 2L_{12} - 3L_{22} + 4L_{44}], \\
L'_{22} &= \frac{1}{16}[9L_{11} + 6L_{12} + L_{22} + 12L_{44}], \\
L'_{24} &= \frac{\sqrt{3}}{16}[-3L_{11} + 2L_{12} + L_{22} + 4L_{44}], \\
L'_{44} &= \frac{1}{16}[3L_{11} - 6L_{12} + 3L_{22} + 4L_{44}], \\
L'_{13} &= \frac{1}{4}[L_{13} + 3L_{23}], \quad L'_{23} = \frac{1}{4}[3L_{13} + L_{23}], \\
L'_{33} &= L_{33}, \quad L'_{34} = \frac{\sqrt{3}}{4}[-L_{13} + L_{23}], \\
L'_{55} &= \frac{1}{4}[L_{55} + 3L_{66}], \quad L'_{56} = \frac{\sqrt{3}}{4}[-L_{55} + 3L_{66}], \\
L'_{66} &= \frac{1}{4}[3L_{55} + L_{66}].
\end{aligned}$$

The equality $\boldsymbol{L}'_{\text{orth}} = \boldsymbol{L}_{\text{orth}}$, considering that $\boldsymbol{L}_{\text{orth}}$ is given by expression (2.15), is satisfied only when

$$L_{22} = L_{11}, \quad L_{23} = L_{13}, \quad L_{66} = L_{55}, \quad L_{12} = L_{11} - 2L_{44}.$$

When the latter relations hold, the elasticity tensor remains unchanged upon rotation at any other angle ϑ inside the plane $x_1 - x_2$. Thus we conclude that:

BOX 2.6 Transversely isotropic material

A transversely isotropic material has five independent nonzero properties. When the plane of symmetry has normal axis parallel to the x_3 direction, its elasticity tensor is written in the Voigt notation as

$$\boldsymbol{L}_{\text{tris}} = \begin{bmatrix} L_{11} & L_{11} - 2L_{44} & L_{13} & 0 & 0 & 0 \\ L_{11} - 2L_{44} & L_{11} & L_{13} & 0 & 0 & 0 \\ L_{13} & L_{13} & L_{33} & 0 & 0 & 0 \\ 0 & 0 & 0 & L_{44} & 0 & 0 \\ 0 & 0 & 0 & 0 & L_{55} & 0 \\ 0 & 0 & 0 & 0 & 0 & L_{55} \end{bmatrix}. \tag{2.18}$$

The elasticity tensor of a transversely isotropic material is often expressed in terms of the transverse bulk modulus K^{tr}, the transverse shear modulus μ^{tr}, the axial shear modulus μ^{ax}, and the stiffness terms $l = L_{13}$ and $n = L_{33}$ in the form [4]

$$\boldsymbol{L}_{\text{tris}} = \begin{bmatrix} K^{\text{tr}} + \mu^{\text{tr}} & K^{\text{tr}} - \mu^{\text{tr}} & l & 0 & 0 & 0 \\ K^{\text{tr}} - \mu^{\text{tr}} & K^{\text{tr}} + \mu^{\text{tr}} & l & 0 & 0 & 0 \\ l & l & n & 0 & 0 & 0 \\ 0 & 0 & 0 & \mu^{\text{tr}} & 0 & 0 \\ 0 & 0 & 0 & 0 & \mu^{\text{ax}} & 0 \\ 0 & 0 & 0 & 0 & 0 & \mu^{\text{ax}} \end{bmatrix}. \tag{2.19}$$

Alternatively, the compliance tensor (inverse of elasticity tensor) can be written in the Voigt form as

$$\widehat{\boldsymbol{M}}_{\text{tris}} = \boldsymbol{L}_{\text{tris}}^{-1} = \begin{bmatrix} \dfrac{1}{E^{\text{tr}}} & -\dfrac{\nu^{\text{tt}}}{E^{\text{tr}}} & -\dfrac{\nu^{\text{ta}}}{E^{\text{tr}}} & 0 & 0 & 0 \\ -\dfrac{\nu^{\text{tt}}}{E^{\text{tr}}} & \dfrac{1}{E^{\text{tr}}} & -\dfrac{\nu^{\text{at}}}{E^{\text{ax}}} & 0 & 0 & 0 \\ -\dfrac{\nu^{\text{ta}}}{E^{\text{tr}}} & -\dfrac{\nu^{\text{at}}}{E^{\text{ax}}} & \dfrac{1}{E^{\text{ax}}} & 0 & 0 & 0 \\ 0 & 0 & 0 & \dfrac{2[1+\nu^{\text{tt}}]}{E^{\text{tr}}} & 0 & 0 \\ 0 & 0 & 0 & 0 & \dfrac{1}{\mu^{\text{ax}}} & 0 \\ 0 & 0 & 0 & 0 & 0 & \dfrac{1}{\mu^{\text{ax}}} \end{bmatrix} \tag{2.20}$$

with

$$\frac{\nu^{\text{ta}}}{E^{\text{tr}}} = \frac{\nu^{\text{at}}}{E^{\text{ax}}}. \tag{2.21}$$

Moreover, E^{ax}, E^{tr}, ν^{at}, ν^{ta}, and ν^{tt} are the axial Young modulus, the transverse Young modulus, the first axial Poisson ratio, the second axial Poisson ratio, and the transverse Poisson ratio, respectively.

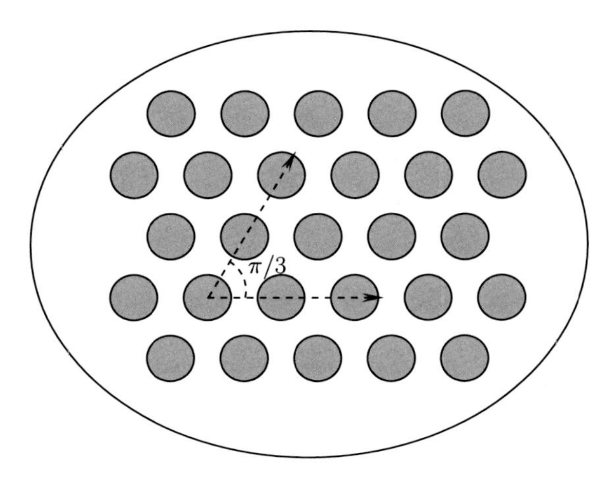

FIGURE 2.3 Schematic representation of unidirectional long fiber composite cross section. The fibers are distributed in hexagonal packing inside the matrix, providing structural material symmetry at angle $\pi/3$.

It is worth noticing that material symmetry in a plane upon rotation at angle $\vartheta = \pi/3$ is sufficient to guarantee the transverse isotropy response. Such symmetry appears in long fiber composites with hexagonal packing of the fibers (Fig. 2.3). When material symmetry is not satisfied at $\vartheta = \pi/3$ but appears at angle $\vartheta = \pi/2$ (as it is the case of long fiber composites with square packing of the fibers, Fig. 2.4), then following analogous steps with the previous derivation, we can conclude that the produced elasticity tensor has six independent nonzero properties and takes the form

$$\boldsymbol{L}_{\text{tetr}} = \begin{bmatrix} L_{11} & L_{12} & L_{13} & 0 & 0 & 0 \\ L_{12} & L_{11} & L_{13} & 0 & 0 & 0 \\ L_{13} & L_{13} & L_{33} & 0 & 0 & 0 \\ 0 & 0 & 0 & L_{44} & 0 & 0 \\ 0 & 0 & 0 & 0 & L_{55} & 0 \\ 0 & 0 & 0 & 0 & 0 & L_{55} \end{bmatrix}. \tag{2.22}$$

As a final remark, it should be mentioned that in several applications the fibers in the unidirectional composites are considered parallel to the

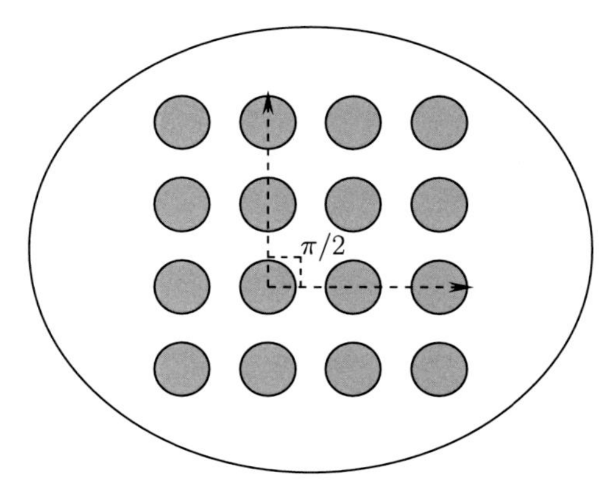

FIGURE 2.4 Schematic representation of unidirectional long fiber composite cross section. The fibers are distributed in square packing inside the matrix, providing structural material symmetry at angle $\pi/2$.

x_1 direction. The transversely isotropic elasticity tensor for these media is expressed through a proper rotation of expression (2.18) and takes the form

$$\boldsymbol{L}^{x_1}_{\text{tris}} = \begin{bmatrix} L_{11} & L_{12} & L_{12} & 0 & 0 & 0 \\ L_{12} & L_{22} & L_{22}-2L_{66} & 0 & 0 & 0 \\ L_{12} & L_{22}-2L_{66} & L_{22} & 0 & 0 & 0 \\ 0 & 0 & 0 & L_{44} & 0 & 0 \\ 0 & 0 & 0 & 0 & L_{44} & 0 \\ 0 & 0 & 0 & 0 & 0 & L_{66} \end{bmatrix}.$$

2.3.2.4 Isotropic materials

One of the most common materials (especially, in engineering structures) is the so–called isotropic material. Isotropic solids present symmetric response at any arbitrary direction. From a mathematical point of view, they can be seen as transversely isotropic materials with three planes of isotropy. If in a transversely isotropic material, it is considered that, in addition, there is a plane of isotropy with normal axis parallel to the x_1 direction, then the elasticity tensor of a material point does not change under an arbitrary rotation φ from the x_1 direction. In other words, a change of the basis

$$\text{from} \quad \begin{bmatrix} \boldsymbol{e}_1 \\ \boldsymbol{e}_2 \\ \boldsymbol{e}_3 \end{bmatrix} \quad \text{to} \quad \begin{bmatrix} \boldsymbol{e}_1 \\ \cos\varphi\, \boldsymbol{e}_2 + \sin\varphi\, \boldsymbol{e}_3 \\ -\sin\varphi\, \boldsymbol{e}_2 + \cos\varphi\, \boldsymbol{e}_3 \end{bmatrix},$$

corresponding to the rotation tensor $\boldsymbol{R}^{\varphi}$ (or the rotator tensor $\check{\boldsymbol{Q}}^{\varphi}$ of "s–s" type in the Voigt notation) with components

$$\boldsymbol{R}^{\varphi} = \begin{bmatrix} 1 & 0 & 0 \\ 0 & c & -s \\ 0 & s & c \end{bmatrix} \quad \text{or} \quad \check{\boldsymbol{Q}}^{\varphi} = \begin{bmatrix} 1 & 0 & 0 & 0 & 0 & 0 \\ 0 & c^2 & s^2 & 0 & 0 & 2cs \\ 0 & s^2 & c^2 & 0 & 0 & -2cs \\ 0 & 0 & 0 & c & s & 0 \\ 0 & 0 & 0 & -s & c & 0 \\ 0 & -cs & cs & 0 & 0 & c^2 - s^2 \end{bmatrix},$$

where $c = \cos\varphi$ and $s = \sin\varphi$, leaves the elasticity tensor unchanged, that is,

$$\boldsymbol{L}'_{\text{tris}} = \check{\boldsymbol{Q}}^{\varphi} \cdot \boldsymbol{L}_{\text{tris}} \cdot \left[\check{\boldsymbol{Q}}^{\varphi}\right]^T = \boldsymbol{L}_{\text{tris}}.$$

Choosing $\varphi = \pi/2$, the latter rotation leads to the form

$$\boldsymbol{L}'_{\text{tris}} = \begin{bmatrix} L_{11} & L_{13} & L_{11} - 2L_{44} & 0 & 0 & 0 \\ L_{13} & L_{33} & L_{13} & 0 & 0 & 0 \\ L_{11} - 2L_{44} & L_{13} & L_{11} & 0 & 0 & 0 \\ 0 & 0 & 0 & L_{55} & 0 & 0 \\ 0 & 0 & 0 & 0 & L_{44} & 0 \\ 0 & 0 & 0 & 0 & 0 & L_{55} \end{bmatrix}.$$

To satisfy the equality $\boldsymbol{L}'_{\text{tris}} = \boldsymbol{L}_{\text{tris}}$, it is sufficient and necessary to have

$$L_{33} = L_{11}, \quad L_{55} = L_{44}, \quad L_{13} = L_{11} - 2L_{44}.$$

For any other choice of angle φ, we obtain the same result. Moreover, assuming as the plane of isotropy the one with normal axis parallel to the x_2 direction yields the same conclusion.

BOX 2.7 Isotropic material

An isotropic material has two independent nonzero properties, and its elasticity tensor is written in the Voigt notation as

$$\boldsymbol{L}_{\text{iso}} = \begin{bmatrix} L_{11} & L_{11} - 2L_{44} & L_{11} - 2L_{44} & 0 & 0 & 0 \\ L_{11} - 2L_{44} & L_{11} & L_{11} - 2L_{44} & 0 & 0 & 0 \\ L_{11} - 2L_{44} & L_{11} - 2L_{44} & L_{11} & 0 & 0 & 0 \\ 0 & 0 & 0 & L_{44} & 0 & 0 \\ 0 & 0 & 0 & 0 & L_{44} & 0 \\ 0 & 0 & 0 & 0 & 0 & L_{44} \end{bmatrix}. \tag{2.23}$$

Isotropic materials are used very often in engineering applications (steel and epoxy resin are two characteristic examples). An isotropic material is usually represented in one of the following three ways:

1. In terms of the bulk modulus K and the shear modulus μ, the elasticity tensor is expressed in the Voigt form as

$$\boldsymbol{L}_{\text{iso}} = \begin{bmatrix} K+\frac{4}{3}\mu & K-\frac{2}{3}\mu & K-\frac{2}{3}\mu & 0 & 0 & 0 \\ K-\frac{2}{3}\mu & K+\frac{4}{3}\mu & K-\frac{2}{3}\mu & 0 & 0 & 0 \\ K-\frac{2}{3}\mu & K-\frac{2}{3}\mu & K+\frac{4}{3}\mu & 0 & 0 & 0 \\ 0 & 0 & 0 & \mu & 0 & 0 \\ 0 & 0 & 0 & 0 & \mu & 0 \\ 0 & 0 & 0 & 0 & 0 & \mu \end{bmatrix}. \tag{2.24}$$

Using the special Hill notation (Example 2 in Section 1.4 of Chapter 1), we can write $\boldsymbol{L}_{\text{iso}}$ in the compact form

$$\boldsymbol{L}_{\text{iso}} = (3K, 2\mu). \tag{2.25}$$

2. In terms of the two Lamé parameters λ and μ, the elasticity tensor is expressed in the Voigt form as

$$\boldsymbol{L}_{\text{iso}} = \begin{bmatrix} \lambda+2\mu & \lambda & \lambda & 0 & 0 & 0 \\ \lambda & \lambda+2\mu & \lambda & 0 & 0 & 0 \\ \lambda & \lambda & \lambda+2\mu & 0 & 0 & 0 \\ 0 & 0 & 0 & \mu & 0 & 0 \\ 0 & 0 & 0 & 0 & \mu & 0 \\ 0 & 0 & 0 & 0 & 0 & \mu \end{bmatrix}. \tag{2.26}$$

3. In terms of the Young modulus E and the Poisson ratio ν, the compliance tensor (inverse of elasticity tensor) is expressed in the Voigt form

as

$$\widehat{\boldsymbol{M}}_{\text{iso}} = \boldsymbol{L}_{\text{iso}}^{-1} = \begin{bmatrix} \frac{1}{E} & -\frac{\nu}{E} & -\frac{\nu}{E} & 0 & 0 & 0 \\ -\frac{\nu}{E} & \frac{1}{E} & -\frac{\nu}{E} & 0 & 0 & 0 \\ -\frac{\nu}{E} & -\frac{\nu}{E} & \frac{1}{E} & 0 & 0 & 0 \\ 0 & 0 & 0 & \frac{2[1+\nu]}{E} & 0 & 0 \\ 0 & 0 & 0 & 0 & \frac{2[1+\nu]}{E} & 0 \\ 0 & 0 & 0 & 0 & 0 & \frac{2[1+\nu]}{E} \end{bmatrix}. \tag{2.27}$$

2.4 Reduction to 2–D problems

In several engineering applications the structures under investigation exhibit characteristics that allow the reduction of the implicated equations from the general 3–D form to a simplified 2–D form. Dams, laminate structures, and thin plates are illustrative examples in which certain stress/strain components can be neglected or the displacement field can be identified through approximate functions.

Ignoring inertia effects and body forces, the governing equation for an anisotropic material reads

$$\frac{\partial}{\partial x_j}\left(L_{ijkl}\varepsilon_{kl}\right) = 0, \tag{2.28}$$

where expressions (2.10) and (2.12) are combined. As already mentioned, the elasticity tensor $\boldsymbol{L}$ connects the stress and strain tensors. In the Voigt notation the above expression takes the form

$$\mathfrak{L}\left(\boldsymbol{L}\cdot\widetilde{\boldsymbol{\varepsilon}}\right) = \boldsymbol{0}, \tag{2.29}$$

where $\boldsymbol{L}$ and $\widetilde{\boldsymbol{\varepsilon}}$ are the Voigt forms of the corresponding tensors, and $\mathfrak{L}$ is the differential operator

$$\mathfrak{L} = \begin{bmatrix} \frac{\partial}{\partial x_1} & 0 & 0 & \frac{\partial}{\partial x_2} & \frac{\partial}{\partial x_3} & 0 \\ 0 & \frac{\partial}{\partial x_2} & 0 & \frac{\partial}{\partial x_1} & 0 & \frac{\partial}{\partial x_3} \\ 0 & 0 & \frac{\partial}{\partial x_3} & 0 & \frac{\partial}{\partial x_1} & \frac{\partial}{\partial x_2} \end{bmatrix}. \tag{2.30}$$

For practical applications, solutions of system (2.29) under appropriate boundary conditions can be obtained through finite element computations. Analytical calculations of the above partial differential equations can be achieved only for particular cases, which are useful for comparison and parameter identification purposes.

Planar approximations of (2.29) are realistic in cases of bodies that can be simulated by two–dimensional idealizations, such as beams and plates. Two of the most common planar approximations include the so–called plane strain and plane stress conditions. Plane strain is often considered for thick structures with applied tractions vertically to the axis of zero normal strain, whereas plane stress is associated with thin plates experiencing stress in their middle plane (Fig. 2.5). Note that a structure is considered thin when its thickness h is much smaller that the other two dimensions l_1 and l_2, that is,

$$h \ll \min(l_1, l_2).$$

Table 2.1 summarizes the planar assumptions in orthotropic media for the two types of conditions, classified according to the variables neglected in each case.

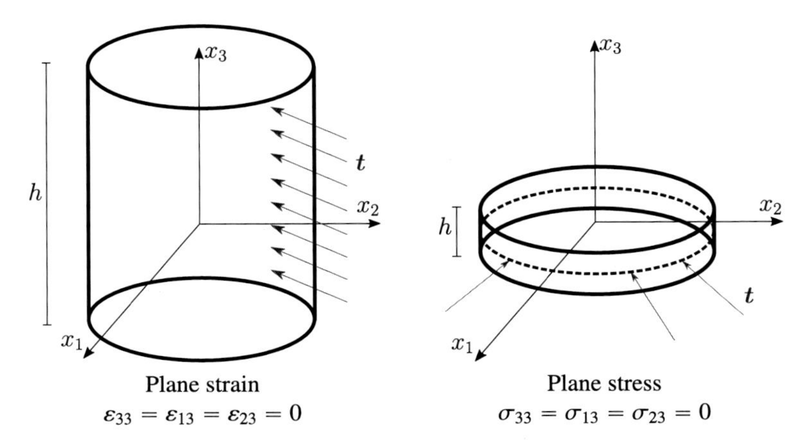

FIGURE 2.5 Plane strain and stress conditions. h denotes the thickness, and $\boldsymbol{t}$ represents the applied traction vector.

2.4.1 Plane strain

Structures like dams (with sufficient length and major loads applied vertically to the length direction) can be studied using simplified theories. If the long axis of such a structure is parallel to the x_3 direction, then it is expected that any cross section lying on the $x_1 - x_2$ plane remains practically unmovable in the x_3 direction, leading to zero normal (ε_{33}) and shear

TABLE 2.1 Stress and strain fields at planar problems for orthotropic media. Zero values declare fields that are ignored, and empty slots denote nonzero fields [5].

	Plane strain	Plane stress
$\varepsilon_{11}, \varepsilon_{22}, \varepsilon_{12}$		
$\varepsilon_{13}, \varepsilon_{23}$	0	0
ε_{33}	0	
$\sigma_{11}, \sigma_{22}, \sigma_{12}$		
σ_{13}, σ_{23}	0	0
σ_{33}		0

($\varepsilon_{13}, \varepsilon_{23}$) strain components. Thus the nonzero strain components form the 2–D type tensor (written in the Voigt notation)

$$\widetilde{\boldsymbol{\varepsilon}} = \begin{bmatrix} \varepsilon_{11} \\ \varepsilon_{22} \\ 2\varepsilon_{12} \end{bmatrix}.$$

It is worth noting that in anisotropic materials, plane strain deformation assumption is applicable only if certain material symmetry requirements are met. These requirements are discussed below.

When plane strain conditions are considered, the displacement field is described by the components

$$u_1 \equiv u_1(x_1, x_2), \quad u_2 \equiv u_2(x_1, x_2). \tag{2.31}$$

Expressing the strains in terms of the displacement vector and using hypothesis (2.31), we write system (2.29) [6] as

$$\begin{aligned} & L_{11}\frac{\partial^2 u_1}{\partial x_1^2} + L_{44}\frac{\partial^2 u_1}{\partial x_2^2} + 2L_{14}\frac{\partial^2 u_1}{\partial x_1 \partial x_2} \\ & + L_{14}\frac{\partial^2 u_2}{\partial x_1^2} + L_{24}\frac{\partial^2 u_2}{\partial x_2^2} + [L_{12} + L_{44}]\frac{\partial^2 u_2}{\partial x_1 \partial x_2} = 0, \end{aligned} \tag{2.32}$$

$$\begin{aligned} & L_{14}\frac{\partial^2 u_1}{\partial x_1^2} + L_{24}\frac{\partial^2 u_1}{\partial x_2^2} + [L_{12} + L_{44}]\frac{\partial^2 u_1}{\partial x_1 \partial x_2} \\ & + L_{44}\frac{\partial^2 u_2}{\partial x_1^2} + L_{22}\frac{\partial^2 u_2}{\partial x_2^2} + 2L_{24}\frac{\partial^2 u_2}{\partial x_1 \partial x_2} = 0, \end{aligned} \tag{2.33}$$

$$\begin{aligned} & L_{15}\frac{\partial^2 u_1}{\partial x_1^2} + L_{46}\frac{\partial^2 u_1}{\partial x_2^2} + [L_{16} + L_{45}]\frac{\partial^2 u_1}{\partial x_1 \partial x_2} \\ & + L_{45}\frac{\partial^2 u_2}{\partial x_1^2} + L_{26}\frac{\partial^2 u_2}{\partial x_2^2} + [L_{25} + L_{46}]\frac{\partial^2 u_2}{\partial x_1 \partial x_2} = 0. \end{aligned} \tag{2.34}$$

This system of three partial differential equations with two unknown functions (u_1 and u_2) has a solution only if one of the equations is an identity. Since L_{11}, L_{22}, and L_{44} cannot be zero for thermodynamics reasons, only the coefficients of the last equation can vanish. Indeed, when

$$L_{15} = L_{16} = L_{25} = L_{26} = L_{45} = L_{46} = 0, \tag{2.35}$$

expression (2.34) is automatically satisfied. Such a requirement is met in the case of monoclinic materials with plane of symmetry normal to the x_3 direction (expression (2.14)).

Under the above simplifications, the stress tensor is computed from Eqs. (2.12)–(2.13) and takes the reduced Voigt form

$$\boldsymbol{\sigma} = \begin{bmatrix} \sigma_{11} \\ \sigma_{22} \\ \sigma_{12} \end{bmatrix} = \begin{bmatrix} L_{11} & L_{12} & L_{14} \\ L_{12} & L_{22} & L_{24} \\ L_{14} & L_{24} & L_{44} \end{bmatrix} \cdot \begin{bmatrix} \varepsilon_{11} \\ \varepsilon_{22} \\ 2\varepsilon_{12} \end{bmatrix}. \tag{2.36}$$

The out–of–plane normal stress component σ_{33} is given by the separate expression

$$\sigma_{33} = L_{13}\varepsilon_{11} + L_{23}\varepsilon_{22} + 2L_{34}\varepsilon_{12}. \tag{2.37}$$

The out–of–plane shear stresses σ_{13} and σ_{23} are usually ignored. For monoclinic materials with plane of symmetry normal to the x_3 direction, these terms are exactly zero, as Eq. (2.14) indicates.

2.4.2 Plane stress

Thin plates subjected to in–plane loading conditions usually present almost zero off plane stresses ($\sigma_{33} = \sigma_{13} = \sigma_{23} = 0$). Under in–plane stress conditions, it is convenient to express the constitutive law by using the compliance tensor $\widehat{\boldsymbol{M}} = \boldsymbol{L}^{-1}$. For anisotropic materials of the form (2.13), the strain–stress relation can be expressed in the reduced Voigt form

$$\begin{bmatrix} \varepsilon_{11} \\ \varepsilon_{22} \\ 2\varepsilon_{12} \end{bmatrix} = \begin{bmatrix} M_{11} & M_{12} & 2M_{14} \\ M_{12} & M_{22} & 2M_{24} \\ 2M_{14} & 2M_{24} & 4M_{44} \end{bmatrix} \cdot \begin{bmatrix} \sigma_{11} \\ \sigma_{22} \\ \sigma_{12} \end{bmatrix}. \tag{2.38}$$

The out–of–plane strain components ε_{33}, ε_{13}, and ε_{23} are given by the separate relations

$$\begin{aligned} \varepsilon_{33} &= M_{13}\sigma_{11} + M_{23}\sigma_{22} + 2M_{34}\sigma_{12}, \\ \varepsilon_{13} &= 2M_{15}\sigma_{11} + 2M_{25}\sigma_{22} + 4M_{45}\sigma_{12}, \\ \varepsilon_{23} &= 2M_{16}\sigma_{11} + 2M_{26}\sigma_{22} + 4M_{46}\sigma_{12}. \end{aligned} \tag{2.39}$$

The last two expressions are usually ignored. For monoclinic materials with plane of symmetry normal to the x_3 direction, these terms are exactly zero, as the inverse of matrix (2.14) indicates.

Under the above simplifications, the equilibrium system (2.29) is reduced to the two equations

$$\frac{\partial \sigma_{11}}{\partial x_1} + \frac{\partial \sigma_{12}}{\partial x_2} = 0, \qquad \frac{\partial \sigma_{12}}{\partial x_1} + \frac{\partial \sigma_{22}}{\partial x_2} = 0. \tag{2.40}$$

Since derivatives with respect to x_3 do not appear, system (2.40) implies that the stress tensor is a function only of the x_1 and x_2 coordinates. Keeping this in mind, only one of the strain compatibility conditions (2.7) is important, specifically,

$$\frac{\partial^2 \varepsilon_{11}}{\partial x_2^2} + \frac{\partial^2 \varepsilon_{22}}{\partial x_1^2} = 2\frac{\partial^2 \varepsilon_{12}}{\partial x_1 \partial x_2}. \tag{2.41}$$

Solving system (2.40)–(2.41) under appropriate boundary conditions is not a trivial task [7]. By introducing the stress function Φ such that

$$\sigma_{11} = \frac{\partial^2 \Phi}{\partial x_2^2}, \qquad \sigma_{22} = \frac{\partial^2 \Phi}{\partial x_1^2}, \qquad \sigma_{12} = -\frac{\partial^2 \Phi}{\partial x_1 \partial x_2}, \tag{2.42}$$

the equilibrium equations (2.40) are satisfied automatically, and the remaining expression (2.41) yields a fourth–order partial differential equation with the only unknown function Φ:

$$\boldsymbol{Z} \cdot \begin{bmatrix} M_{11} & M_{12} & 2M_{14} \\ M_{12} & M_{22} & 2M_{24} \\ 2M_{14} & 2M_{24} & 4M_{44} \end{bmatrix} \cdot \boldsymbol{Z}^T(\Phi) = 0, \tag{2.43}$$

where $\boldsymbol{Z}$ is the differential operator

$$\boldsymbol{Z} = \begin{bmatrix} \dfrac{\partial^2}{\partial x_2^2} & \dfrac{\partial^2}{\partial x_1^2} & -\dfrac{\partial^2}{\partial x_1 \partial x_2} \end{bmatrix}. \tag{2.44}$$

Expression (2.43) is supplemented by boundary conditions for the derivatives of Φ. The error by this approximation is considered to be small.

2.5 Examples

Example 1: known displacement field in 3–D body

Consider a body made of linear elastic solid with Lamé constants λ and μ, which is deformed under mechanical loading. If the displacement field on the body is given by

$$\begin{bmatrix} u_1 \\ u_2 \\ u_3 \end{bmatrix} = \begin{bmatrix} a\left[x_1^2 - x_2^2\right] - 2bx_1x_2 \\ b\left[x_2^2 - x_1^2\right] - 2ax_1x_2 \\ cx_3 \end{bmatrix}$$

and no body forces are applied, then i) identify the strain tensor, ii) identify the stress tensor, and iii) illustrate that the body is in equilibrium.

Solution:

For the first question, use of (2.6) yields

$$\begin{bmatrix} \varepsilon_{11} \\ \varepsilon_{22} \\ \varepsilon_{33} \\ 2\varepsilon_{12} \\ 2\varepsilon_{13} \\ 2\varepsilon_{23} \end{bmatrix} = \begin{bmatrix} 2ax_1 - 2bx_2 \\ -2ax_1 + 2bx_2 \\ c \\ -4bx_1 - 4ax_2 \\ 0 \\ 0 \end{bmatrix}.$$

For the second question, the Hooke law (2.12) and the elasticity tensor form (2.26) give

$$\begin{bmatrix} \sigma_{11} \\ \sigma_{22} \\ \sigma_{33} \\ \sigma_{12} \\ \sigma_{13} \\ \sigma_{23} \end{bmatrix} = \begin{bmatrix} 4a\mu x_1 - 4b\mu x_2 + c\lambda \\ -4a\mu x_1 + 4b\mu x_2 + c\lambda \\ c[\lambda + 2\mu] \\ -4b\mu x_1 - 4a\mu x_2 \\ 0 \\ 0 \end{bmatrix}.$$

Concerning the last question, the equilibrium equations (2.10) read

$$\frac{\partial \sigma_{11}}{\partial x_1} + \frac{\partial \sigma_{12}}{\partial x_2} + \frac{\partial \sigma_{13}}{\partial x_3} = 4a\mu - 4a\mu = 0,$$

$$\frac{\partial \sigma_{12}}{\partial x_1} + \frac{\partial \sigma_{22}}{\partial x_2} + \frac{\partial \sigma_{23}}{\partial x_3} = 4b\mu - 4b\mu = 0,$$

$$\frac{\partial \sigma_{13}}{\partial x_1} + \frac{\partial \sigma_{23}}{\partial x_2} + \frac{\partial \sigma_{33}}{\partial x_3} = 0 - 0 = 0.$$

Thus the body is in equilibrium.

Example 2: thin cantilever beam under plane stress conditions

Consider a cantilever beam of length L, width b, and height h, made of an isotropic material with Young modulus E and Poisson ratio ν, which

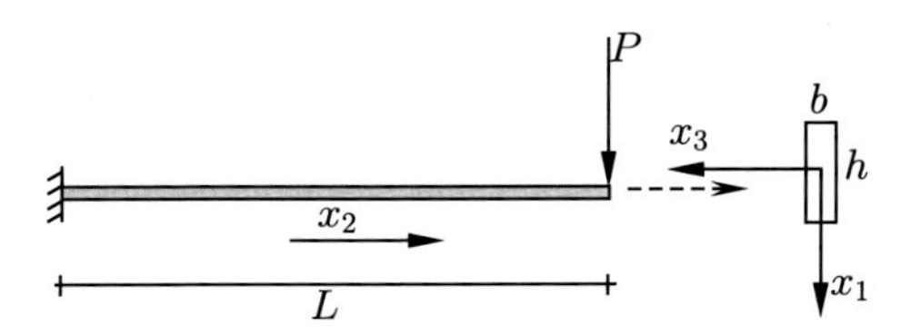

FIGURE 2.6 Schematic of cantilever beam and its cross section. The beam is loaded with force at the free end. Hypothesis of plane stress conditions.

is loaded with force P at the free end (Fig. 2.6). If $b \ll h < L$, then we can assume plane stress conditions. The Airy stress function corresponding to this boundary value problem is [8]

$$\Phi = \frac{Px_1}{2bh^3}\left[4[x_2 - L]x_1^2 - 3h^2x_2\right].$$

Noted that, according to Fig. 2.6,

$$-\frac{h}{2} \leqslant x_1 \leqslant \frac{h}{2}, \quad 0 \leqslant x_2 \leqslant L, \quad -\frac{b}{2} \leqslant x_3 \leqslant \frac{b}{2}.$$

Compute the stress and strain tensors and demonstrate that Φ satisfies the strain compatibility equation (2.41).

Solution:

For the isotropic material, the 2–D compliance tensor is written as

$$\widehat{\boldsymbol{M}} = \begin{bmatrix} \frac{1}{E} & -\frac{\nu}{E} & 0 \\ -\frac{\nu}{E} & \frac{1}{E} & 0 \\ 0 & 0 & \frac{2[1+\nu]}{E} \end{bmatrix}.$$

Due to the plane stress conditions, the nonzero stress components are given by (2.42). Thus

$$\sigma_{11} = \frac{\partial^2\Phi}{\partial x_2^2} = 0,$$
$$\sigma_{22} = \frac{\partial^2\Phi}{\partial x_1^2} = -\frac{12P}{bh^3}x_1[L - x_2],$$
$$\sigma_{12} = -\frac{\partial^2\Phi}{\partial x_1\partial x_2} = \frac{3P}{2bh^3}[h^2 - 4x_1^2].$$

The constitutive law (2.38) yields the strain components

$$\varepsilon_{11} = \frac{12P\nu}{Ebh^3}x_1[L - x_2],$$
$$\varepsilon_{22} = -\frac{12P}{Ebh^3}x_1[L - x_2],$$
$$\varepsilon_{12} = \frac{3P[1+\nu]}{Ebh^3}[h^2 - 4x_1^2].$$

The last normal strain component ε_{33} is obtained from Eq. $(2.39)_1$:

$$\varepsilon_{33} = \varepsilon_{11} = \frac{12P\nu}{Ebh^3}x_1[L - x_2].$$

Concerning (2.41), we have

$$\frac{\partial^2 \varepsilon_{11}}{\partial x_2^2} + \frac{\partial^2 \varepsilon_{22}}{\partial x_1^2} = 0,$$
$$2\frac{\partial^2 \varepsilon_{12}}{\partial x_1 \partial x_2} = 0.$$

Consequently, the strain compatibility condition is satisfied.

Remark:

The above study has been extended to the case of transversely isotropic materials [8]. It has been revealed that the closed–form solution for the Φ works well for the interior of the beam but violates the essential boundary condition at the clamped end. This violation increases with increasing anisotropy of the material. Thus the plane stress simplification of the problem should be utilized with caution.

References

[1] W.M. Lai, D. Rubin, E. Krempl, Introduction to continuum mechanics, 4th ed., Elsevier, Burlington, 2010.
[2] G. Chatzigeorgiou, N. Charalambakis, Y. Chemisky, F. Meraghni, Thermomechanical Behavior of Dissipative Composite Materials, ISTE Press – Elsevier, London, 2018.
[3] L.E. Malvern, Introduction to the Mechanics of a Continuous Medium, Prentice Hall, Hertfordshire, 1969.
[4] Z. Hashin, Thermoelastic properties of fiber composites with imperfect interface, Mechanics of Materials 8 (1990) 333–348.
[5] O. Rand, V. Rovenski, Analytical Methods in Anisotropic Elasticity, Birkhäuser, Basel, 2005.
[6] A.F. Bower, Applied Mechanics of Solids, CRC Press, Boca Raton, FL, 2010.
[7] B.A. Jayne, M.O. Hunt, Plane stress and plane strain in orthotropic and anisotropic media, Wood and Fiber Science (3) (1969) 236–247.
[8] G. Kress, C. Thurnherr, Bending stiffness of transversal isotropic materials, Composite Structures 176 (2017) 692–701.

PART II

Micromechanics for composite media

CHAPTER

3

General concepts of micromechanics

OUTLINE

3.1 Heterogeneous media

Solid materials present the internal structure (called microstructure), which often presents significant complexity, including spatial variation. In the microstructure, local variations (called heterogeneities) of the matter appear. The existence of heterogeneities induce physical fields, like strain and temperature, highly non–uniform inside the matter.

An illustrative example of material heterogeneities can be found in the case of metals. A metal is composed by crystals (or grains) that coexist in various orientations, constructing a polycrystal. Certain material prop-

Multiscale Modeling Approaches for Composites
https://doi.org/10.1016/B978-0-12-823143-2.00013-8

erties depend on the interatomic distance of the lattice structure. In the general case, this distance varies inside the lattice. For instance, the components of the elasticity tensor L_{ijkl} depend on the interatomic distance d.

Polycrystalline materials is another case of heterogeneous media. For a polycrystal, it has been shown that the components of the elasticity tensor L_{ijkl} depend on the interatomic distance d. Considering a material with cubic structure for example, its constitutive law can be written in the Voigt notation as

$$\begin{bmatrix} \sigma_{11} \\ \sigma_{22} \\ \sigma_{33} \\ \sigma_{12} \\ \sigma_{13} \\ \sigma_{23} \end{bmatrix} = \begin{bmatrix} L_{11} & L_{12} & L_{12} & 0 & 0 & 0 \\ L_{12} & L_{11} & L_{12} & 0 & 0 & 0 \\ L_{12} & L_{12} & L_{11} & 0 & 0 & 0 \\ 0 & 0 & 0 & L_{44} & 0 & 0 \\ 0 & 0 & 0 & 0 & L_{44} & 0 \\ 0 & 0 & 0 & 0 & 0 & L_{44} \end{bmatrix} \cdot \begin{bmatrix} \varepsilon_{11} \\ \varepsilon_{22} \\ \varepsilon_{33} \\ 2\varepsilon_{12} \\ 2\varepsilon_{13} \\ 2\varepsilon_{23} \end{bmatrix}. \tag{3.1}$$

This relation is established in the local coordinates of the crystal assuming that the axis 1 lies on the [1 0 0] direction. If the coordinate systems of the laboratory frame (where the boundary conditions are imposed) and the crystal are not the same, then the elasticity tensor of the crystal should be rotated properly. This indicates that the elastic stiffness, as expressed in the laboratory coordinate system, will vary between grains, depending on the grain orientation. Consequently, the grain orientation can be seen as one source of heterogeneity in the microstructure.

There are of course additional sources of heterogeneities. In steel, for instance, a perlitic phase can be formed in the grain boundaries. This phase consists of cementite lamellae (Fe_3C phase) and ferrite lamellae (phase α). In such a case, even if the crystals possess the same orientation, the difference in composition results in a variation of the elastic constants between perlite and ferrite lamellae. In addition, with regard to steel, phenomena like the formation of martensitic phase, the residual austenite, and so on also contribute in the heterogeneous nature of the microstructure.

Nowadays there exist many ways to observe the microstructure of the material. The most recent methods allow us to observe it in three dimensions. One way is combining EBSD analysis (electron backscatter diffraction) with FIB (focused ion beam) to perform a layer–by–layer texture analysis. Another possibility is analyzing the materials by using a microtomograph. Many materials can be analyzed with this method (composite materials with random/continuous reinforcement, polycrystalline materials, architectured materials, porous media, etc.).

3.2 Homogenization

Composites with random or periodic microstructure frequently appear in engineering applications (characteristic examples are shown in Fig. 3.1). Considering such type of materials, with generally complex and spatially dependent microstructure, the following question arises: How can someone obtain the overall behavior of a material exhibiting heterogeneous response? The answer to this question relies on the theory of micromechanics.

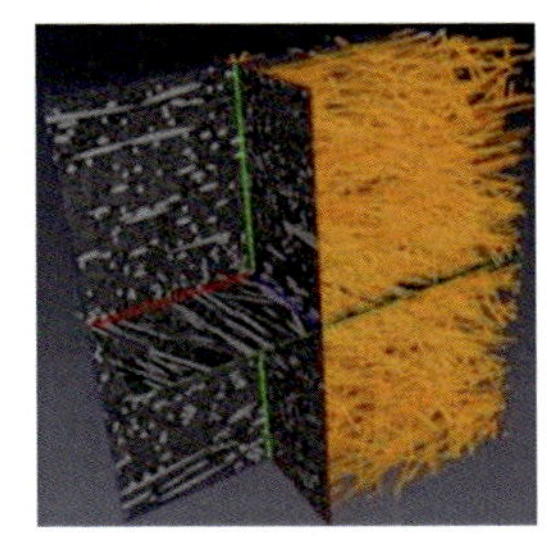

short fiber composite
(a)

woven composite
(b)

FIGURE 3.1 Microtomography of media with random (a) and periodic (b) microstructure.

The homogenization of composite materials (Fig. 3.2) can be defined as follows:

BOX 3.1 Homogenization

Given a composite material with known heterogeneous microstructure (random or periodic), the homogenization is the mathematical process that attempts to substitute its mechanical response with that of an equivalent homogeneous medium.

To achieve the complicated task of homogenization, we first need to specify the characteristic sizes of the heterogeneities with respect to the studied structure. Indeed, an equivalent medium can be considered as a structure of a material point only if there is a difference of large scale between the characteristic lengths of the structure and microstructure. In a metallic material, for example, several characteristic scales can be considered (Fig. 3.3):

- The scale of the structure.
- The scale of the crystalline structure.

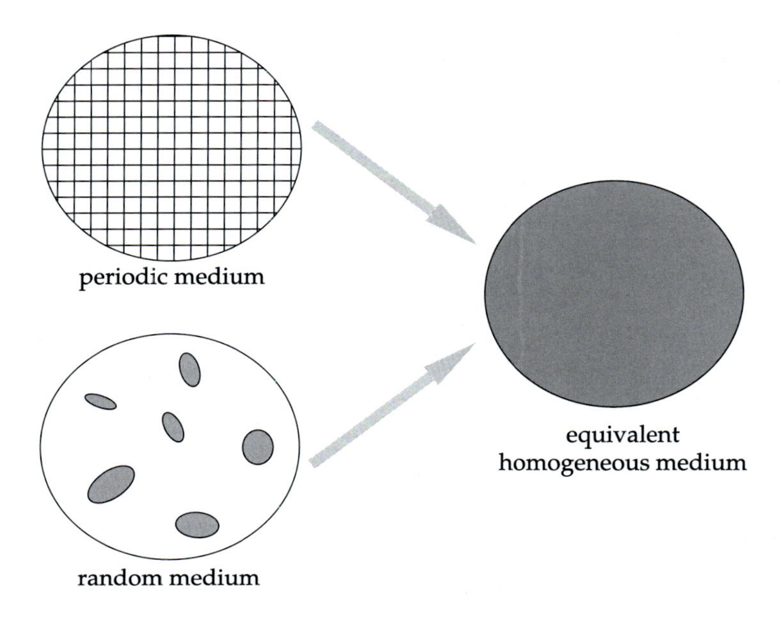

FIGURE 3.2 The problem of homogenization.

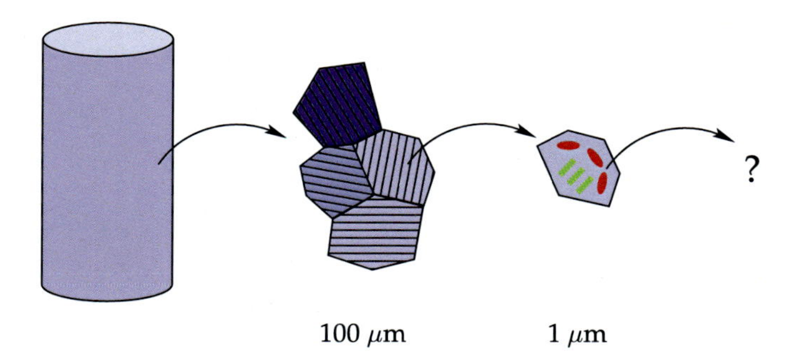

FIGURE 3.3 Different characteristic scales in the study of polycrystals.

- The scale of the intragranular structure (precipitates, lamellae perlite/martensite, etc.).
- The scale of the crystallographic lattice.

It becomes clear that an appropriate choice of the characteristic scale of heterogeneities is not obvious. Finer scale under consideration corresponds to finer way of translating the material behavior. Consequently, the representative volume of the medium needs to be smaller. Furthermore, under a certain scale, the continuum mechanics are not able to describe the material response, and interatomic interactions must be considered discretely. These methods will not be described here.

An important factor in the modeling of composite materials is choosing the most appropriate scale with regard to the problem under investigation. The modeling approaches discussed in this book have an important restriction: The characteristic size of the studied material must be several orders of magnitude lower than the characteristic size of the structure, or, in other terms, the characteristic length d of heterogeneities should be much smaller than the characteristic size of the structure L,

$$d \ll L. \tag{3.2}$$

BOX 3.2 Representative volume element

A characteristic volume of the studied heterogeneous medium should be sufficiently large to be representative of the microstructure, but at the same time, it should be considered as a material point at the higher scale. This notion leads to define a representative volume element (RVE). Such consideration is very important in choosing an appropriate modeling strategy.

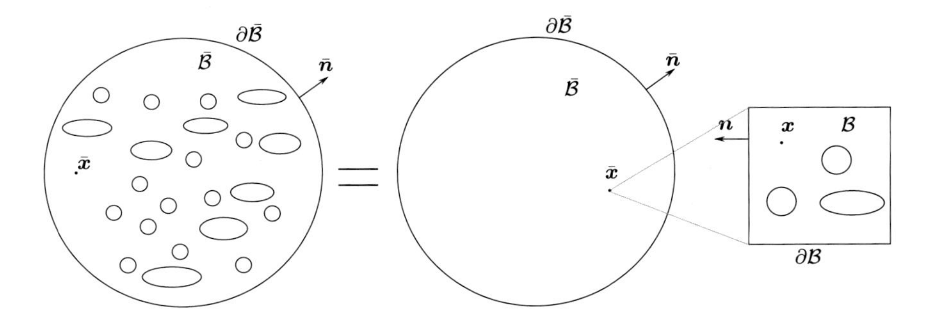

FIGURE 3.4 Homogenization method: the actual composite is split in two scales.

Let us consider the composite of Fig. 3.4. The homogenization is a procedure that allows us to split the actual problem of a composite material into two parts: i) a macroscopic one, where the composite is treated in total, considering that is described by a homogeneous equivalent medium, and ii) a microscopic problem (RVE of the composite), where the different phases and geometrical characteristics of the microstructure are taken into account. Various fields (strain, stress, etc.) appear in both scales: In the macroscopic scale, each point $\bar{\boldsymbol{x}}$ is connected with an RVE. The position vector in the RVE is denoted as $\boldsymbol{x}$. The fictitious material at the macroscopic scale should provide strain and stress fields representing the real heterogeneous medium.

All the homogenization methods are based on a crucial assumption: the composite under investigation needs to be statistically homogeneous, that is, under uniform boundary conditions, the material response is homogeneous in an average sense, and averaging of stresses or strains in an arbitrary RVE of the composite provides essentially the same result. The overall properties of a composite, obtained by homogenization, are meaningful only when appropriate averaging taken over the RVE gives the same mechanical response as that of the overall composite [1,2].

In homogenization theories the following definition is of vital importance:

BOX 3.3 Macroscopic stress and strain

The macroscopic strain $\bar{\boldsymbol{\varepsilon}}$ and stress $\bar{\boldsymbol{\sigma}}$ at a macroscopic point $\bar{\boldsymbol{x}}$ are equal to the volume average of their microscopic counterparts $\boldsymbol{\varepsilon}$ and $\boldsymbol{\sigma}$, respectively, over the RVE (with volume V) attached to the point $\bar{\boldsymbol{x}}$, that is,

$$\bar{\boldsymbol{\sigma}} = \frac{1}{V}\int_{\mathcal{B}} \sigma_{ij}\,\mathrm{d}V, \quad \bar{\boldsymbol{\varepsilon}} = \frac{1}{V}\int_{\mathcal{B}} \varepsilon_{ij}\,\mathrm{d}V. \tag{3.3}$$

The aim of this book is to discuss various existing homogenization theories, which, given specific microstructural geometric and material characteristics, provide information about the overall behavior of the composite material. This chapter is devoted to the discussion of the general principles of homogenization that a proper micromechanics theory should respect.

3.3 Homogenization principles

Before discussing the various theorems and concepts that govern all micromechanics theories, the following definitions are important to be mentioned:

BOX 3.4 Admissible stress and strain

Consider a heterogeneous medium that lies in the space $\mathcal{B}$ with volume V and is bounded by the surface $\partial\mathcal{B}$. Each material point on this volume possesses a position vector $\boldsymbol{x}$.

A statically admissible stress $\boldsymbol{\sigma}$ in the domain $\mathcal{B}$ is any symmetric second–order tensor that satisfies the equilibrium equation (ignoring body forces)

$$\frac{\partial \sigma_{ij}}{\partial x_j} = 0.$$

A kinematically admissible strain $\boldsymbol{\varepsilon}$ in the domain $\mathcal{B}$ is any symmetric second–order tensor related to the displacement vector $\boldsymbol{u}$ through the relation

$$\varepsilon_{ij} = \frac{1}{2}\left[\frac{\partial u_i}{\partial x_j} + \frac{\partial u_j}{\partial x_i}\right].$$

In addition, the Green–Ostrogradski theorem (also usually referred to as the divergence theorem) needs to be introduced in its generalized form. This theorem can be written for an arbitrary tensor field χ, and it connects the volume integral of a partial derivative of the quantity $\frac{\partial \chi_{i_1 i_2 \ldots i_q \ldots i_n}}{\partial x_{i_q}}$ with the surface integral of the flux of this quantity, $\phi_{i_1 i_2 \ldots i_n} = \chi_{i_1 i_2 \ldots i_q \ldots i_n} n_{i_q}$. It states that

$$\int_{\mathcal{B}} \frac{\partial \chi_{i_1 i_2 \ldots i_q \ldots i_n}}{\partial x_{i_q}} \, \mathrm{d}V = \int_{\partial \mathcal{B}} \chi_{i_1 i_2 \ldots i_q \ldots i_n} n_{i_q} \, \mathrm{d}S. \tag{3.4}$$

In this section, we present three very important concepts in micromechanics: i) the average stress and strain theorems, ii) the Hill–Mandel (or average energy) theorem, and iii) the notion of the concentration tensors.

3.3.1 Average theorems

The average theorems are powerful tools in the theory of homogenization that provide a link between boundary conditions and average value of fields inside a body.

3.3.1.1 Average stress theorem

The average stress theorem connects the average stress developed inside a body with the applied traction at the boundary of the body.

BOX 3.5 Average stress theorem

Assume the body of Fig. 3.5. At each point M of the boundary surface $\partial\mathcal{B}$ (with normal vector $\boldsymbol{n}$) a traction vector $\mathbf{t}^0$ is applied such that

$$\mathbf{t}^0(M, \mathbf{n}) = \boldsymbol{\sigma}^0 \cdot \boldsymbol{n}. \tag{3.5}$$

If the applied traction vector at the boundary surface of the domain is such that $\boldsymbol{\sigma}^0$ is constant, then for a statically admissible stress,

$$\langle \sigma \rangle = \frac{1}{V} \int_{\mathcal{B}} \sigma(\boldsymbol{x}) \, \mathrm{d}V = \boldsymbol{\sigma}^0. \tag{3.6}$$

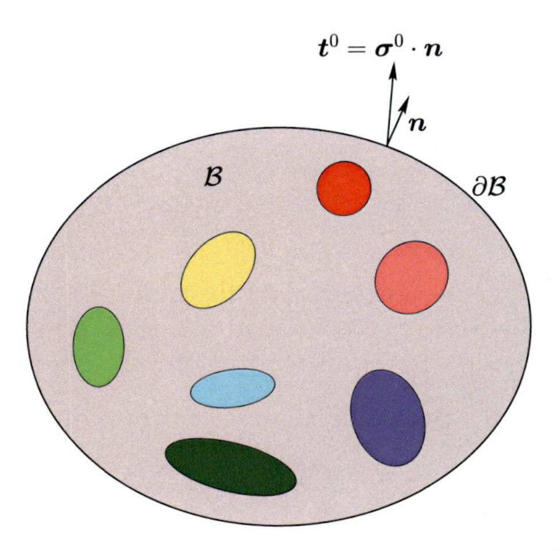

FIGURE 3.5 Boundary conditions: applied stress vector on the boundary surface of the volume.

Proof. The proof of the average stress theorem involves the use of the Green–Ostrogradski theorem. Recall that in the following calculations the Einstein convention for tensors considers summation over repeated indices. We have

$$\frac{1}{V}\int_{\mathcal{B}} \sigma_{ij}\,\mathrm{d}V = \frac{1}{V}\int_{\mathcal{B}} \sigma_{ik}\delta_{jk}\,\mathrm{d}V = \frac{1}{V}\int_{\mathcal{B}} \sigma_{ik}\frac{\partial x_j}{\partial x_k}\,\mathrm{d}V.$$

Integration by parts and the Green–Ostrogradski theorem yield

$$\begin{aligned}\frac{1}{V}\int_{\mathcal{B}} \sigma_{ij}\,\mathrm{d}V &= \frac{1}{V}\int_{\mathcal{B}} \left[\frac{\partial}{\partial x_k}\left(\sigma_{ik}x_j\right) - \frac{\partial \sigma_{ik}}{\partial x_k}x_j\right]\mathrm{d}V \\ &= \frac{1}{V}\int_{\partial\mathcal{B}} \sigma_{ik}x_j n_k\,\mathrm{d}S - \frac{1}{V}\int_{\mathcal{B}} \frac{\partial \sigma_{ik}}{\partial x_k}x_j\,\mathrm{d}V.\end{aligned}$$

Considering the equilibrium equation, the last expression is simplified to

$$\frac{1}{V}\int_{\mathcal{B}} \sigma_{ij}\,\mathrm{d}V = \frac{1}{V}\int_{\partial\mathcal{B}} \sigma_{ik}x_j n_k\,\mathrm{d}S = \frac{1}{V}\int_{\partial\mathcal{B}} t_i^0 x_j\,\mathrm{d}S = \frac{1}{V}\int_{\partial\mathcal{B}} \sigma_{ik}^0 n_k x_j\,\mathrm{d}S.$$

Applying once more the Green–Ostrogradski theorem and using that σ_{ik}^0 is constant give

$$\frac{1}{V}\int_{\mathcal{B}} \sigma_{ij}\,\mathrm{d}V = \frac{1}{V}\int_{\mathcal{B}} \sigma_{ik}^0\frac{\partial x_j}{\partial x_k}\,\mathrm{d}V = \frac{1}{V}\int_{\mathcal{B}} \sigma_{ij}^0\,\mathrm{d}V = \sigma_{ij}^0.$$

The latter concludes the proof. □

3.3.1.2 *Average strain theorem*

In a complete analogy with the average stress theorem, the average strain theorem connects the average strain developed inside a body with the applied displacement at the boundary of the body.

BOX 3.6 Average strain theorem

Assume the body of Fig. 3.6. At each point M of the boundary surface $\partial\mathcal{B}$ (with normal vector $\boldsymbol{n}$) a displacement vector $\mathbf{u}^0$ is applied such that

$$\mathbf{u}^0(M, \boldsymbol{x}) = \boldsymbol{\varepsilon}^0 \cdot \boldsymbol{x} + \mathbf{u}^c. \tag{3.7}$$

If the applied displacement vector at the boundary surface of the domain $\mathcal{B}$ is such that $\boldsymbol{\varepsilon}^0$ and $\mathbf{u}^c$ are constant, then for a kinematically admissible strain,

$$\langle \boldsymbol{\varepsilon} \rangle = \frac{1}{V} \int_{\mathcal{B}} \boldsymbol{\varepsilon}(\boldsymbol{x}) \, \mathrm{d}V = \boldsymbol{\varepsilon}^0. \tag{3.8}$$

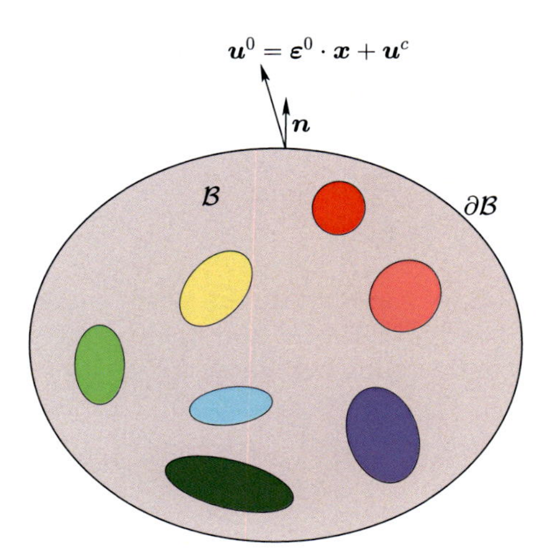

FIGURE 3.6 Boundary conditions: applied displacement vector on the boundary surface of the volume.

Proof. The proof of the statement also involves the use of the Green–Ostrogradski theorem. We have

$$\frac{1}{V} \int_{\mathcal{B}} \varepsilon_{ij} \, \mathrm{d}V = \frac{1}{2V} \int_{\mathcal{B}} \left[\frac{\partial u_i}{\partial x_j} + \frac{\partial u_j}{\partial x_i} \right] \mathrm{d}V = \frac{1}{2V} \int_{\partial\mathcal{B}} [u_i n_j + u_j n_i] \, \mathrm{d}S.$$

Due to the boundary conditions, the last integral is written as

$$\begin{aligned}\frac{1}{V}\int_{\mathcal{B}}\varepsilon_{ij}\,\mathrm{d}V &= \frac{1}{2V}\int_{\partial\mathcal{B}}\left[\varepsilon^0_{ik}x_k n_j + \varepsilon^0_{jk}x_k n_i\right]\mathrm{d}S + \frac{1}{2V}\int_{\partial\mathcal{B}}\left[u^c_i n_j + u^c_j n_i\right]\mathrm{d}S\\ &= \frac{\varepsilon^0_{ik}}{2V}\int_{\partial\mathcal{B}}x_k n_j\,\mathrm{d}S + \frac{\varepsilon^0_{jk}}{2V}\int_{\partial\mathcal{B}}x_k n_i\,\mathrm{d}S + \frac{1}{2V}\int_{\mathcal{B}}\left[\frac{\partial u^c_i}{\partial x_j} + \frac{\partial u^c_j}{\partial x_i}\right]\mathrm{d}V\\ &= \frac{\varepsilon^0_{ik}}{2V}\int_{\mathcal{B}}\frac{\partial x_k}{\partial x_j}\,\mathrm{d}V + \frac{\varepsilon^0_{jk}}{2V}\int_{\mathcal{B}}\frac{\partial x_k}{\partial x_i}\,\mathrm{d}V = \frac{1}{2}\left[\varepsilon^0_{ij} + \varepsilon^0_{ji}\right] = \varepsilon^0_{ij}.\end{aligned}$$

This concludes the proof. □

3.3.2 Hill–Mandel principle

Consider a composite material and an arbitrary RVE occupying the space $\mathcal{B}$ with boundary surface $\partial\mathcal{B}$ that is attached to a macroscopic point $\bar{\boldsymbol{x}}$. Let us assume that this representative volume element is subjected to a set of boundary conditions that produce a kinematically admissible strain $\boldsymbol{\varepsilon}(\boldsymbol{x})$. The same RVE can be subjected to the same, or different, boundary conditions, which produce a statically admissible stress $\boldsymbol{\sigma}(\boldsymbol{x})$. The average strain energy inside $\mathcal{B}$ produced by the kinematically admissible strain and the statically admissible stress is given by

$$\begin{aligned}W_c &= \frac{1}{2}\langle\sigma_{ij}\varepsilon_{ij}\rangle = \frac{1}{2V}\int_{\mathcal{B}}\sigma_{ij}\varepsilon_{ij}\,\mathrm{d}V\\ &= \frac{1}{2V}\int_{\mathcal{B}}\frac{\partial u_i}{\partial x_j}\sigma_{ij}\,\mathrm{d}V = \frac{1}{2V}\int_{\partial\mathcal{B}}u_i\sigma_{ij}n_j\,\mathrm{d}S.\end{aligned} \tag{3.9}$$

On the other hand, we can identify the strain energy W'_h produced by the product of the average strain $\langle\boldsymbol{\varepsilon}\rangle$ and the average stress $\langle\boldsymbol{\sigma}\rangle$. This energy can be written in the following ways:

$$W'_h = \frac{1}{2}\langle\sigma_{ij}\rangle\langle\varepsilon_{ij}\rangle = \frac{1}{2V}\langle\sigma_{ij}\rangle\int_{\mathcal{B}}\frac{\partial u_i}{\partial x_j}\,\mathrm{d}V = \frac{1}{2V}\int_{\partial\mathcal{B}}u_i\langle\sigma_{ij}\rangle n_j\,\mathrm{d}S, \tag{3.10}$$

$$W'_h = \frac{1}{2}\langle\sigma_{ij}\rangle\langle\varepsilon_{ij}\rangle = \frac{1}{2V}\langle\varepsilon_{ij}\rangle\int_{\mathcal{B}}\sigma_{ij}\,\mathrm{d}V = \frac{1}{2V}\int_{\partial\mathcal{B}}\langle\varepsilon_{ij}\rangle\sigma_{ik}n_k x_j\,\mathrm{d}S, \tag{3.11}$$

$$W'_h = \frac{1}{2V}\langle\varepsilon_{ij}\rangle\langle\sigma_{ik}\rangle\int_{\mathcal{B}}\delta_{jk}\,\mathrm{d}V = \frac{1}{2V}\int_{\partial\mathcal{B}}\langle\varepsilon_{ij}\rangle\langle\sigma_{ik}\rangle n_k x_j\,\mathrm{d}S. \tag{3.12}$$

Regrouping all the above terms leads to the Hill lemma: In an RVE a kinematically admissible strain $\boldsymbol{\varepsilon}$ and a statically admissible stress $\boldsymbol{\sigma}$ produce average strain energy that obeys the relation

$$\langle\sigma_{ij}\varepsilon_{ij}\rangle - \langle\sigma_{ij}\rangle\langle\varepsilon_{ij}\rangle = \frac{1}{V}\int_{\partial\mathcal{B}}\left[u_i - x_j\langle\varepsilon_{ij}\rangle\right]\left[\sigma_{ik}n_k - \langle\sigma_{ik}\rangle n_k\right]\mathrm{d}S. \tag{3.13}$$

We can easily observe that the right–hand side of Eq. (3.13) vanishes under three types of boundary conditions:

- In the case of homogeneous stresses on the boundary

$$t_i = \sigma^0_{ij} n_j \tag{3.14}$$

with constant σ^0_{ij}. Indeed, due to the average stress theorem, $\langle \sigma_{ij} \rangle = \sigma^0_{ij}$, and the right–hand side of (3.13) vanishes.
- In the case of homogeneous deformation on the boundary

$$u_i = \varepsilon^0_{ij} x_j + u^c_i \tag{3.15}$$

with constant ε^0_{ij} and u^c_i. Indeed, due to the average strain theorem, $\langle \varepsilon_{ij} \rangle = \varepsilon^0_{ij}$, and the right–hand side of (3.13) is written as

$$\frac{1}{V} \int_{\partial \mathcal{B}} \left[\varepsilon^0_{ij} x_j + u^c_i - \varepsilon^0_{ij} x_j \right] \left[\sigma_{ik} n_k - \langle \sigma_{ik} \rangle n_k \right] \mathrm{d}S$$
$$= \frac{u^c_i}{V} \left[\int_{\mathcal{B}} \frac{\partial \sigma_{ik}}{\partial x_k} \mathrm{d}V - \int_{\mathcal{B}} \frac{\partial \langle \sigma_{ik} \rangle}{\partial x_k} \mathrm{d}V \right] = 0,$$

since the divergence of stress is zero due to the statically admissible stress and $\langle \sigma_{ik} \rangle$ is constant inside $\mathcal{B}$.
- In the case of periodic media, we can consider the following displacement field:

$$u_i = \langle \varepsilon_{ij} \rangle x_j + u^{\mathrm{per}}_i + u^c_i, \tag{3.16}$$

where $\boldsymbol{u}^{\mathrm{per}}$ is a periodic displacement field, and $\boldsymbol{u}^c$ is a constant displacement. The traction vector field $t_i = \sigma_{ij} n_j$ is assumed antiperiodic. In such a case the difference $\langle \sigma_{ij} \varepsilon_{ij} \rangle - \langle \sigma_{ij} \rangle \langle \varepsilon_{ij} \rangle$ is expressed as

$$\langle \sigma_{ij} \varepsilon_{ij} \rangle - \langle \sigma_{ij} \rangle \langle \varepsilon_{ij} \rangle = \frac{1}{V} \int_{\partial \mathcal{B}} \left[u_i - x_j \langle \varepsilon_{ij} \rangle \right] \left[\sigma_{ik} n_k - \langle \sigma_{ik} \rangle n_k \right] \, \mathrm{d}S$$
$$= \frac{1}{V} \int_{\partial \mathcal{B}} \left[u^{\mathrm{per}}_i + u^c_i \right] \left[\sigma_{ik} n_k - \langle \sigma_{ik} \rangle n_k \right] \, \mathrm{d}S = 0,$$

since the product between periodic and antiperiodic terms vanishes when integrated on the total boundary.

At this point, it is necessary to introduce a very important principle in micromechanics:

BOX 3.7 Hill–Mandel principle

In an arbitrary RVE of a composite material the average strain energy produced by a kinematically admissible strain and a statically admissible stress must be equal to the strain energy produced by the average strain and stress fields. In mathematical terms, this can be expressed by the relation

$$\langle \boldsymbol{\sigma} : \boldsymbol{\varepsilon} \rangle = \langle \boldsymbol{\sigma} \rangle : \langle \boldsymbol{\varepsilon} \rangle \quad \text{or} \quad \langle \sigma_{ij} \varepsilon_{ij} \rangle = \langle \sigma_{ij} \rangle \langle \varepsilon_{ij} \rangle . \tag{3.17}$$

The Hill–Mandel principle is satisfied under the following boundary conditions:

1. uniform tractions,
2. linear displacements,
3. periodic displacements and antiperiodic tractions.

If the kinematically admissible strains and statically admissible stresses are produced under the same type of boundary conditions, then we can consider that their average values represent macroscopic quantities.

Concerning the fields at the macroscopic scale, the divergence theorem and the symmetry of the stress and strain tensors allow us to write the macroscopic stress and strain given by Eqs. (3.3) in the alternative forms

$$\bar{\sigma}_{ij} = \frac{1}{V} \int_{\mathcal{B}} \sigma_{ij} \, \mathrm{d}V = \frac{1}{V} \int_{\partial \mathcal{B}} \sigma_{ik} n_k x_j \, \mathrm{d}S = \frac{1}{2V} \int_{\partial \mathcal{B}} [\sigma_{ik} n_k x_j + x_i \sigma_{jk} n_k] \, \mathrm{d}S \tag{3.18}$$

and

$$\bar{\varepsilon}_{ij} = \frac{1}{V} \int_{\mathcal{B}} \varepsilon_{ij} \, \mathrm{d}V = \frac{1}{2V} \int_{\partial \mathcal{B}} [u_i n_j + u_j n_i] \, \mathrm{d}S. \tag{3.19}$$

With regard to the produced mechanical energies, the Hill–Mandel principle states that the strain energy of the equivalent homogeneous medium at the macroscopic position $\bar{\boldsymbol{x}}$, identified as

$$W_h = \frac{1}{2} \bar{\sigma}_{ij} \bar{\varepsilon}_{ij}, \tag{3.20}$$

should be equal to the average strain energy W_c inside the RVE of $\bar{\boldsymbol{x}}$. From the Hill lemma it becomes clear that when the RVE is subjected to one of the three types of boundary conditions (3.14), (3.15), and (3.16), the Hill–Mandel principle is automatically satisfied.

As a final remark, note that the mechanical properties of the equivalent homogeneous medium can be defined from the macroscopic strain energy through the relation

$$W_h = \frac{1}{2}\bar{\varepsilon}_{ij}\bar{L}_{ijkl}\bar{\varepsilon}_{kl}. \tag{3.21}$$

In this case, $\bar{\boldsymbol{L}}$ represents the elasticity tensor of the equivalent homogeneous medium.

3.3.3 Concentration tensors

As has been illustrated earlier, the knowledge of the microscale fields (strains or stresses) is sufficient to identify their macroscopic counterparts by volume averaging in the RVE. In most of the cases though, only a direct access to the macroscopic fields can be achieved. Upon loading of a composite structure, it is usually much easier experimentally to obtain values of the fields at the observable scale (macroscale).

In light of the above, it becomes of vital importance to establish appropriate theoretical tools that allow us to obtain the microscopic fields using as information their macroscopic values. The latter procedure in the micromechanics terminology is usually referred to as localization. The localization (or concentration) tensors play a very crucial role in homogenization schemes:

BOX 3.8 Concentration tensors

In the case of heterogeneous materials with linear elastic behavior, microscopic quantities (strain, stress) can be connected with their macroscopic counterparts through relations of the form

$$\boldsymbol{\varepsilon}(\boldsymbol{x}) = \boldsymbol{A}(\boldsymbol{x}) : \bar{\boldsymbol{\varepsilon}}, \quad \boldsymbol{\sigma}(\boldsymbol{x}) = \boldsymbol{B}(\boldsymbol{x}) : \bar{\boldsymbol{\sigma}}. \tag{3.22}$$

In these expressions, $\boldsymbol{A}(\boldsymbol{x})$ is called the strain concentration (or localization) tensor, and $\boldsymbol{B}(\boldsymbol{x})$ is called the stress concentration (or localization) tensor.

This general definition of the concentration tensors is convenient for all types of heterogeneous materials, random or periodic (Fig. 3.7). It takes into account that the fields in the medium are heterogeneous from one point to another. Both stress and strain concentration components are unitless.

When a composite can be seen as a medium with specific number of phases, then the following statement holds:

BOX 3.9 Concentration tensors per phase

When the medium has distinct phases (Fig. 3.7_a), it is possible to define average strains and stresses per phase r:

$$\boldsymbol{\varepsilon}_r = \frac{1}{V_r}\int_{\mathcal{B}_r} \boldsymbol{\varepsilon}(\boldsymbol{x})\,\mathrm{d}V, \quad \boldsymbol{\sigma}_r = \frac{1}{V_r}\int_{\mathcal{B}_r} \boldsymbol{\sigma}(\boldsymbol{x})\,\mathrm{d}V, \tag{3.23}$$

where V_r denotes the volume of space $\mathcal{B}_r$ that the phase r occupies in the medium. Thus the equations of localization can be re–expressed in the form

$$\boldsymbol{\varepsilon}_r = \boldsymbol{A}_r : \bar{\boldsymbol{\varepsilon}}, \quad \boldsymbol{\sigma}_r = \boldsymbol{B}_r : \bar{\boldsymbol{\sigma}}, \tag{3.24}$$

where $\boldsymbol{A}_r$ and $\boldsymbol{B}_r$ are constant fourth–order tensors per phase.

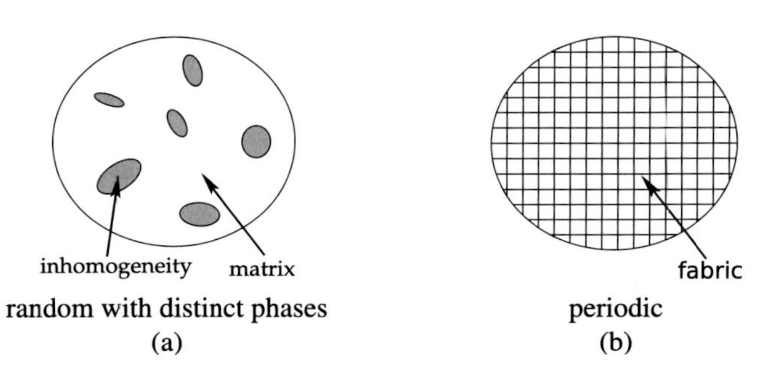

FIGURE 3.7 Different types of heterogeneous media.

In periodic media (Fig. 3.7_b), we can often define separate phases. However, periodic media homogenization theory considers heterogeneous fields of strain and stress, and the average quantities per phase do not appear. Thus the equations of localization can be expressed by relations (3.22).

3.3.3.1 *Properties of concentration tensors*

From the definitions of the macroscopic strain (3.8) and the strain concentration tensor $(3.22)_1$ we obtain

$$\bar{\boldsymbol{\varepsilon}} = \langle \boldsymbol{\varepsilon} \rangle = \frac{1}{V}\int_{\mathcal{B}} \boldsymbol{\varepsilon}(\boldsymbol{x})\,\mathrm{d}V = \frac{1}{V}\int_{\mathcal{B}} \boldsymbol{A}(\boldsymbol{x})\,\mathrm{d}V : \bar{\boldsymbol{\varepsilon}}, \tag{3.25}$$

which leads to the conclusion that

$$\langle \boldsymbol{A} \rangle = \frac{1}{V}\int_{\mathcal{B}} \boldsymbol{A}(\boldsymbol{x})\,\mathrm{d}V = \boldsymbol{I}. \tag{3.26}$$

Recall that $\boldsymbol{I}$ denotes the fourth–order symmetric identity tensor. In a similar manner the definitions of the macroscopic stress (3.6) and the stress concentration tensor $(3.22)_2$ allow us to obtain

$$\bar{\boldsymbol{\sigma}} = \langle \boldsymbol{\sigma} \rangle = \frac{1}{V} \int_{\mathcal{B}} \boldsymbol{\sigma}(\boldsymbol{x})\, \mathrm{d}V = \frac{1}{V} \int_{\mathcal{B}} \boldsymbol{B}(\boldsymbol{x})\, \mathrm{d}V : \bar{\boldsymbol{\sigma}}, \tag{3.27}$$

which leads to the conclusion that

$$\langle \boldsymbol{B} \rangle = \frac{1}{V} \int_{\mathcal{B}} \boldsymbol{B}(\boldsymbol{x})\, \mathrm{d}V = \boldsymbol{I}. \tag{3.28}$$

When the medium has $N + 1$ distinct phases (a matrix 0 and N types of fibers and/or particles), then expressions (3.26) and (3.28) can be written as

$$\sum_{r=0}^{N} c_r \boldsymbol{A}_r = \boldsymbol{I}, \quad c_r = \frac{V_r}{V}, \tag{3.29}$$

and

$$\sum_{r=0}^{N} c_r \boldsymbol{B}_r = \boldsymbol{I}, \quad c_r = \frac{V_r}{V}, \tag{3.30}$$

respectively, where c_r denotes the volume fraction of the rth constituent. Moreover, the constitutive law between average stresses and strains in the elastic phase r is

$$\boldsymbol{\sigma}_r = \boldsymbol{L}_r : \boldsymbol{\varepsilon}_r, \tag{3.31}$$

with $\boldsymbol{L}_r$ denoting the elasticity tensor of the phase. If the composite obeys a homogenized law of a similar form

$$\bar{\boldsymbol{\sigma}} = \bar{\boldsymbol{L}} : \bar{\boldsymbol{\varepsilon}}, \tag{3.32}$$

then we have the following relation between the stress and strain concentration tensors:

$$\boldsymbol{B}_r : \bar{\boldsymbol{L}} = \boldsymbol{L}_r : \boldsymbol{A}_r. \tag{3.33}$$

3.4 Bounds in the overall response

Obtaining the overall properties of a composite material is not always a trivial task. In many occasions, microstructural characteristics such as distribution or shape of reinforcement are unknown. In these cases, we can search for bounds on the macroscopic response using all the available information. Through variational principles we can identify proper bounds on the elastic energy of a composite. We further present the two most famous cases of bounds.

3.4.1 Voigt and Reuss bounds

Voigt has identified an upper bound in the response of a composite by considering that the macroscopic strain is distributed uniformly in all the RVE constituents. This hypothetical situation causes the RVE to behave like a system of different phases connected in parallel, and the overall elasticity tensor is given by

$$\bar{\boldsymbol{L}}^V = \frac{1}{V}\int_{\mathcal{B}} \boldsymbol{L}(\boldsymbol{x})\,\mathrm{d}V. \tag{3.34}$$

On the other hand, Reuss has identified a lower bound in the response of a composite by considering that the macroscopic stress is distributed uniformly in all the RVE constituents. This hypothetical situation causes the RVE to behave like a system of different phases connected in series, and the overall elasticity tensor is given by

$$\left[\bar{\boldsymbol{L}}^R\right]^{-1} = \frac{1}{V}\int_{\mathcal{B}} \boldsymbol{L}^{-1}(\boldsymbol{x})\,\mathrm{d}V. \tag{3.35}$$

Using these two results, it can be shown through variational principles [2] that the macroscopic energy of the composite under any type of boundary conditions must be bounded by the energies produced by the two extreme cases, that is,

$$W_h^R \leqslant W_h \leqslant W_h^V \quad \Rightarrow \quad \bar{\varepsilon}_{ij}\bar{L}^R_{ijkl}\bar{\varepsilon}_{kl} \leqslant \bar{\varepsilon}_{ij}\bar{L}_{ijkl}\bar{\varepsilon}_{kl} \leqslant \bar{\varepsilon}_{ij}\bar{L}^V_{ijkl}\bar{\varepsilon}_{kl}. \tag{3.36}$$

Note that the latter is valid for any symmetric tensor $\bar{\varepsilon}_{ij}$. This inequality allows us to identify bounds in the eigenvalues of the effective elasticity tensor $\bar{\boldsymbol{L}}$.

Whereas for generally anisotropic materials, these bounds are not directly linked with physical material properties, in isotropic media the elasticity tensor eigenvalues are connected with the bulk and the shear modulus. For a composite consisting of two isotropic elastic materials with bulk moduli K_1, K_2, shear moduli μ_1, μ_2, and volume fractions c_1, c_2 ($c_1 + c_2 = 1$), we obtain the following bounds for the macroscopic bulk ($\bar{K}$) and shear ($\bar{\mu}$) moduli:

$$\bar{K}^R \leqslant \bar{K} \leqslant \bar{K}^V \quad \text{and} \quad \bar{\mu}^R \leqslant \bar{\mu} \leqslant \bar{\mu}^V, \tag{3.37}$$

where

$$\bar{K}^R = \frac{K_1 K_2}{c_1 K_2 + c_2 K_1}, \quad \bar{K}^V = c_1 K_1 + c_2 K_2,$$

$$\bar{\mu}^R = \frac{\mu_1 \mu_2}{c_1 \mu_2 + c_2 \mu_1}, \quad \bar{\mu}^V = c_1 \mu_1 + c_2 \mu_2. \tag{3.38}$$

The Reuss and Voigt bounds are computed very easily, since they depend only on the properties of the different phases and their volume fractions inside the composite. The drawback is that they are very general and quite far apart. Tighter bounds are derived by Hashin and Shtrikman [3].

3.4.2 Hashin–Shtrikman bounds

The Hashin–Shtrikman bounds are tighter than the Voigt–Reuss bounds for describing the macroscopic response of a composite material. They are derived through proper variational principles, and theoretical details can be found in [4,3] and in the third part of this book. In case a bi–phase material with isotropic phases behaves isotropically, the final form of these bounds is presented below.

Consider a composite consisting of two isotropic elastic materials with bulk moduli K_1, K_2, shear moduli μ_1, μ_2, and volume fractions c_1, c_2 ($c_1 + c_2 = 1$). Assuming that material 2 is stiffer than material 1, that is, $K_2 > K_1$ and $\mu_2 > \mu_1$, Hashin and Shtrikman have identified the following bounds for the macroscopic bulk ($\bar{K}$) and shear ($\bar{\mu}$) moduli:

$$\bar{K}^L_{HS} \leqslant \bar{K} \leqslant \bar{K}^U_{HS} \quad \text{and} \quad \bar{\mu}^L_{HS} \leqslant \bar{\mu} \leqslant \bar{\mu}^U_{HS}, \tag{3.39}$$

where

$$\bar{K}^L_{HS} = K_1 + \frac{c_2}{\dfrac{1}{K_2 - K_1} + \dfrac{3c_1}{3K_1 + 4\mu_1}},$$

$$\bar{K}^U_{HS} = K_2 + \frac{c_1}{\dfrac{1}{K_1 - K_2} + \dfrac{3c_2}{3K_2 + 4\mu_2}}, \tag{3.40}$$

and

$$\bar{\mu}^L_{HS} = \mu_1 + \frac{c_2}{\dfrac{1}{\mu_2 - \mu_1} + \dfrac{6}{5}\dfrac{c_1[K_1 + 2\mu_1]}{[3K_1 + 4\mu_1]\mu_1}},$$

$$\bar{\mu}^U_{HS} = \mu_2 + \frac{c_1}{\dfrac{1}{\mu_1 - \mu_2} + \dfrac{6}{5}\dfrac{c_2[K_2 + 2\mu_2]}{[3K_2 + 4\mu_2]\mu_2}}. \tag{3.41}$$

3.5 Examples

Example 1

For an elastic composite:

1. Prove that $\bar{L} = \langle L\colon A\rangle$ and $\bar{L}^{-1} = \left\langle L^{-1}\colon B\right\rangle$, where L, A, and B depend on the position inside the RVE.
2. Generalizing expression (3.33), identify the relation between A and B.
3. Using the Hill–Mandel principle and the major symmetry of the elasticity tensor of the material phases, show that $\bar{L}$ preserves the major symmetry.

Solution:

1. For the first question, the stress–strain relation at every point inside an RVE is given by the expression

$$\boldsymbol{\sigma}(\boldsymbol{x}) = \boldsymbol{L}(\boldsymbol{x})\colon \boldsymbol{\varepsilon}(\boldsymbol{x}).$$

On the other hand, the macroscopic stress–strain behavior corresponding to this RVE is expected to satisfy a similar relation, that is,

$$\bar{\boldsymbol{\sigma}} = \bar{\boldsymbol{L}}\colon \bar{\boldsymbol{\varepsilon}}.$$

a) Using $(3.22)_1$ and the definition of the macroscopic stress yield

$$\bar{\boldsymbol{\sigma}} = \langle \boldsymbol{\sigma}(\boldsymbol{x})\rangle = \langle \boldsymbol{L}(\boldsymbol{x})\colon \boldsymbol{\varepsilon}(\boldsymbol{x})\rangle = \langle \boldsymbol{L}(\boldsymbol{x})\colon \boldsymbol{A}(\boldsymbol{x})\rangle \colon \bar{\boldsymbol{\varepsilon}}.$$

Comparing with the macroscopic constitutive law leads to the conclusion that

$$\bar{\boldsymbol{L}} = \langle \boldsymbol{L}\colon \boldsymbol{A}\rangle.$$

b) Using $(3.22)_2$ and the definition of the macroscopic strain yield

$$\bar{\boldsymbol{\varepsilon}} = \langle \boldsymbol{\varepsilon}(\boldsymbol{x})\rangle = \left\langle \boldsymbol{L}^{-1}(\boldsymbol{x})\colon \boldsymbol{\sigma}(\boldsymbol{x})\right\rangle = \left\langle \boldsymbol{L}^{-1}(\boldsymbol{x})\colon \boldsymbol{B}(\boldsymbol{x})\right\rangle \colon \bar{\boldsymbol{\sigma}}.$$

Comparing with the macroscopic constitutive law leads to the conclusion that

$$\bar{\boldsymbol{L}}^{-1} = \left\langle \boldsymbol{L}^{-1}\colon \boldsymbol{B}\right\rangle.$$

2. For the second question, the stress–strain relation at every point inside an RVE

$$\boldsymbol{\sigma}(\boldsymbol{x}) = \boldsymbol{L}(\boldsymbol{x})\colon \boldsymbol{\varepsilon}(\boldsymbol{x}),$$

the macroscopic stress–strain behavior corresponding to this RVE

$$\bar{\boldsymbol{\sigma}} = \bar{\boldsymbol{L}} : \bar{\boldsymbol{\varepsilon}},$$

and relations (3.22) allow us to write

$$\boldsymbol{B}(\boldsymbol{x}) : \bar{\boldsymbol{\sigma}} = \boldsymbol{L}(\boldsymbol{x}) : \boldsymbol{A}(\boldsymbol{x}) : \bar{\boldsymbol{\varepsilon}}$$

or

$$\boldsymbol{B}(\boldsymbol{x}) : \bar{\boldsymbol{L}} : \bar{\boldsymbol{\varepsilon}} = \boldsymbol{L}(\boldsymbol{x}) : \boldsymbol{A}(\boldsymbol{x}) : \bar{\boldsymbol{\varepsilon}}.$$

The last expression should hold for arbitrary choice of macroscopic strain, and thus the formula

$$\boldsymbol{B}(\boldsymbol{x}) : \bar{\boldsymbol{L}} = \boldsymbol{L}(\boldsymbol{x}) : \boldsymbol{A}(\boldsymbol{x})$$

defines a general relation between the strain and stress concentration tensors at any microscopic point of the RVE.

3. For the third question, the indicial notation is more convenient. According to the Hill–Mandel principle,

$$\langle \sigma_{ij} \varepsilon_{ij} \rangle = \bar{\sigma}_{ij} \bar{\varepsilon}_{ij},$$

where both σ_{ij} and ε_{ij} are functions of the position vector $\boldsymbol{x}$ inside the RVE. Using the elastic microscopic stress–strain relations, we have

$$\langle \varepsilon_{ij} L_{ijkl} \varepsilon_{kl} \rangle = \bar{\varepsilon}_{ij} \bar{L}_{ijkl} \bar{\varepsilon}_{kl}.$$

Expression $(3.22)_1$ allows us to write

$$\begin{aligned} \langle \varepsilon_{ij} L_{ijkl} \varepsilon_{kl} \rangle &= \langle A_{ijmn} \bar{\varepsilon}_{mn} L_{ijkl} A_{klpq} \bar{\varepsilon}_{pq} \rangle \\ &= \bar{\varepsilon}_{mn} \langle A_{ijmn} L_{ijkl} A_{klpq} \rangle \bar{\varepsilon}_{pq} \\ &= \bar{\varepsilon}_{ij} \langle A_{mnij} L_{mnpq} A_{pqkl} \rangle \bar{\varepsilon}_{kl}. \end{aligned}$$

Comparing the latter expression with its macroscopic analog yields

$$\bar{L}_{ijkl} = \langle A_{mnij} L_{mnpq} A_{pqkl} \rangle.$$

Using the major symmetry of $\boldsymbol{L}$, the latter expression gives

$$\begin{aligned} \bar{L}_{ijkl} &= \langle A_{mnij} L_{mnpq} A_{pqkl} \rangle = \langle A_{mnij} L_{pqmn} A_{pqkl} \rangle \\ &= \langle A_{pqkl} L_{pqmn} A_{mnij} \rangle = \bar{L}_{klij}, \end{aligned}$$

that is, the macroscopic elasticity tensor preserves the major symmetry.

Example 2: Levin's formula

Consider a bi–phase composite material whose RVE occupies the space $\mathcal{B}$ and contains a matrix (phase 0, elasticity tensor $\boldsymbol{L}_0$) and reinforcement (phase 1, elasticity tensor $\boldsymbol{L}_1$). The reinforcement occupies the space Ω. Each phase is subjected to a constant eigenstrain, denoted by $\boldsymbol{\varepsilon}_0^*$ and $\boldsymbol{\varepsilon}_1^*$ for the matrix and the particle, respectively. The constitutive relation as a function of the position inside the RVE is expressed as

$$\boldsymbol{\varepsilon}(\boldsymbol{x}) = \boldsymbol{L}^{-1}(\boldsymbol{x}):\boldsymbol{\sigma}(\boldsymbol{x}) + \boldsymbol{\varepsilon}^*(\boldsymbol{x}),$$

where

$$\boldsymbol{L}^{-1}(\boldsymbol{x}) = \boldsymbol{L}_0^{-1} + \left[\boldsymbol{L}_1^{-1} - \boldsymbol{L}_0^{-1}\right]\theta(\boldsymbol{x}), \qquad \boldsymbol{\varepsilon}^*(\boldsymbol{x}) = \boldsymbol{\varepsilon}_0^* + [\boldsymbol{\varepsilon}_1^* - \boldsymbol{\varepsilon}_0^*]\theta(\boldsymbol{x}),$$

$$\theta(\boldsymbol{x}) = \begin{cases} 1, & \forall \boldsymbol{x} \in \Omega, \\ 0, & \forall \boldsymbol{x} \in \mathcal{B} - \Omega. \end{cases}$$

If the RVE is subjected to arbitrary uniform traction $\boldsymbol{t} = \overline{\boldsymbol{\sigma}} \cdot \boldsymbol{n}$ (Fig. 3.8), then prove, using the Hill–Mandel principle, that the macroscopic constitutive law

$$\overline{\boldsymbol{\varepsilon}} = \overline{\boldsymbol{L}}^{-1}:\overline{\boldsymbol{\sigma}} + \overline{\boldsymbol{\varepsilon}}^*$$

has a macroscopic eigenstrain given by the formula

$$\overline{\boldsymbol{\varepsilon}}^* = \langle \boldsymbol{\varepsilon}^* : \boldsymbol{B} \rangle$$

with $\boldsymbol{B}$ denoting the stress concentration tensor.

Remark:

The above result can be extended to any number of material phases and shape of inhomogeneities inside the RVE. For the simplicity of the example, here only two phases are considered.

Solution:

Since the eigenstrains in the RVE are constant, the microscopic constitutive law remains linear. The superposition principle allows us to split the boundary value problem of Fig. 3.8 into two simpler ones:

1. The first problem has only the external boundary conditions and no eigenstrains inside the RVE, that is,

$$\boldsymbol{t} = \overline{\boldsymbol{\sigma}} \cdot \boldsymbol{n} \quad \text{on} \quad \partial\mathcal{B},$$
$$\boldsymbol{\varepsilon}^* = \boldsymbol{0} \quad \text{in} \quad \mathcal{B}.$$

Under these conditions, the classical micromechanics provide the solution

$$\boldsymbol{\varepsilon}(\boldsymbol{x}) = \boldsymbol{L}^{-1}(\boldsymbol{x}):\boldsymbol{\sigma}(\boldsymbol{x}),$$

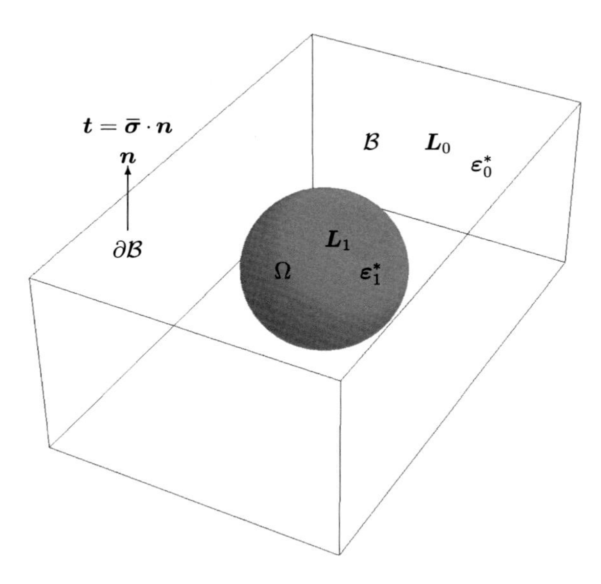

FIGURE 3.8 RVE of a bi–phase composite in which the matrix and the reinforcement are under constant eigenstrains. The RVE is subjected to arbitrary uniform traction.

$$\overline{\boldsymbol{\varepsilon}} = \overline{\boldsymbol{L}}^{-1} : \overline{\boldsymbol{\sigma}},$$
$$\boldsymbol{\sigma}(\boldsymbol{x}) = \boldsymbol{B}(\boldsymbol{x}) : \overline{\boldsymbol{\sigma}}.$$

2. The second problem has only eigenstrains and the RVE is traction free, that is,

$$\boldsymbol{t}' = \boldsymbol{0} \quad \text{on} \quad \partial\mathcal{B},$$
$$\boldsymbol{\varepsilon}^* \neq \boldsymbol{0} \quad \text{in} \quad \mathcal{B}.$$

Under these conditions, thanks to the average stress theorem, we obtain

$$\overline{\boldsymbol{\sigma}}' = \langle \boldsymbol{\sigma}'(\boldsymbol{x}) \rangle = \boldsymbol{0},$$
$$\boldsymbol{\varepsilon}'(\boldsymbol{x}) = \boldsymbol{L}^{-1}(\boldsymbol{x}) : \boldsymbol{\sigma}'(\boldsymbol{x}) + \boldsymbol{\varepsilon}^*(\boldsymbol{x}),$$
$$\overline{\boldsymbol{\varepsilon}}' = \overline{\boldsymbol{\varepsilon}}^*.$$

In both problems the strains are kinematically admissible, and the stresses are statically admissible. Moreover, in both problems the traction is uniform on the boundary. Thus the Hill–Mandel principle can be applied in a mixed manner, that is, combining the strains of one problem with the stresses of the other.

Considering the strains of the first problem and the stresses of the second problem, the Hill–Mandel principle yields

$$\left\langle \boldsymbol{\sigma}' : \boldsymbol{\varepsilon} \right\rangle = \bar{\boldsymbol{\sigma}}' : \bar{\boldsymbol{\varepsilon}} = 0, \quad \text{or} \quad \left\langle \boldsymbol{\sigma}' : \boldsymbol{L}^{-1} : \boldsymbol{\sigma} \right\rangle = 0, \quad \text{or} \quad \left\langle \boldsymbol{\sigma} : \boldsymbol{L}^{-1} : \boldsymbol{\sigma}' \right\rangle = 0,$$

due to the symmetry[1] of $\boldsymbol{L}^{-1}$. Considering the strains of the second problem and the stresses of the first problem, the Hill–Mandel principle yields

$$\begin{aligned}
\bar{\boldsymbol{\sigma}} : \bar{\boldsymbol{\varepsilon}}' &= \left\langle \boldsymbol{\sigma} : \boldsymbol{\varepsilon}' \right\rangle, \quad \text{or} \\
\bar{\boldsymbol{\sigma}} : \bar{\boldsymbol{\varepsilon}}^* &= \left\langle \boldsymbol{\sigma} : \boldsymbol{L}^{-1} : \boldsymbol{\sigma}' \right\rangle + \left\langle \boldsymbol{\sigma} : \boldsymbol{\varepsilon}^* \right\rangle, \quad \text{or} \\
\bar{\boldsymbol{\varepsilon}}^* : \bar{\boldsymbol{\sigma}} &= \left\langle \boldsymbol{\varepsilon}^* : \boldsymbol{B} \right\rangle : \bar{\boldsymbol{\sigma}}.
\end{aligned}$$

Since the applied traction is arbitrary, the above relation should hold for any choice of $\bar{\boldsymbol{\sigma}}$. This leads to the conclusion that

$$\bar{\boldsymbol{\varepsilon}}^* = \left\langle \boldsymbol{\varepsilon}^* : \boldsymbol{B} \right\rangle.$$

Example 3

Consider a composite made of two material phases. The bulk and shear moduli of these phases are as follows:

phase 1	phase 2
$K_1 = 2.5$ GPa	$K_2 = 44$ MPa
$\mu_1 = 1.15$ GPa	$\mu_2 = 30$ GPa

Identify the Voigt–Reuss and Hashin–Shtrikman bounds of the composite moduli in terms of the phase 1 volume fraction.

Solution:

For the Voigt and Reuss bounds, expressions (3.38) are utilized, whereas formulas (3.40) and (3.41) are considered for the Hashin–Shtrikman bounds. Figs. 3.9 and 3.10 demonstrate the bounds on the macroscopic bulk and shear moduli as functions of the phase 1 volume fraction.

Example 4

Consider a bi–phase elastic material in which an isotropic matrix (phase 0) is reinforced with isotropic spherical particles (phase 1). The Young modulus and Poisson ratio of the two phases are as follows:

matrix	particles
$E_0 = 3000$ MPa	$E_1 = 30000$ MPa
$v_0 = 0.3$	$v_1 = 0.2$

[1] If $\boldsymbol{L}$ can be expressed as a 6×6 matrix, then linear algebra theorems show that the symmetry of $\boldsymbol{L}$ and the uniqueness of its inverse ensure the symmetry of $\boldsymbol{L}^{-1}$.

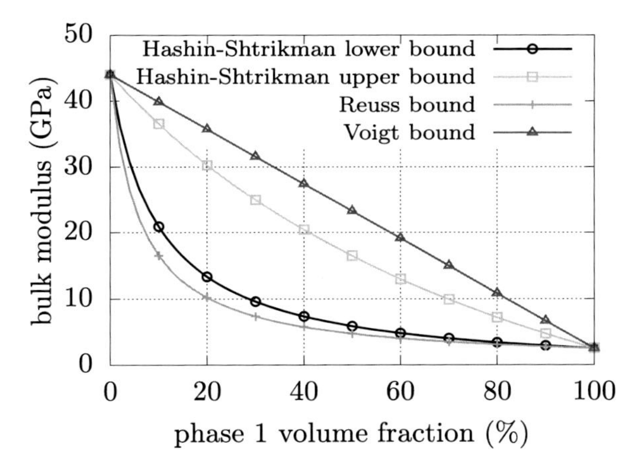

FIGURE 3.9 Voigt–Reuss and Hashin–Shtrikman bounds on macroscopic bulk modulus vs function of phase 1 volume fraction.

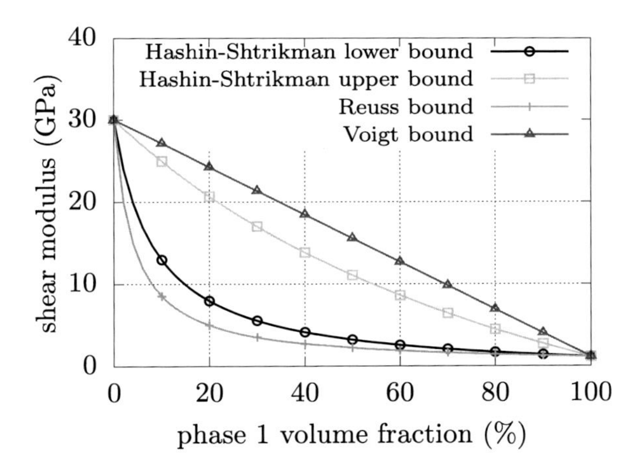

FIGURE 3.10 Voigt–Reuss and Hashin–Shtrikman bounds on macroscopic shear modulus vs function of phase 1 volume fraction.

The volume fraction of the spherical particles is $c_1 = 30\%$.

Such a composite presents an isotropic behavior. The strain concentration of the particles in the Hill notation (Example 2 in Section 1.4 of Chapter 1) is equal to

$$\boldsymbol{A}_1 = (3A_1^b, 2A_1^s) = (0.2894, 0.2338).$$

1. Calculate the Young modulus $\bar{E}$ and Poisson ratio $\bar{\upsilon}$ of the composite.

2. Assuming the macroscopic deformation

$$\bar{\boldsymbol{\varepsilon}} = \begin{bmatrix} 0.001 & 0 & 0 \\ 0 & 0.001 & 0 \\ 0 & 0 & 0 \end{bmatrix},$$

calculate the average stresses and strains in the particles and in the matrix.
3. For the same macroscopic deformation, compute the macroscopic stress of the composite.

Solution:

In the Hill notation the elasticity tensors of the two phases (matrix and particles) are expressed through the bulk and shear moduli (see Chapter 2, Subsection 2.3.2). For the two materials, these moduli are equal to

$$K_0 = \frac{E_0}{3[1-2\nu_0]} = 2500 \text{ MPa}, \quad \mu_0 = \frac{E_0}{2[1+\nu_0]} = 1153.846 \text{ MPa},$$
$$K_1 = \frac{E_1}{3[1-2\nu_1]} = 16666.667 \text{ MPa}, \quad \mu_1 = \frac{E_0}{2[1+\nu_1]} = 12500 \text{ MPa}.$$

Thus

$$\boldsymbol{L}_0 = (3K_0, 2\mu_0) = (7500, 2307.692) \text{ MPa},$$
$$\boldsymbol{L}_1 = (3K_1, 2\mu_1) = (50000, 25000) \text{ MPa}.$$

The consistency condition (3.29) allows us to identify the strain concentration tensor for the matrix, that is,

$$\boldsymbol{A}_0 = \frac{1}{c_0}[\boldsymbol{I} - c_1\boldsymbol{A}_1] = \frac{1}{1-0.3}[(1,1) - 0.3 \cdot (0.2894, 0.2338)]$$
$$= (1.3045, 1.3284).$$

1. The relation $\bar{\boldsymbol{L}} = \langle \boldsymbol{L} : \boldsymbol{A} \rangle$ established in the first example of this chapter is written for a bi–phase composite as

$$\bar{\boldsymbol{L}} = (3\bar{K}, 2\bar{\mu}) = c_0\boldsymbol{L}_0 : \boldsymbol{A}_0 + c_1\boldsymbol{L}_1 : \boldsymbol{A}_1.$$

For the two materials,

$$\boldsymbol{L}_0 : \boldsymbol{A}_0 = (7500, 2307.692) \cdot (1.3045, 1.3284)$$
$$= (9783.75, 3065.538) \text{ MPa},$$
$$\boldsymbol{L}_1 : \boldsymbol{A}_1 = (50000, 25000) \cdot (0.2894, 0.2338)$$
$$= (14470, 5845) \text{ MPa}.$$

Consequently,

$$\bar{\boldsymbol{L}} = [1 - 0.3]\cdot(9783.75, 3065.538) + 0.3\cdot(14470, 5845)$$
$$= (11189.625, 3899.377) \text{ MPa},$$

which means that

$$\bar{K} = 3729.875 \text{ MPa}, \qquad \bar{\mu} = 1949.689 \text{ MPa},$$

and

$$\bar{E} = \frac{9\bar{K}\bar{\mu}}{3\bar{K} + \bar{\mu}} = 4981.148 \text{ MPa}, \qquad \bar{\nu} = \frac{3\bar{K} - 2\bar{\mu}}{6\bar{K} + 2\bar{\mu}} = 0.2774.$$

2. The average strains in the matrix and particles are expressed as

$$\boldsymbol{\varepsilon}_0 = \boldsymbol{A}_0 : \bar{\boldsymbol{\varepsilon}}, \qquad \boldsymbol{\varepsilon}_1 = \boldsymbol{A}_1 : \bar{\boldsymbol{\varepsilon}}.$$

On the other hand, the average stresses in the same phases are given by the expressions

$$\boldsymbol{\sigma}_0 = \boldsymbol{L}_0 : \boldsymbol{\varepsilon}_0 = \boldsymbol{L}_0 : \boldsymbol{A}_0 : \bar{\boldsymbol{\varepsilon}}, \qquad \boldsymbol{\sigma}_1 = \boldsymbol{L}_1 : \boldsymbol{\varepsilon}_1 = \boldsymbol{L}_1 : \boldsymbol{A}_1 : \bar{\boldsymbol{\varepsilon}}.$$

When a fourth–order isotropic tensor $\boldsymbol{C} = (3C^b, 2C^s)$ is multiplied with a second–order symmetric tensor

$$\boldsymbol{c} = \begin{bmatrix} c_{11} & c_{12} & c_{13} \\ c_{12} & c_{22} & c_{23} \\ c_{13} & c_{23} & c_{33} \end{bmatrix},$$

then the tensor algebra yields

$$\boldsymbol{C} : \boldsymbol{c} = \begin{bmatrix} D_1 & D_4 & D_5 \\ D_4 & D_2 & D_6 \\ D_5 & D_6 & D_3 \end{bmatrix}$$

with

$$D_1 = \left[3C^b - 2C^s\right] \frac{c_{11} + c_{22} + c_{33}}{3} + 2C^s c_{11},$$
$$D_2 = \left[3C^b - 2C^s\right] \frac{c_{11} + c_{22} + c_{33}}{3} + 2C^s c_{22},$$
$$D_3 = \left[3C^b - 2C^s\right] \frac{c_{11} + c_{22} + c_{33}}{3} + 2C^s c_{33},$$
$$D_4 = 2C^s c_{12}, \quad D_5 = 2C^s c_{13}, \quad D_6 = 2C^s c_{23}.$$

So

$$\boldsymbol{\varepsilon}_0 = \begin{bmatrix} 1.312 \cdot 10^{-3} & 0 & 0 \\ 0 & 1.312 \cdot 10^{-3} & 0 \\ 0 & 0 & -1.593 \cdot 10^{-5} \end{bmatrix},$$

$$\boldsymbol{\varepsilon}_1 = \begin{bmatrix} 2.709 \cdot 10^{-4} & 0 & 0 \\ 0 & 2.709 \cdot 10^{-4} & 0 \\ 0 & 0 & 3.707 \cdot 10^{-5} \end{bmatrix},$$

$$\boldsymbol{\sigma}_0 = \begin{bmatrix} 7.544 & 0 & 0 \\ 0 & 7.544 & 0 \\ 0 & 0 & 4.479 \end{bmatrix} \text{MPa},$$

$$\boldsymbol{\sigma}_1 = \begin{bmatrix} 11.595 & 0 & 0 \\ 0 & 11.595 & 0 \\ 0 & 0 & 5.75 \end{bmatrix} \text{MPa}.$$

3. The macroscopic stress tensor can be computed in two ways:

- from the macroscopic constitutive law

$$\bar{\boldsymbol{\sigma}} = \bar{\boldsymbol{L}} : \bar{\boldsymbol{\varepsilon}},$$

- from volume averaging the phases stress fields, that is,

$$\bar{\boldsymbol{\sigma}} = c_0 \boldsymbol{\sigma}_0 + c_1 \boldsymbol{\sigma}_1.$$

Both methods lead to

$$\bar{\boldsymbol{\sigma}} = \begin{bmatrix} 8.759 & 0 & 0 \\ 0 & 8.759 & 0 \\ 0 & 0 & 4.86 \end{bmatrix} \text{MPa}.$$

References

[1] S. Nemat–Nasser, M. Hori, Micromechanics: Overall Properties of Heterogeneous Materials, 2nd ed., North–Holland, Amsterdam, 1999.
[2] J. Qu, M. Cherkaoui, Fundamentals of Micromechanics of Solids, Wiley, New Jersey, 2006.
[3] Z. Hashin, S. Shtrikman, A variational approach to the theory of the elastic behaviour of multiphase materials, Journal of the Mechanics and Physics of Solids 11 (2) (1963) 127–140.
[4] Z. Hashin, S. Shtrikman, On some variational principles in anisotropic and nonhomogeneous elasticity, Journal of the Mechanics and Physics of Solids 10 (1962) 335–342.

CHAPTER 4

Voigt and Reuss bounds

OUTLINE

4.1 Theory

Both Voigt and Reuss developed their theories for the study of aggregates of crystals. Voigt [1] assumed that the strain remains uniform throughout an aggregate and the overall stress can be estimated by averaging the stresses over all possible lattice orientations. Reuss [2] considered the opposite procedure: assuming that the stress remains uniform throughout an aggregate, the overall strain can be estimated by averaging the strains over all possible lattice orientations. These two approaches allow us to obtain the macroscopic moduli of an aggregate. However, experimental evidences showed that the actual macroscopic properties lie between the predictions of the two methods. Indeed, the Voigt approach provides an upper bound estimate, whereas the Reuss approach yields a lower bound estimate.

Multiscale Modeling Approaches for Composites
https://doi.org/10.1016/B978-0-12-823143-2.00014-X

4.1.1 Voigt upper bound

Let us consider a heterogeneous body $\mathcal{B}$ with elasticity tensor $\boldsymbol{L}(\boldsymbol{x})$ subjected to the linear displacement $\boldsymbol{u}^0 = \bar{\boldsymbol{\varepsilon}} \cdot \boldsymbol{x}$ on the boundary $\partial\mathcal{B}$. To identify the Voigt upper bound, it is essential to employ the minimum potential energy theorem. The latter states that for all kinematically admissible strains $\boldsymbol{\varepsilon}_a$, the actual strain $\boldsymbol{\varepsilon}$ minimizes the mechanical strain energy

$$W = \frac{1}{2V} \int_{\mathcal{B}} \boldsymbol{\varepsilon} : \boldsymbol{L} : \boldsymbol{\varepsilon} \, \mathrm{d}V = \inf_{\boldsymbol{\varepsilon}_a} \frac{1}{2V} \int_{\mathcal{B}} \boldsymbol{\varepsilon}_a : \boldsymbol{L} : \boldsymbol{\varepsilon}_a \, \mathrm{d}V. \tag{4.1}$$

According to the average strain theorem and the Hill–Mandel principle (Definitions 3.6 and 3.7 of Chapter 3), W can be expressed as

$$W = \frac{1}{2V} \int_{\mathcal{B}} \boldsymbol{\varepsilon} : \boldsymbol{\sigma} \, \mathrm{d}V = \frac{1}{2} \left[\frac{1}{V} \int_{\mathcal{B}} \boldsymbol{\varepsilon} \, \mathrm{d}V \right] : \left[\frac{1}{V} \int_{\mathcal{B}} \boldsymbol{\sigma} \, \mathrm{d}V \right] = \frac{1}{2} \bar{\boldsymbol{\varepsilon}} : \bar{\boldsymbol{\sigma}},$$

where $\bar{\boldsymbol{\sigma}}$ can be written as the product of the hypothetical homogenized medium elasticity tensor $\bar{\boldsymbol{L}}$ and the applied macroscopic strain $\bar{\boldsymbol{\varepsilon}}$,

$$\bar{\boldsymbol{\sigma}} = \bar{\boldsymbol{L}} : \bar{\boldsymbol{\varepsilon}}.$$

Thus the minimum potential energy theorem allows us to write

$$\bar{\boldsymbol{\varepsilon}} : \bar{\boldsymbol{L}} : \bar{\boldsymbol{\varepsilon}} \leqslant \frac{1}{V} \int_{\mathcal{B}} \boldsymbol{\varepsilon}_a : \boldsymbol{L} : \boldsymbol{\varepsilon}_a \, \mathrm{d}V \tag{4.2}$$

for all kinematically admissible strain fields $\boldsymbol{\varepsilon}_a$. Voigt proposed to consider that the strain in all phases of the $\mathcal{B}$ are the same,

$$\boldsymbol{\varepsilon}_a = \bar{\boldsymbol{\varepsilon}} \quad \text{in } \mathcal{B}.$$

With such simplification we obtain

$$\bar{\boldsymbol{\varepsilon}} : \bar{\boldsymbol{L}} : \bar{\boldsymbol{\varepsilon}} \leqslant \bar{\boldsymbol{\varepsilon}} : \frac{1}{V} \int_{\mathcal{B}} \boldsymbol{L} \, \mathrm{d}V : \bar{\boldsymbol{\varepsilon}} = \bar{\boldsymbol{\varepsilon}} : \langle \boldsymbol{L} \rangle : \bar{\boldsymbol{\varepsilon}}. \tag{4.3}$$

Since $\boldsymbol{\varepsilon}_a$ is arbitrary, relation (4.3) holds for every macroscopic strain $\bar{\boldsymbol{\varepsilon}}$. The above result is called the Voigt upper bound estimate.

4.1.2 Reuss lower bound

Let us consider again the heterogeneous body $\mathcal{B}$ with elasticity tensor $\boldsymbol{L}(\boldsymbol{x})$, but this time subjected to the uniform traction $\boldsymbol{t}^0 = \bar{\boldsymbol{\sigma}} \cdot \boldsymbol{n}$ on the boundary $\partial\mathcal{B}$ ($\boldsymbol{n}$ being the normal vector to the boundary surface). To identify the Reuss lower bound, we can employ the minimum complementary energy theorem. The latter states that for all statically admissible stresses $\boldsymbol{\sigma}_a$, the actual stress $\boldsymbol{\sigma}$ minimizes the mechanical complementary energy

$$\Psi = \frac{1}{2V}\int_{\mathcal{B}} \boldsymbol{\sigma} : \boldsymbol{L}^{-1} : \boldsymbol{\sigma}\, \mathrm{d}V = \inf_{\boldsymbol{\sigma}_a} \frac{1}{2V}\int_{\mathcal{B}} \boldsymbol{\sigma}_a : \boldsymbol{L}^{-1} : \boldsymbol{\sigma}_a\, \mathrm{d}V. \tag{4.4}$$

According to the average stress theorem and the Hill–Mandel principle (Definitions 3.5 and 3.7 of Chapter 3), Ψ can be written as

$$\Psi = \frac{1}{2V}\int_{\mathcal{B}} \boldsymbol{\varepsilon} : \boldsymbol{\sigma}\, \mathrm{d}V = \frac{1}{2}\left[\frac{1}{V}\int_{\mathcal{B}} \boldsymbol{\varepsilon}\, \mathrm{d}V\right] : \left[\frac{1}{V}\int_{\mathcal{B}} \boldsymbol{\sigma}\, \mathrm{d}V\right] = \frac{1}{2}\bar{\boldsymbol{\varepsilon}} : \bar{\boldsymbol{\sigma}},$$

where $\bar{\boldsymbol{\varepsilon}}$ is expressed as the product of the hypothetical homogenized medium compliance tensor $\bar{\boldsymbol{L}}^{-1}$ and the applied macroscopic stress $\bar{\boldsymbol{\sigma}}$,

$$\bar{\boldsymbol{\varepsilon}} = \bar{\boldsymbol{L}}^{-1} : \bar{\boldsymbol{\sigma}}.$$

Thus the minimum potential energy theorem allows us to write

$$\bar{\boldsymbol{\sigma}} : \bar{\boldsymbol{L}}^{-1} : \bar{\boldsymbol{\sigma}} \leqslant \frac{1}{V}\int_{\mathcal{B}} \boldsymbol{\sigma}_a : \boldsymbol{L}^{-1} : \boldsymbol{\sigma}_a\, \mathrm{d}V \tag{4.5}$$

for all statically admissible stress fields $\boldsymbol{\sigma}_a$. Reuss proposed to consider that the stress in all phases of the $\mathcal{B}$ are the same,

$$\boldsymbol{\sigma}_a = \bar{\boldsymbol{\sigma}} \quad \text{in } \mathcal{B}.$$

With such simplification we obtain

$$\bar{\boldsymbol{\sigma}} : \bar{\boldsymbol{L}}^{-1} : \bar{\boldsymbol{\sigma}} \leqslant \bar{\boldsymbol{\sigma}} : \frac{1}{V}\int_{\mathcal{B}} \boldsymbol{L}^{-1}\, \mathrm{d}V : \bar{\boldsymbol{\sigma}} = \bar{\boldsymbol{\sigma}} : \left\langle \boldsymbol{L}^{-1} \right\rangle : \bar{\boldsymbol{\sigma}}. \tag{4.6}$$

Since $\boldsymbol{\sigma}_a$ is arbitrary, relation (4.6) holds for every macroscopic stress $\bar{\boldsymbol{\sigma}}$. The $\left\langle \boldsymbol{L}^{-1} \right\rangle$ minimizes the mechanical complementary energy, and, consequently, its inverse[1] necessarily maximizes the mechanical strain energy:

$$\bar{\boldsymbol{\varepsilon}} : \bar{\boldsymbol{L}} : \bar{\boldsymbol{\varepsilon}} \geqslant \bar{\boldsymbol{\varepsilon}} : \left\langle \boldsymbol{L}^{-1} \right\rangle^{-1} : \bar{\boldsymbol{\varepsilon}}. \tag{4.7}$$

This result is called the Reuss lower bound estimate.

The Voigt and Reuss estimates nowadays have been shown to be very conservative predictions for the overall properties of the majority of composites. Nevertheless, in certain cases, where the actual characteristics of the microstructure are unknown, these approaches are still utilized.

[1] This property for the inverse tensor holds due to the positive definite nature of $\boldsymbol{L}$. More details about positive definiteness are provided in Chapter 10.

To illustrate the use of the Voigt and Reuss estimates, the sections that follow are dedicated to the study of aligned fiber composites and composite beams under bending.

4.2 Simple methods for fiber composites

The fiber composites are material systems that have been extensively studied in the literature. In the following chapters, we discuss several methods for treating these media. In this section, we present a simplistic approach that provides a rough estimation for the overall behavior of a unidirectional fiber composite. This approach is based on the Voigt and Reuss bounds.

The unidirectional long cylindrical fiber composites exhibit transversely isotropic effective behavior. The compliance tensor (2.20) in Chapter 2 allows us to express the strain in terms of the stress (in the Voigt notation), thus allowing for imposing simple loading cases as uniaxial stress or simple shearing to compute the macroscopic moduli. This process can be combined with additional assumptions and some known results of the two–phase material behavior.

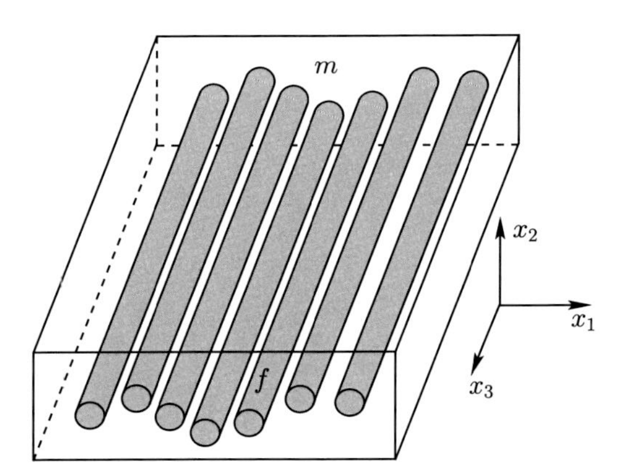

FIGURE 4.1 Composite consisting of a matrix phase (m) and unidirectional long cylindrical fibers (f).

Fig. 4.1 illustrates the heterogeneous body consisting of two isotropic materials, the matrix (index m), and the unidirectional fibers (index f). In these types of composites, usually the fiber is stiffer than the matrix phase. The constitutive relations between the strains and stresses for these media are written in the Voigt notation as

$$\begin{bmatrix} \varepsilon_{11}^{(r)} \\ \varepsilon_{22}^{(r)} \\ \varepsilon_{33}^{(r)} \\ 2\varepsilon_{12}^{(r)} \\ 2\varepsilon_{13}^{(r)} \\ 2\varepsilon_{23}^{(r)} \end{bmatrix} = \begin{bmatrix} \frac{1}{E_r} & -\frac{\nu_r}{E_r} & -\frac{\nu_r}{E_r} & 0 & 0 & 0 \\ -\frac{\nu_r}{E_r} & \frac{1}{E_r} & -\frac{\nu_r}{E_r} & 0 & 0 & 0 \\ -\frac{\nu_r}{E_r} & -\frac{\nu_r}{E_r} & \frac{1}{E_r} & 0 & 0 & 0 \\ 0 & 0 & 0 & \frac{1}{\mu_r} & 0 & 0 \\ 0 & 0 & 0 & 0 & \frac{1}{\mu_r} & 0 \\ 0 & 0 & 0 & 0 & 0 & \frac{1}{\mu_r} \end{bmatrix} \cdot \begin{bmatrix} \sigma_{11}^{(r)} \\ \sigma_{22}^{(r)} \\ \sigma_{33}^{(r)} \\ \sigma_{12}^{(r)} \\ \sigma_{13}^{(r)} \\ \sigma_{23}^{(r)} \end{bmatrix}, \tag{4.8}$$

where

$$\mu_r = \frac{E_r}{2[1+\nu_r]}. \tag{4.9}$$

The index r stands for matrix m or fiber f. Moreover, the stress and strain fields represent average quantities over the matrix or the fiber following the concept of composites with distinct phases discussed in Chapter 3. The volume fraction of the fibers is denoted as c_f, whereas the volume fraction of the matrix is denoted as $c_m = 1 - c_f$. The constitutive relations for the transversely isotropic composite (x_3 being the normal direction of the plane of isotropy) are given in the Voigt notation form

$$\begin{bmatrix} \bar{\varepsilon}_{11} \\ \bar{\varepsilon}_{22} \\ \bar{\varepsilon}_{33} \\ 2\bar{\varepsilon}_{12} \\ 2\bar{\varepsilon}_{13} \\ 2\bar{\varepsilon}_{23} \end{bmatrix} = \begin{bmatrix} \frac{1}{\bar{E}^{\rm tr}} & -\frac{\bar{\nu}^{\rm tt}}{\bar{E}^{\rm tr}} & -\frac{\bar{\nu}^{\rm at}}{\bar{E}^{\rm ax}} & 0 & 0 & 0 \\ -\frac{\bar{\nu}^{\rm tt}}{\bar{E}^{\rm tr}} & \frac{1}{\bar{E}^{\rm tr}} & -\frac{\bar{\nu}^{\rm at}}{\bar{E}^{\rm ax}} & 0 & 0 & 0 \\ -\frac{\bar{\nu}^{\rm at}}{\bar{E}^{\rm ax}} & -\frac{\bar{\nu}^{\rm at}}{\bar{E}^{\rm ax}} & \frac{1}{\bar{E}^{\rm ax}} & 0 & 0 & 0 \\ 0 & 0 & 0 & \frac{1}{\bar{\mu}^{\rm tr}} & 0 & 0 \\ 0 & 0 & 0 & 0 & \frac{1}{\bar{\mu}^{\rm ax}} & 0 \\ 0 & 0 & 0 & 0 & 0 & \frac{1}{\bar{\mu}^{\rm ax}} \end{bmatrix} \cdot \begin{bmatrix} \bar{\sigma}_{11} \\ \bar{\sigma}_{22} \\ \bar{\sigma}_{33} \\ \bar{\sigma}_{12} \\ \bar{\sigma}_{13} \\ \bar{\sigma}_{23} \end{bmatrix} \tag{4.10}$$

with

$$\bar{\mu}^{\rm tr} = \frac{\bar{E}^{\rm tr}}{2[1+\bar{\nu}^{\rm tt}]}. \tag{4.11}$$

The above expressions indicate that parallel and transverse to the fiber direction the composite behaves differently. Indeed, along the fiber direction the fibers support the matrix, providing a significant enhancement in the stiffness of the final medium. In contrast, transversely to the fibers the response of the composite is similar to that of the matrix. In light of these observations, the Voigt and Reuss assumptions can be used to identify the macroscopic properties of the medium.

Along the x_3 direction, we can reasonably assume that the fibers and the matrix experience the same normal strain $\bar{\varepsilon}_{33}$. However, in the $x_1 - x_2$ plane direction a more realistic approach is to consider that the stresses (normal and shear) of the two phases are equal. A similar assumption can be also made for the axial shear stresses. These main assumptions are written formally as [3]

$$\begin{aligned}
\bar{\sigma}_{11} &= \sigma_{11}^{(f)} = \sigma_{11}^{(m)}, \\
\bar{\sigma}_{22} &= \sigma_{22}^{(f)} = \sigma_{22}^{(m)}, \\
\bar{\varepsilon}_{33} &= \varepsilon_{33}^{(f)} = \varepsilon_{33}^{(m)}, \\
\bar{\sigma}_{12} &= \sigma_{12}^{(f)} = \sigma_{12}^{(m)}, \\
\bar{\sigma}_{13} &= \sigma_{13}^{(f)} = \sigma_{13}^{(m)}, \\
\bar{\sigma}_{23} &= \sigma_{23}^{(f)} = \sigma_{23}^{(m)}.
\end{aligned} \tag{4.12}$$

Taking into account the latter relations, rearrangement of system (4.8) yields

$$\begin{bmatrix} \varepsilon_{11}^{(r)} \\ \varepsilon_{22}^{(r)} \\ \sigma_{33}^{(r)} \\ 2\varepsilon_{12}^{(r)} \\ 2\varepsilon_{13}^{(r)} \\ 2\varepsilon_{23}^{(r)} \end{bmatrix} = \begin{bmatrix} \dfrac{1-\nu_r^2}{E_r} & -\dfrac{\nu_r+\nu_r^2}{E_r} & -\nu_r & 0 & 0 & 0 \\ -\dfrac{\nu_r+\nu_r^2}{E_r} & \dfrac{1-\nu_r^2}{E_r} & -\nu_r & 0 & 0 & 0 \\ \nu_r & \nu_r & E_r & 0 & 0 & 0 \\ 0 & 0 & 0 & \dfrac{1}{\mu_r} & 0 & 0 \\ 0 & 0 & 0 & 0 & \dfrac{1}{\mu_r} & 0 \\ 0 & 0 & 0 & 0 & 0 & \dfrac{1}{\mu_r} \end{bmatrix} \cdot \begin{bmatrix} \bar{\sigma}_{11} \\ \bar{\sigma}_{22} \\ \bar{\varepsilon}_{33} \\ \bar{\sigma}_{12} \\ \bar{\sigma}_{13} \\ \bar{\sigma}_{23} \end{bmatrix}. \tag{4.13}$$

In addition, a similar rearrangement for the macroscopic relations (4.10) gives

$$\begin{bmatrix} \bar{\varepsilon}_{11} \\ \bar{\varepsilon}_{22} \\ \bar{\sigma}_{33} \\ 2\bar{\varepsilon}_{12} \\ 2\bar{\varepsilon}_{13} \\ 2\bar{\varepsilon}_{23} \end{bmatrix} = \begin{bmatrix} \frac{1}{\bar{E}^{\mathrm{tr}}} - \frac{[\bar{\nu}^{\mathrm{at}}]^2}{\bar{E}^{\mathrm{ax}}} & -\frac{\bar{\nu}^{\mathrm{tt}}}{\bar{E}^{\mathrm{tr}}} - \frac{[\bar{\nu}^{\mathrm{at}}]^2}{\bar{E}^{\mathrm{ax}}} & -\bar{\nu}^{\mathrm{at}} & 0 & 0 & 0 \\ -\frac{\bar{\nu}^{\mathrm{tt}}}{\bar{E}^{\mathrm{tr}}} - \frac{[\bar{\nu}^{\mathrm{at}}]^2}{\bar{E}^{\mathrm{ax}}} & \frac{1}{\bar{E}^{\mathrm{tr}}} - \frac{[\bar{\nu}^{\mathrm{at}}]^2}{\bar{E}^{\mathrm{ax}}} & -\bar{\nu}^{\mathrm{at}} & 0 & 0 & 0 \\ \bar{\nu}^{\mathrm{at}} & \bar{\nu}^{\mathrm{at}} & \bar{E}^{\mathrm{ax}} & 0 & 0 & 0 \\ 0 & 0 & 0 & \frac{1}{\bar{\mu}^{\mathrm{tr}}} & 0 & 0 \\ 0 & 0 & 0 & 0 & \frac{1}{\bar{\mu}^{\mathrm{ax}}} & 0 \\ 0 & 0 & 0 & 0 & 0 & \frac{1}{\bar{\mu}^{\mathrm{ax}}} \end{bmatrix} \cdot \begin{bmatrix} \bar{\sigma}_{11} \\ \bar{\sigma}_{22} \\ \bar{\varepsilon}_{33} \\ \bar{\sigma}_{12} \\ \bar{\sigma}_{13} \\ \bar{\sigma}_{23} \end{bmatrix}. \tag{4.14}$$

According to the basic concepts of the homogenization theory, the remaining macroscopic strains and stress must be equal to the volume average in the RVE of their counterparts. This leads to the expressions

$$\begin{aligned} \bar{\varepsilon}_{11} &= c_f \varepsilon_{11}^{(f)} + c_m \varepsilon_{11}^{(m)}, \\ \bar{\varepsilon}_{22} &= c_f \varepsilon_{22}^{(f)} + c_m \varepsilon_{22}^{(m)}, \\ \bar{\sigma}_{33} &= c_f \sigma_{33}^{(f)} + c_m \sigma_{33}^{(m)}, \\ \bar{\varepsilon}_{12} &= c_f \varepsilon_{12}^{(f)} + c_m \varepsilon_{12}^{(m)}, \\ \bar{\varepsilon}_{13} &= c_f \varepsilon_{13}^{(f)} + c_m \varepsilon_{13}^{(m)}, \\ \bar{\varepsilon}_{23} &= c_f \varepsilon_{23}^{(f)} + c_m \varepsilon_{23}^{(m)}. \end{aligned} \tag{4.15}$$

Substituting (4.13) into (4.15) and comparing with (4.14), we can compute the macroscopic properties of the unidirectional composite.

BOX 4.1 Unidirectional composite (Voigt–Reuss approach)

For unidirectional long cylindrical fiber composites (Fig. 4.1), the Voigt and Reuss approximations for the strain and stress fields yield the following macroscopic properties formulas:

$$\begin{aligned} \bar{E}^{\mathrm{ax}} &= c_m E_m + c_f E_f, \\ \bar{\nu}^{\mathrm{at}} &= c_m \nu_m + c_f \nu_f, \\ \bar{E}^{\mathrm{tr}} &= \left[\frac{c_m}{E_m} + \frac{c_f}{E_f} - \frac{c_f c_m [E_m \nu_f - E_f \nu_m]^2}{E_f E_m [c_m E_m + c_f E_f]} \right]^{-1}, \\ \bar{\mu}^{\mathrm{tr}} = \bar{\mu}^{\mathrm{ax}} &= \left[\frac{c_m}{\mu_m} + \frac{c_f}{\mu_f} \right]^{-1}. \end{aligned} \tag{4.16}$$

The transverse Poisson ratio $\bar{\nu}^{tt}$ can be computed using the generic expression

$$\bar{\nu}^{tt} = \frac{\bar{E}^{tr}}{2\bar{\mu}^{tr}} - 1. \tag{4.17}$$

Note that a more conservative prediction for the transverse Young modulus $\bar{E}^{tr}$ commonly appears in the literature. If the third term in the right–hand side of $(4.16)_3$ is ignored, then the remaining terms provide the simpler formula

$$\bar{E}^{tr} = \left[\frac{c_m}{E_m} + \frac{c_f}{E_f} \right]^{-1}. \tag{4.18}$$

4.3 Composite beams

Beams are structural elements that appear frequently in engineering applications. Especially in civil engineering structures, bending is a critical loading condition that drives the material selection and the characteristics of the beams. In concrete–based structures the requirements of proper design lead to the use of reinforced concrete and the construction of composite beams, that is, beams consisting of more than one material.

The Euler–Bernoulli theory is a well–established methodology for beams under bending. The following subsections present some key ingredients of the classical beam bending theory and its application to composite beams by considering homogenization techniques like the Voigt and Reuss bounds.

4.3.1 Essential elements of beam bending theory

This part is a brief recall of the key elements from the beam bending theory that allow us to pass to the composite beams study. A more thorough discussion and presentation of the beam bending theory can be found in [4].

Let us consider a typical simply supported beam of length L, Young modulus E, and rectangular cross–section (width b, height h), subjected to the distributed load $q(x)$ (Fig. 4.2). Due to this loading, the beam is deformed and normal strains parallel to the x direction are developed. The main assumption in the classical beam theory is that the deformed cross–sections remain plane, independently of the material properties. Under this condition, the longitudinal strains are linear and related to the local radius of curvature $R(x)$ of the beam by

$$\varepsilon_{xx}(x, z) = \frac{z}{R(x)}, \tag{4.19}$$

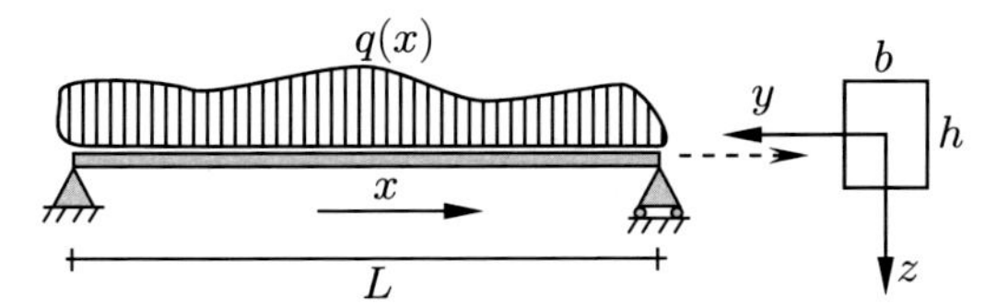

FIGURE 4.2 Simply supported beam with rectangular cross–section under distributed loading.

where x is the coordinate measuring the position of the cross–section along the axis of the beam, and z denotes the distance from the neutral axis (Fig. 4.3$_a$). Recall that in rectangular cross–sections the neutral axis coincides with the axis that passes from the centroid of the beam cross–section.

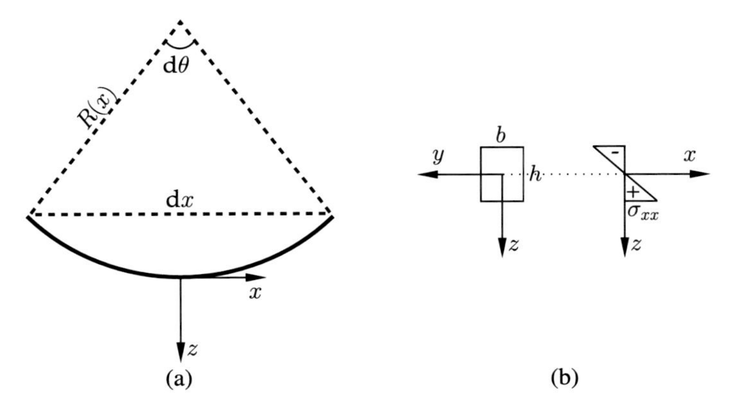

FIGURE 4.3 Bending of simply supported beam: (a) Deflection of beam and (b) distribution of longitudinal stresses across the cross–section.

From the above expression of the longitudinal strain component the corresponding stress component is expressed as

$$\sigma_{xx}(x,z) = E\varepsilon_{xx} = E\frac{z}{R(x)}. \tag{4.20}$$

The distribution of σ_{xx} in the z direction of the cross–section is schematically illustrated in Fig. 4.3$_b$. Since the beam is in equilibrium, the average stress over the cross–section must be zero, that is,

$$\int_A \sigma_{xx} \mathrm{d}A = 0, \tag{4.21}$$

where A is the area of the cross–section. On the other hand, the average moment M in the cross–section is computed by the formula

$$M = \int_A z\sigma_{xx} \mathrm{d}A. \tag{4.22}$$

Eqs. (4.21) and (4.22) are general and hold also for composite beams. For the specific beam of Fig. 4.2, the relation

$$\int_A \sigma_{xx} \mathrm{d}A = \frac{E}{R(x)} \int_{-b/2}^{b/2} \int_{-z/2}^{z/2} z \mathrm{d}y \mathrm{d}z = 0 \tag{4.23}$$

holds due to the symmetric structure of the cross–section, confirming that the neutral axis x coincides with the centroid axis [4]. In addition,

$$M = \int_A z\sigma_{xx} \mathrm{d}A = \frac{E}{R(x)} \int_{-b/2}^{b/2} \int_{-z/2}^{z/2} z^2 \mathrm{d}y \mathrm{d}z = \frac{EI}{R(x)}, \tag{4.24}$$

where

$$I = \int_A z^2 \mathrm{d}A \tag{4.25}$$

denotes the moment of inertia. For a rectangular cross–section, $I = \frac{bh^3}{12}$. If the cross–section consists of multiple rectangles with different heights and widths, then the moment of inertia can be evaluated through the Steiner theorem.

4.3.2 Beam made of two materials

We can apply the basic principles discussed in the previous subsection to composite beams. Consider a beam of length L with rectangular cross–section (width b, height h), made of two distinct material phases, the matrix and the reinforcement (Fig. 4.4). The reinforcement occupies the space Ω_1 and has Young modulus E_1 and the volume fraction c, whereas the matrix occupies the space Ω_2 and has Young modulus $E_2 < E_1$ and the volume fraction $1 - c$. If the beam is simply supported and is subjected to distributed load, then the cross–sections remain plane after deformation according to the main beam theory hypothesis. Such an assumption leads to a Voigt–type approach of equal longitudinal strains in the two phases, that is,

$$\begin{aligned}
\varepsilon_{xx}^{(1)}(x, y, z) &= \bar{\varepsilon}_{xx}(x, z) = \frac{z}{R(x)}, \qquad \forall y, z \in \Omega_1, \ \ \forall x \in [0, L], \\
\varepsilon_{xx}^{(2)}(x, y, z) &= \bar{\varepsilon}_{xx}(x, z) = \frac{z}{R(x)}, \qquad \forall y, z \in \Omega_2, \ \ \forall x \in [0, L],
\end{aligned} \tag{4.26}$$

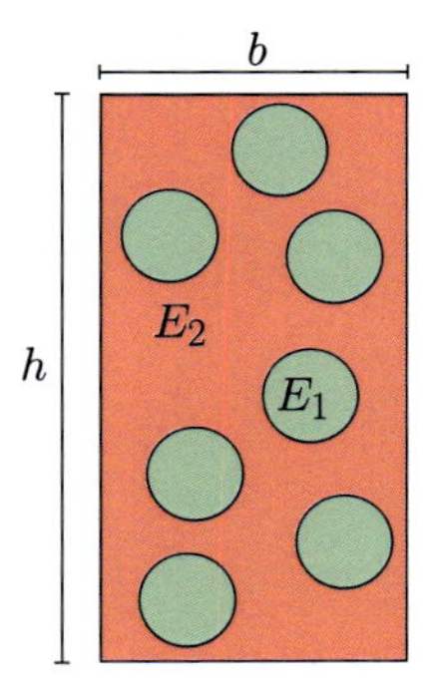

FIGURE 4.4 Composite beam made of a matrix phase and reinforcement. The volume fraction of reinforcement is equal to c.

where indices (1) and (2) denote quantities in the reinforcement and the matrix, respectively. As usual, a bar above a symbol denotes the overall (i.e., macroscopic) quantity. The resulting longitudinal stresses are expressed as

$$\begin{aligned}\sigma_{xx}^{(1)}(x,y,z) &= \frac{E_1 z}{R(x)}, \qquad \forall y,z \in \Omega_1, \quad \forall x \in [0,L],\\ \sigma_{xx}^{(2)}(x,y,z) &= \frac{E_2 z}{R(x)}, \qquad \forall y,z \in \Omega_2, \quad \forall x \in [0,L].\end{aligned} \tag{4.27}$$

The Voigt approach leads to an overall Young modulus given by the rule of mixtures:

$$\bar{E} = cE_1 + [1-c]E_2. \tag{4.28}$$

In addition, the overall longitudinal stress is given by

$$\bar{\sigma}_{xx} = c\sigma_{xx}^{(1)} + [1-c]\sigma_{xx}^{(2)} = \frac{\bar{E}z}{R(x)}. \tag{4.29}$$

The presence of the reinforcement alters the neutral axis position, which no longer coincides with the axis that passes from the centroid of the beam cross–section. To identify the new position of the neutral axis, we have to proceed to further assumptions. A simplified approach considers that the reinforcement can be seen as one solid phase of height ch that occupies the upper part of the cross–section (Fig. 4.5_a). In the "new beam", the neutral axis has a distance z_1 from the centroid of phase 1 and a distance z_2 from the centroid of phase 2. The two distances are connected with the relation

$$z_1 + z_2 = h/2. \tag{4.30}$$

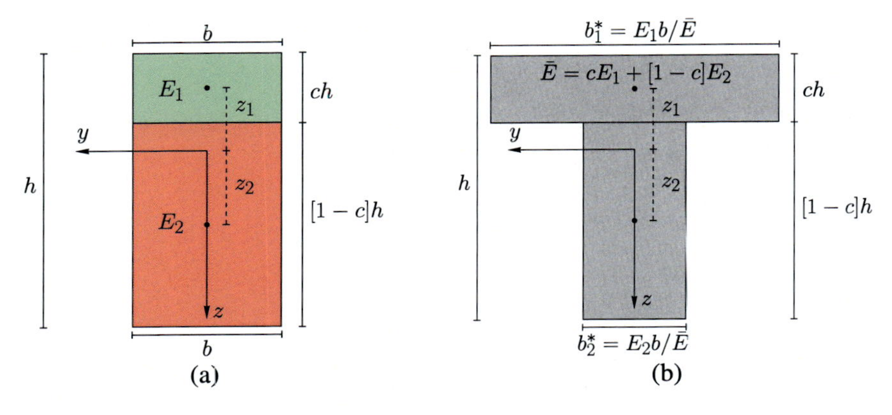

FIGURE 4.5 (a) Simplified model of the beam, where the reinforcement occupies the upper part of the cross–section. (b) Equivalent cross–section for the macroscopic medium with Young modulus $\bar{E}$.

For the clarity of the subsequent calculations, we define the following distances:

$$
\begin{aligned}
y_a &= \frac{b}{2}, \\
z_a &= z_1 + \frac{ch}{2}, \\
z_b &= z_2 + \frac{[1-c]h}{2}, \\
z_c &= z_1 - \frac{ch}{2} = \frac{[1-c]h}{2} - z_2.
\end{aligned} \tag{4.31}
$$

The equilibrium equation (4.21) for the composite beam reads

$$
\begin{aligned}
\int_{-y_a}^{y_a} \int_{-z_a}^{z_b} \sigma_{xx} \mathrm{d}z\mathrm{d}y &= \frac{E_1 b}{R(x)} \int_{-z_a}^{-z_c} z\mathrm{d}z + \frac{E_2 b}{R(x)} \int_{-z_c}^{z_b} z\mathrm{d}z \\
&= -E_1 bchz_1 + E_2 b[1-c]hz_2 = 0.
\end{aligned} \tag{4.32}
$$

Setting

$$
b_1^* = \frac{E_1 b}{\bar{E}}, \quad b_2^* = \frac{E_2 b}{\bar{E}}, \tag{4.33}
$$

we get the following relation:

$$
cb_1^* z_1 = [1-c]b_2^* z_2. \tag{4.34}
$$

Combining the last expression with (4.30), we obtain the actual distance of the neutral axis from the bottom plane of the cross–section:

$$z_2 = \frac{h}{2}\frac{cb_1^*}{cb_1^* + [1-c]b_2^*}. \tag{4.35}$$

Concerning the average moment at the cross–section, this is given by

$$\begin{aligned} M(x) &= \int_{-y_a}^{y_a}\int_{-z_a}^{z_b} z\sigma_{xx}\mathrm{d}z\mathrm{d}y \\ &= \frac{E_1}{R(x)}\int_{-y_a}^{y_a}\int_{-z_a}^{-z_c} z^2\mathrm{d}z\mathrm{d}y + \frac{E_2}{R(x)}\int_{-y_a}^{y_a}\int_{-z_c}^{z_b} z^2\mathrm{d}z\mathrm{d}y. \end{aligned} \tag{4.36}$$

The two integrals of the last expression represent the moments of inertia of the two parts of the cross–section (the matrix and the reinforcement). According to the Steiner theorem,

$$\begin{aligned} \int_{-y_a}^{y_a}\int_{-z_a}^{-z_c} z^2\mathrm{d}z\mathrm{d}y &= \frac{bc^3h^3}{12} + bchz_1^2, \\ \int_{-y_a}^{y_a}\int_{-z_c}^{z_b} z^2\mathrm{d}z\mathrm{d}y &= \frac{b[1-c]^3h^3}{12} + b[1-c]hz_2^2. \end{aligned} \tag{4.37}$$

Thus

$$M(x) = \frac{\bar{E}\bar{I}}{R(x)}, \tag{4.38}$$

where

$$\bar{I} = \left[\frac{b_1^*c^3h^3}{12} + b_1^*chz_1^2 + \frac{b_2^*[1-c]^3h^3}{12} + b_2^*[1-c]hz_2^2\right] \tag{4.39}$$

is the moment of inertia of the equivalent beam, which consists of the overall (macroscopic) medium (Fig. 4.5_b). Consequently:

BOX 4.2 Composite beam (Voigt approach)

For a composite beam made of a matrix phase and reinforcement subjected to bending, the Voigt approach yields substitution of the actual cross–section of Fig. 4.4 with the equivalent cross–section of Fig. 4.5_b, whose Young modulus is equal to $\bar{E}$ given by the rule of mixtures formula. The distances z_1 and z_2 are provided by the expressions

$$z_1 = \frac{h}{2}\frac{[1-c]b_2^*}{cb_1^* + [1-c]b_2^*}, \quad z_2 = \frac{h}{2}\frac{cb_1^*}{cb_1^* + [1-c]b_2^*}. \tag{4.40}$$

4.4 Examples

Example 1: glass fiber reinforced epoxy

Assuming a bi–phase composite made of aligned glass fibers and epoxy matrix, identify its overall Young and shear moduli as a function of the fibers volume fraction using the approach presented in Section 4.2. The glass is assumed to have Young modulus $E_f = 72$ GPa and Poisson's ratio $\nu_f = 0.2$, whereas the epoxy is characterized by Young modulus $E_m = 3$ GPa and Poisson ratio $\nu_m = 0.3$.

Solution:

With the help of expressions (4.16), the five properties of the fiber composite can be estimated as functions of c_f, given that $c_m = 1 - c_f$. The obtained Young and shear moduli are summarized in Fig. 4.6.

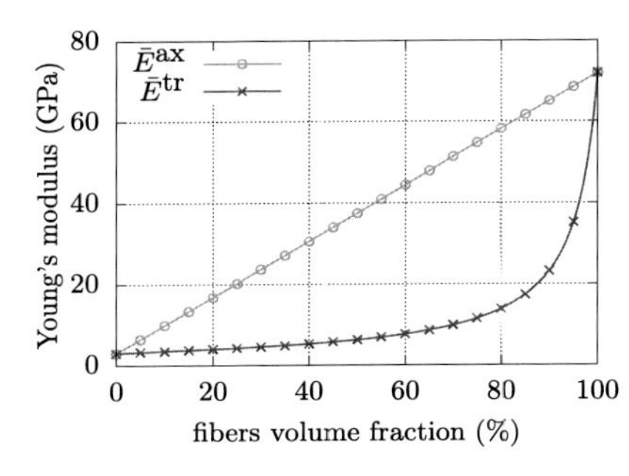

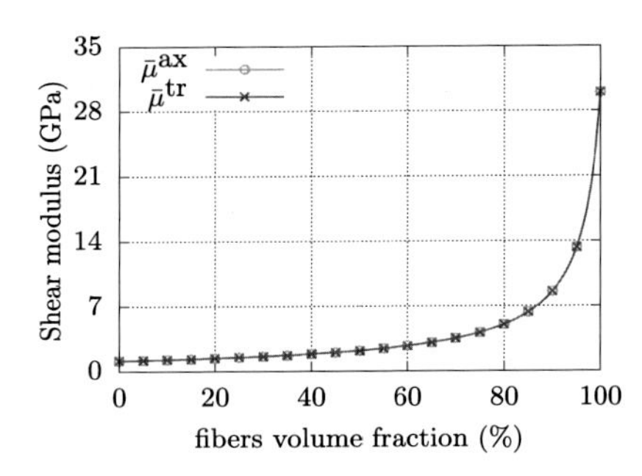

FIGURE 4.6 (a) Macroscopic Young moduli and (b) macroscopic shear moduli vs fibers volume fraction. Voigt and Reuss approach estimates.

Example 2: beam made of two periodic laminae

Consider a simply supported elastic beam of length L with doubly symmetric composite cross–section (width b, height $2h$) that is transversely loaded in the $x - z$ plane with the distributed load $q(x)$ (Fig. 4.7). The cross–section consists of two laminae with Young moduli E_1 and E_2, which are repeated $2k$ times (in the schematic of Fig. 4.7, $k = 3$). The total volume fractions of the two materials are c for the lamina 1 and $1 - c$ for the lamina 2. Identify the equivalent beam consisting of the equivalent medium as $k \to \infty$.

Solution:

This example can be seen as an extension of the classical technological beam theory to the case of composite beams with perfectly bonded constituents in the form of laminae. The basic assumptions of this approach are summarized in Section 4.3.

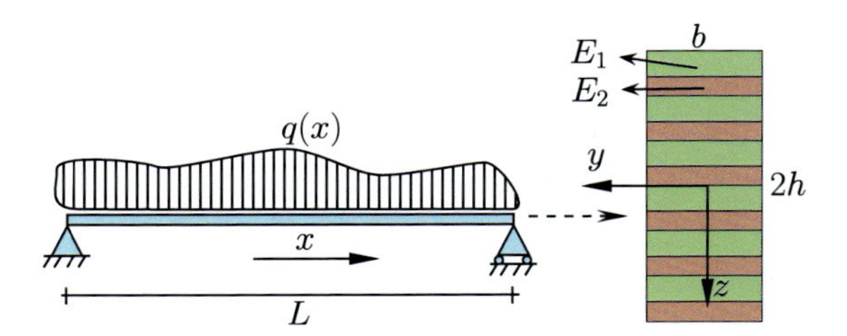

FIGURE 4.7 Simply supported beam with rectangular cross–section under distributed loading. The cross–section consists of two laminae that are repeated $2k$ times (in this schematic, $k = 3$).

For writing convenience, the rth lamina with Young modulus E_1 is considered to have a z coordinate from the neutral axis that belongs to the space $Z^{(1r)}$, and its arbitrary field ϕ is denoted as $\phi^{(1r)}$. Similarly, the rth lamina with Young modulus E_2 is considered to have a z coordinate from the neutral axis that belongs to the space $Z^{(2r)}$, and its arbitrary field ϕ is denoted as $\phi^{(2r)}$. The laminae have volume fraction (i.e., height) ch/k or $[1-c]h/k$ if they have Young modulus E_1 or E_2, respectively (Fig. 4.8).

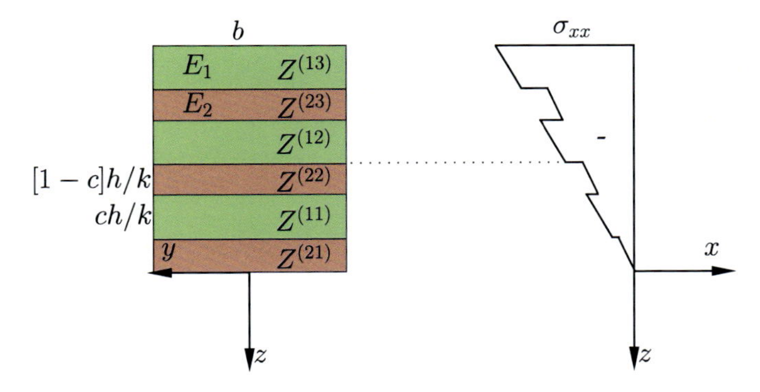

FIGURE 4.8 Schematic of longitudinal stress distribution at the upper half of a specific cross–section of the composite beam. At the lower half the stress is antisymmetric.

According to the Voigt approximation, the microscopic longitudinal strain at every material phase coincides with the macroscopic longitudinal strain

$$\varepsilon_{xx}^{(1r)}(x,z) = \bar{\varepsilon}_{xx}(x,z) = \frac{z}{R(x)}, \qquad \forall z \in Z^{(1r)}, \quad \forall x \in [0,L],$$
$$\varepsilon_{xx}^{(2r)}(x,z) = \bar{\varepsilon}_{xx}(x,z) = \frac{z}{R(x)}, \qquad \forall z \in Z^{(2r)}, \quad \forall x \in [0,L], \tag{4.41}$$

where x and z ares the coordinate measuring the position of the cross–section along the axis of the beam and from the neutral axis, respectively.

Due to the double symmetry of the section structure, the neutral axis coincides with the axis that passes from the centroid of the beam. The longitudinal stress at each lamina is given by

$$\begin{aligned}\sigma_{xx}^{(1r)}(x,z) &= \frac{E_1 z}{R(x)}, \qquad \forall z \in Z^{(1r)}, \quad \forall x \in [0,L],\\ \sigma_{xx}^{(2r)}(x,z) &= \frac{E_2 z}{R(x)}, \qquad \forall z \in Z^{(2r)}, \quad \forall x \in [0,L].\end{aligned} \tag{4.42}$$

Clearly, the stress field exhibits discontinuities at the interfaces (Fig. 4.8).

The Voigt approximation leads to a macroscopic Young modulus that obeys the rule of mixtures

$$\bar{E} = cE_1 + [1-c]E_2, \tag{4.43}$$

leading to a macroscopic longitudinal stress–strain relation of the form

$$\bar{\sigma}_{xx}(x,z) = \bar{E}\bar{\varepsilon}_{xx}(x,z) = \frac{\bar{E}z}{R(x)}. \tag{4.44}$$

The equilibrium of forces along the axis x leads to an identity due to the symmetric structure of the section. On the other hand, the equilibrium of moments is expressed as

$$M(x) = \int_{-b/2}^{b/2}\int_{-h}^{h} z\sigma_{xx}(x,z)\mathrm{d}z\mathrm{d}y. \tag{4.45}$$

Using Eqs. (4.42) and the symmetric structure of the cross–section, we write the moment (4.45) as

$$M(x) = \frac{E_1}{R(x)}\int_{-b/2}^{b/2} 2\int_{Z_1} z^2\mathrm{d}z\mathrm{d}y + 2\frac{E_2}{R(x)}\int_{-b/2}^{b/2} 2\int_{Z_2} z^2\mathrm{d}z\mathrm{d}y, \tag{4.46}$$

where

$$\begin{aligned}Z_1 &= Z^{(11)} \cup Z^{(12)} \cup Z^{(13)} \cup \cdots \cup Z^{(1k)},\\ Z_2 &= Z^{(21)} \cup Z^{(22)} \cup Z^{(23)} \cup \cdots \cup Z^{(2k)}.\end{aligned} \tag{4.47}$$

The integrals in (4.46) represent the moments of inertia occupied by the two constituents. To compute these moments of inertia, use of the Steiner theorem is needed. Considering the cross–section part shown in Fig. 4.8, we have

$$\int_{-b/2}^{b/2}\int_{Z^{(11)}} z^2\mathrm{d}z\mathrm{d}y = \frac{bch}{k}\left[\frac{h}{k}\left[1-\frac{c}{2}\right]\right]^2 + \frac{bc^3h^3}{12k^3},$$

$$\int_{-b/2}^{b/2}\int_{Z^{(12)}} z^2 \mathrm{d}z\mathrm{d}y = \frac{bch}{k}\left[\frac{h}{k}\left[2-\frac{c}{2}\right]\right]^2 + \frac{bc^3h^3}{12k^3},$$
$$\int_{-b/2}^{b/2}\int_{Z^{(13)}} z^2 \mathrm{d}z\mathrm{d}y = \frac{bch}{k}\left[\frac{h}{k}\left[3-\frac{c}{2}\right]\right]^2 + \frac{bc^3h^3}{12k^3}. \tag{4.48}$$

Thus, for k laminae in the upper half of the cross–section, the first integral of (4.46) is expressed as

$$\begin{aligned}\int_{-b/2}^{b/2} 2\int_{Z_1} z^2 \mathrm{d}z\mathrm{d}y &= 2\sum_{n=1}^{k}\left[\frac{bch^3}{k^3}[n^2+\frac{c^2}{4}-nc]+\frac{bc^3h^3}{12k^3}\right] \\ &= 2\frac{bch^3}{k^3}\sum_{k=1}^{n} n^2 + O(k^{-1}) \\ &= 2\frac{bch^3}{k^3}\frac{k(k+1)(2k+1)}{6} + O(k^{-1}) \\ &= 2\frac{bch^3}{3} + O(k^{-1}),\end{aligned} \tag{4.49}$$

where $O(k^{-1})$ denotes terms of order k^{-1} or lower. In a similar manner, working on the bottom half of the cross–section, we can obtain by analogous computations the following result for the second integral of (4.46):

$$\int_{-b/2}^{b/2} 2\int_{Z_2} z^2 \mathrm{d}z\mathrm{d}y = 2\frac{b[1-c]h^3}{3} + O(k^{-1}). \tag{4.50}$$

As $k \to \infty$, the $O(k^{-1})$ vanish. Then substituting (4.49), (4.50), and (4.43) into (4.46) yields

$$M(x) = \frac{cE_1 + [1-c]E_2}{R(x)}\frac{b[2h]^3}{12} = \frac{\bar{E}I}{R(x)}, \tag{4.51}$$

where I is the moment of inertia of the rectangular cross–section of width b and height $2h$.

Concluding, we observe in this example that the equivalent beam, in which Young modulus is given by the rule of mixtures formula, has a cross–section whose geometrical characteristics are the same as those of the initial composite beam cross–section.

References

[1] W. Voigt, Lehrbuch der Kristallphysik, B.G. Teulner, Leipzig, Germany, 1928.

[2] A. Reuss, Berechnung der Fließgrenze von Mischkristallen auf Grund der Plastizitätsbedingung für Einkristalle, Zeitschrift für Angewandte Mathematik und Mechanik (Journal of Applied Mathematics and Mechanics) 9 (1929) 49–58.

[3] M.C. Ray, Micromechanics of piezoelectric composites with improved effective piezoelectric constant, International Journal of Mechanics and Materials in Design 3 (2006) 361–371.
[4] F.P. Beer, E.R. Johnston, J.T. DeWolf, D.E. Mazurek, Mechanics of Materials, sixth ed., McGraw Hill, New York, 2012.

CHAPTER 5

Eshelby solution–based mean–field methods

OUTLINE

5.1 Inclusion problems

Amongst numerous developed homogenization theories for composite media, the "mean–field" approaches are considered as the most popular. In these methodologies the microscopic (RVE) problem is not solved explicitly. Instead, only average fields per phase are identified and utilized for obtaining the overall response of the composite. The pioneering work of Eshelby is the cornerstone of the mean–field theories.

The Eshelby inclusion problems [1] are revolutionary for the development of the micromechanics of media with distinct phases. Eshelby identified an analytical expression for the strain disturbance in an infinite elastic medium caused by the existence of an arbitrary inclusion of ellipsoidal shape. This problem constitutes the necessary background for the study of

Multiscale Modeling Approaches for Composites
https://doi.org/10.1016/B978-0-12-823143-2.00015-1

composites consisting of a matrix phase and randomly distributed ellipsoidal particles.

Here the Eshelby inclusion problem is briefly presented. More theoretical details are provided by Mura [2]. Before proceeding further, the following definitions are necessary [3]:

- An inclusion Ω of a body $\mathcal{B}$ is a region where uniform eigenstrains $\boldsymbol{\varepsilon}^*$ appear. These eigenstrains are generally considered inelastic, and they disturb the total strains $\boldsymbol{\varepsilon}$ in the body.
- An inhomogeneity Ω of a body $\mathcal{B}$ is a region where the material properties are different with regard to the rest of the body.

5.1.1 Eshelby's inclusion problem

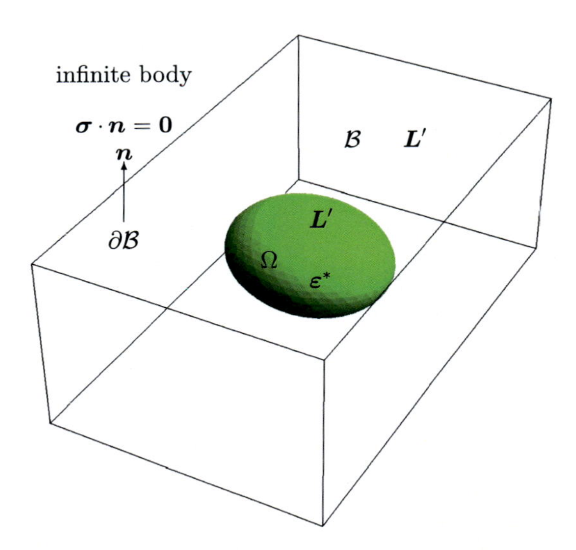

FIGURE 5.1 Eshelby's inclusion problem.

The inclusion problem studied by Eshelby assumes that an inclusion Ω is embedded in an infinite elastic body. Practically, this is equivalent to the case of a very small inclusion Ω inside a finite elastic body $\mathcal{B}$ with elasticity tensor $\boldsymbol{L}'$ (Fig. 5.1). The tractions and displacements at the boundary of the body are assumed zero, and in addition the body forces are ignored. In mathematical formalism the problem is described by the following set of equations:

$$\begin{aligned} \mathrm{div}\boldsymbol{\sigma} &= \mathbf{0} \quad \text{in } \mathcal{B}, \\ \boldsymbol{\sigma} &= \boldsymbol{L}' : [\boldsymbol{\varepsilon} - \boldsymbol{\varepsilon}^*] \quad \text{in } \mathcal{B}, \\ \boldsymbol{\sigma}\cdot\boldsymbol{n} &= \mathbf{0} \quad \text{in } \partial\mathcal{B}, \end{aligned}$$

$$\boldsymbol{\varepsilon}^* = \begin{cases} \text{constant} \neq \mathbf{0} & \text{in } \Omega\,, \\ \mathbf{0} & \text{in } \mathcal{B} - \Omega\,. \end{cases} \tag{5.1}$$

BOX 5.1 Eshelby tensor

Eshelby has proven that the total strain in an ellipsoidal inclusion, which is embedded in an infinite body and has an eigenstrain $\boldsymbol{\varepsilon}^*$, is uniform and is given by the expression

$$\boldsymbol{\varepsilon} = \text{constant} = \boldsymbol{S} : \boldsymbol{\varepsilon}^* \quad \text{in } \Omega\,, \tag{5.2}$$

where $\boldsymbol{S}$ is the fourth–order Eshelby tensor that depends on the geometry of the inclusion and the material properties $\boldsymbol{L}'$.

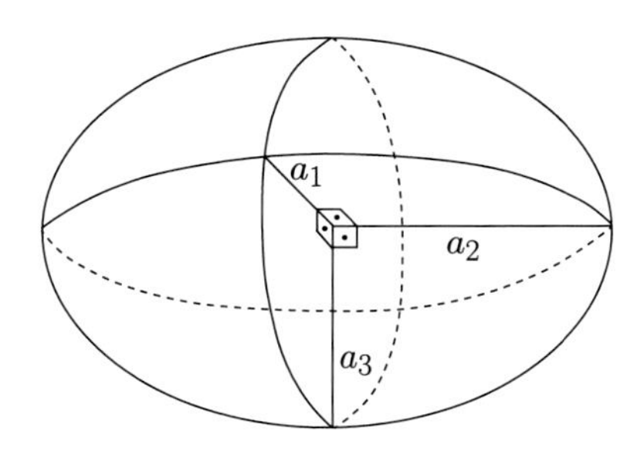

FIGURE 5.2 Ellipsoidal inclusion.

For an ellipsoidal inclusion with semiaxes a_1, a_2, a_3 (Fig. 5.2), the Eshelby tensor is given by the expression [2]

$$\begin{aligned} S_{ijkl} &= P_{ijmn} L'_{mnkl}, \\ P_{ijkl} &= \frac{1}{16\pi} \int_{-1}^{1} \int_{0}^{2\pi} \left[Z_{ik}^{-1} \widehat{\zeta}_j \widehat{\zeta}_l + Z_{jk}^{-1} \widehat{\zeta}_i \widehat{\zeta}_l + Z_{jl}^{-1} \widehat{\zeta}_i \widehat{\zeta}_k + Z_{il}^{-1} \widehat{\zeta}_j \widehat{\zeta}_k \right] \mathrm{d}\omega \mathrm{d}\zeta_3, \\ Z_{ij} &= L'_{ikjl} \widehat{\zeta}_k \widehat{\zeta}_l. \end{aligned} \tag{5.3}$$

The exponent -1 on the second–order tensor $\boldsymbol{Z}$ denotes the usual 3×3 matrix inverse. In the above expression, $\widehat{\boldsymbol{\zeta}}$ is the vector with components

$$\widehat{\zeta}_1 = \sqrt{1 - \zeta_3^2}\, \frac{\cos \omega}{a_1}, \quad \widehat{\zeta}_2 = \sqrt{1 - \zeta_3^2}\, \frac{\sin \omega}{a_2}, \quad \widehat{\zeta}_3 = \frac{\zeta_3}{a_3}. \tag{5.4}$$

For special inclusion geometries and isotropic materials, this tensor has analytical form [2]. For anisotropic materials, the Eshelby tensor can be computed numerically [4]. We need to be aware that the Eshelby tensor for a general ellipsoidal inclusion is a fourth–order tensor without major symmetries. Two special examples are the following:

- For isotropic elastic materials with spherical inclusions ($a_1 = a_2 = a_3$), the Eshelby tensor is isotropic and in the Voigt notation takes the "e–e" form

$$\widetilde{\boldsymbol{S}} = \begin{bmatrix} \gamma' + \frac{4}{3}\delta' & \gamma' - \frac{2}{3}\delta' & \gamma' - \frac{2}{3}\delta' & 0 & 0 & 0 \\ \gamma' - \frac{2}{3}\delta' & \gamma' + \frac{4}{3}\delta' & \gamma' - \frac{2}{3}\delta' & 0 & 0 & 0 \\ \gamma' - \frac{2}{3}\delta' & \gamma' - \frac{2}{3}\delta' & \gamma' + \frac{4}{3}\delta' & 0 & 0 & 0 \\ 0 & 0 & 0 & 2\delta' & 0 & 0 \\ 0 & 0 & 0 & 0 & 2\delta' & 0 \\ 0 & 0 & 0 & 0 & 0 & 2\delta' \end{bmatrix} \tag{5.5}$$

with

$$\gamma' = \frac{K'}{3K' + 4\mu'}, \quad \delta' = \frac{3K' + 6\mu'}{15K' + 20\mu'}. \tag{5.6}$$

In the above expressions, K' is the bulk modulus, and μ' is the shear modulus of the isotropic elastic body. Note that the multiplier 2 for the shear terms of $\boldsymbol{S}$ in the Voigt notation arises from the nature of the Eshelby tensor to connect two strain tensors (Chapter 1, Section 1.5). Due to its isotropy in the specific case, the Eshelby tensor can also be written in the special form of the Hill notation as (Example 2 in Section 1.4 of Chapter 1)

$$\boldsymbol{S} = (3\gamma', 2\delta').$$

- For isotropic elastic materials with infinitely long elliptical inclusions ($a_3 \to \infty$), the Eshelby tensor is non–symmetric and in the Voigt notation takes the "e–e" form

$$\widetilde{\boldsymbol{S}} = \begin{bmatrix} S_{1111} & S_{1122} & S_{1133} & 0 & 0 & 0 \\ S_{2211} & S_{2222} & S_{2233} & 0 & 0 & 0 \\ 0 & 0 & 0 & 0 & 0 & 0 \\ 0 & 0 & 0 & 2S_{1212} & 0 & 0 \\ 0 & 0 & 0 & 0 & 2S_{3131} & 0 \\ 0 & 0 & 0 & 0 & 0 & 2S_{2323} \end{bmatrix} \tag{5.7}$$

with

$$S_{1111} = \frac{1}{2[1-\nu']}\left[\frac{a_2^2 + 2a_1a_2}{[a_1+a_2]^2} + [1-2\nu']\frac{a_2}{a_1+a_2}\right],$$

$$S_{1122} = \frac{1}{2[1-\nu']}\left[\frac{a_2^2}{[a_1+a_2]^2} - [1-2\nu']\frac{a_2}{a_1+a_2}\right],$$

$$S_{2211} = \frac{1}{2[1-\nu']}\left[\frac{a_1^2}{[a_1+a_2]^2} - [1-2\nu']\frac{a_1}{a_1+a_2}\right],$$

$$S_{2222} = \frac{1}{2[1-\nu']}\left[\frac{a_1^2 + 2a_1a_2}{[a_1+a_2]^2} + [1-2\nu']\frac{a_1}{a_1+a_2}\right],$$

$$S_{1133} = \frac{\nu'}{1-\nu'}\frac{a_2}{a_1+a_2}, \qquad S_{2233} = \frac{\nu'}{1-\nu'}\frac{a_1}{a_1+a_2},$$

$$S_{1212} = \frac{1}{4[1-\nu']}\left[\frac{a_1^2 + a_2^2}{[a_1+a_2]^2} + 1 - 2\nu'\right],$$

$$S_{3131} = \frac{a_2}{2[a_1+a_2]}, \qquad S_{2323} = \frac{a_1}{2[a_1+a_2]},$$

and ν' denotes the Poisson ratio of the isotropic elastic body. When the two free semiaxes are equal ($a_1 = a_2$), the inclusions have cylindrical shape, and the Eshelby tensor reads

$$\widetilde{\boldsymbol{S}} = \begin{bmatrix} \frac{5-4\nu'}{8-8\nu'} & \frac{4\nu'-1}{8-8\nu'} & \frac{\nu'}{2-2\nu'} & 0 & 0 & 0 \\ \frac{4\nu'-1}{8-8\nu'} & \frac{5-4\nu'}{8-8\nu'} & \frac{\nu'}{2-2\nu'} & 0 & 0 & 0 \\ 0 & 0 & 0 & 0 & 0 & 0 \\ 0 & 0 & 0 & \frac{3-4\nu'}{4-4\nu'} & 0 & 0 \\ 0 & 0 & 0 & 0 & \frac{1}{2} & 0 \\ 0 & 0 & 0 & 0 & 0 & \frac{1}{2} \end{bmatrix}. \tag{5.8}$$

5.1.2 Inhomogeneity problem

Let us now assume a small inhomogeneity Ω with elasticity tensor $\boldsymbol{L}_r$ embedded in an infinite elastic body $\mathcal{B}$ with elasticity tensor $\boldsymbol{L}'$. A stress $\boldsymbol{\sigma}'$ and the corresponding strain $\boldsymbol{\varepsilon}'$ are applied at the far field (Fig. 5.3).

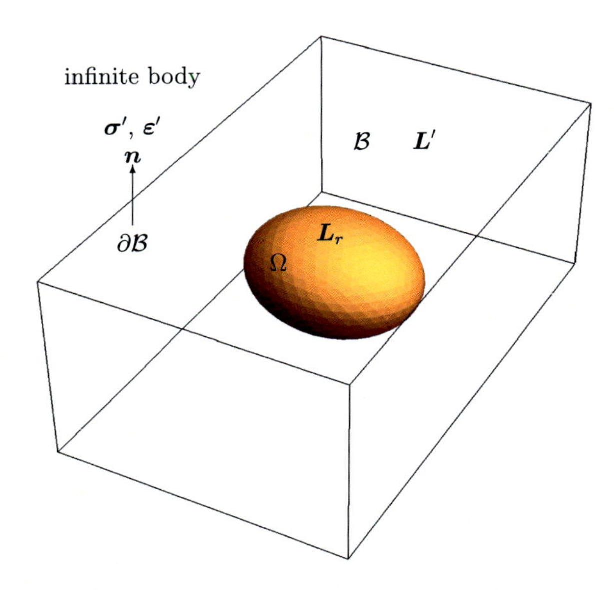

FIGURE 5.3 Inhomogeneity problem.

Mathematically speaking, this problem is expressed by the following set of equations:

$$
\begin{aligned}
&\operatorname{div}\boldsymbol{\sigma} = \mathbf{0} \quad \text{in } \mathcal{B}, \\
&\boldsymbol{\sigma} = \begin{cases} \boldsymbol{L}' : \boldsymbol{\varepsilon} & \text{in } \mathcal{B} - \Omega, \\ \boldsymbol{L}_r : \boldsymbol{\varepsilon} & \text{in } \Omega, \end{cases} \\
&\boldsymbol{\sigma} = \boldsymbol{\sigma}' \quad \text{and} \quad \boldsymbol{\varepsilon} = \boldsymbol{\varepsilon}' \quad \text{at far field.}
\end{aligned}
\tag{5.9}
$$

Eshelby [1] solves the above system by splitting it into two simpler cases:

1. In the first problem, the constant stresses $\boldsymbol{\sigma}^a = \boldsymbol{\sigma}'$ and strains $\boldsymbol{\varepsilon}^a = \boldsymbol{\sigma}'$ are applied at the far field, whereas the region Ω has the same properties with the rest of the material (Fig. 5.4_a). In mathematical formalism, this is written as

$$
\operatorname{div}\boldsymbol{\sigma}^a = \operatorname{div}\left(\boldsymbol{L}' : \boldsymbol{\varepsilon}^a\right) = \mathbf{0} \quad \text{in } \mathcal{B}, \qquad \boldsymbol{\sigma}^a = \boldsymbol{\sigma}' \text{ at far field.} \tag{5.10}
$$

The solution of the latter is trivial and is expressed as

$$
\boldsymbol{\sigma}^a = \boldsymbol{\sigma}' \quad \text{and} \quad \boldsymbol{\varepsilon}^a = \boldsymbol{\varepsilon}' \quad \text{in } \mathcal{B}. \tag{5.11}
$$

2. In the second problem, zero surface tractions are applied, whereas the inhomogeneity is substituted by a mechanically equivalent inclusion. Such equivalence implies that the region Ω possesses the same properties with the rest of the body, and the disturbance caused by the

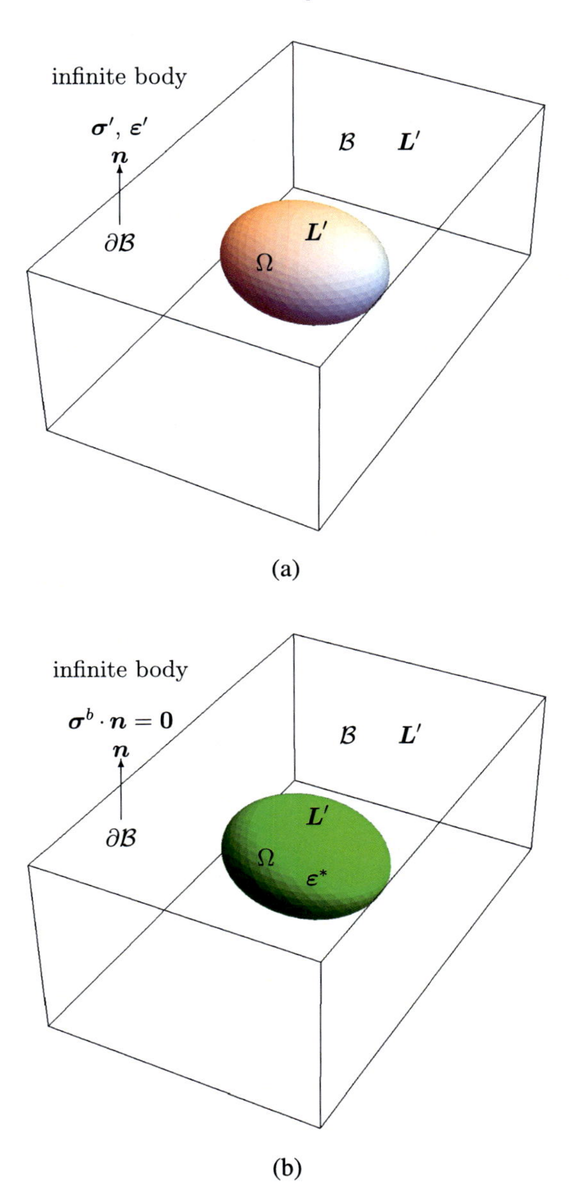

FIGURE 5.4 Inhomogeneity problem split into two problems.

inhomogeneity is recovered by an imaginary eigenstrain (Fig. 5.4$_b$). In mathematical terms, this problem is expressed by the equations

$$\mathrm{div}\boldsymbol{\sigma}^b = \mathbf{0} \quad \text{in } \mathcal{B},$$

$$\begin{aligned} \boldsymbol{\sigma}^b &= \boldsymbol{L}' : [\boldsymbol{\varepsilon}^b - \boldsymbol{\varepsilon}^*] \quad \text{in } \mathcal{B}, \\ \boldsymbol{\sigma}^b \cdot \boldsymbol{n} &= \boldsymbol{0} \quad \text{in } \partial\mathcal{B}, \\ \boldsymbol{\varepsilon}^* &= \begin{cases} \text{constant} \neq \boldsymbol{0} & \text{in } \Omega, \\ \boldsymbol{0} & \text{in } \mathcal{B} - \Omega. \end{cases} \end{aligned} \tag{5.12}$$

This system is the same as that of Subsection 5.1.1. Consequently,

$$\boldsymbol{\varepsilon}^b = \text{constant} = \boldsymbol{S} : \boldsymbol{\varepsilon}^* \quad \text{in } \Omega. \tag{5.13}$$

Due to the elastic nature of the media, the principle of superposition holds. Thus the solution of (5.9) is obtained by the sum of the solutions of two simpler problems. At the region inside the inclusion the superposition principle yields

$$\boldsymbol{\sigma} = \boldsymbol{\sigma}^a + \boldsymbol{\sigma}^b \quad \text{and} \quad \boldsymbol{\varepsilon} = \boldsymbol{\varepsilon}^a + \boldsymbol{\varepsilon}^b \quad \text{in } \Omega. \tag{5.14}$$

Eq. $(5.14)_1$, combined with $(5.9)_2$, (5.11), $(5.12)_2$, and $(5.14)_2$, is written as

$$\boldsymbol{L}_r : [\boldsymbol{\varepsilon}' + \boldsymbol{\varepsilon}^b] = \boldsymbol{L}' : \boldsymbol{\varepsilon}' + \boldsymbol{L}' : [\boldsymbol{\varepsilon}^b - \boldsymbol{\varepsilon}^*] \quad \text{in } \Omega. \tag{5.15}$$

Substituting $\boldsymbol{\varepsilon}^b$ with $\boldsymbol{\varepsilon}^*$ from Eq. (5.13), we obtain the following result:

$$\left[[\boldsymbol{L}_r - \boldsymbol{L}'] : \boldsymbol{S} + \boldsymbol{L}'\right] : \boldsymbol{\varepsilon}^* = -[\boldsymbol{L}_r - \boldsymbol{L}'] : \boldsymbol{\varepsilon}' \quad \text{in } \Omega. \tag{5.16}$$

The last relation allows us to connect the eigenstrain $\boldsymbol{\varepsilon}^*$ with the far field constant strain $\boldsymbol{\varepsilon}'$:

$$\boldsymbol{\varepsilon}^* = -\left[[\boldsymbol{L}_r - \boldsymbol{L}'] : \boldsymbol{S} + \boldsymbol{L}'\right]^{-1} : [\boldsymbol{L}_r - \boldsymbol{L}'] : \boldsymbol{\varepsilon}' \quad \text{in } \Omega. \tag{5.17}$$

Substituting the latter into Eq. $(5.14)_2$, we conclude that:

BOX 5.2 Interaction tensor

The total strain inside an inhomogeneity embedded in an infinite body is expressed as

$$\boldsymbol{\varepsilon} = \text{constant} = \boldsymbol{T} : \boldsymbol{\varepsilon}' \quad \text{in } \Omega, \tag{5.18}$$

where $\boldsymbol{\varepsilon}'$ is the far field strain, and

$$\begin{aligned} \boldsymbol{T} &= \boldsymbol{I} - \boldsymbol{S} : \left[[\boldsymbol{L}_r - \boldsymbol{L}'] : \boldsymbol{S} + \boldsymbol{L}'\right]^{-1} : [\boldsymbol{L}_r - \boldsymbol{L}'] \\ &= \boldsymbol{I} - \boldsymbol{S} : \left[\boldsymbol{S} + [\boldsymbol{L}_r - \boldsymbol{L}']^{-1} : \boldsymbol{L}'\right]^{-1} \\ &= \left[\boldsymbol{I} + \boldsymbol{S} : [\boldsymbol{L}']^{-1} : [\boldsymbol{L}_r - \boldsymbol{L}']\right]^{-1}. \end{aligned} \tag{5.19}$$

In the last expression, $\boldsymbol{L}'$ and $\boldsymbol{L}_r$ are the elasticity tensors of the body and the inhomogeneity, respectively, $\boldsymbol{S}$ is the Eshelby tensor (depending on $\boldsymbol{L}'$ and the shape of the inhomogeneity), and $\boldsymbol{I}$ is the fourth-order symmetric identity tensor. $\boldsymbol{T}$ is called the interaction tensor and plays an important role in the Eshelby–based approaches for obtaining the overall mechanical properties of composites with distinct phases.

Both the Eshelby tensor and the interaction tensor components are unitless.

5.2 Eshelby–based homogenization approaches

Consider a composite material composed by an elastic matrix material (index 0) and N elastic inhomogeneities of ellipsoidal shape. In the RVE of this composite the matrix has volume V_0, whereas the rth inhomogeneity occupies the space $\mathcal{B}_r$ and has volume V_r (Fig. 5.5).

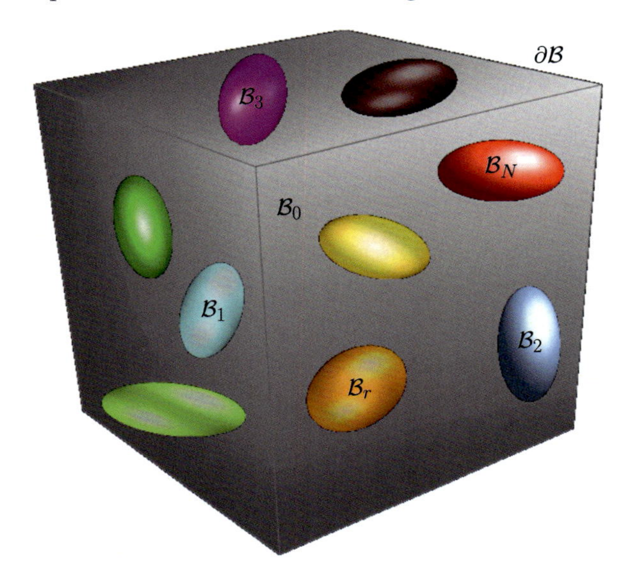

FIGURE 5.5 Composite with multiple inclusions (random medium).

As already mentioned, when the medium is constructed by $N+1$ separate phases, the average strains and stresses per phase are given by the expressions

$$\boldsymbol{\varepsilon}_r = \frac{1}{V_r}\int_{\mathcal{B}_r} \boldsymbol{\varepsilon}\,\mathrm{d}V, \quad \boldsymbol{\sigma}_r = \frac{1}{V_r}\int_{\mathcal{B}_r} \boldsymbol{\sigma}\,\mathrm{d}V \tag{5.20}$$

for $r = 0, 1, 2, ..., N$. In this formalism the constitutive law per phase that connects the stress with the strain is written as

$$\boldsymbol{\sigma}_r = \boldsymbol{L}_r : \boldsymbol{\varepsilon}_r. \tag{5.21}$$

From these definitions and equations we can easily show that

$$\bar{\boldsymbol{\varepsilon}} = \sum_{r=0}^{N} c_r \boldsymbol{\varepsilon}_r, \quad \bar{\boldsymbol{\sigma}} = \sum_{r=0}^{N} c_r \boldsymbol{\sigma}_r, \quad c_r = \frac{V_r}{V}. \tag{5.22}$$

As discussed in Chapter 3, the strain concentration tensors that are independent of the position vector can be identified per phase,

$$\boldsymbol{\varepsilon}_r = \boldsymbol{A}_r : \bar{\boldsymbol{\varepsilon}}. \tag{5.23}$$

The consistency condition (3.29) of Chapter 3 allows us to express the matrix concentration tensor in terms of the other phase concentration tensors:

BOX 5.3 Concentration tensor of matrix

Once $\boldsymbol{A}_r$ is identified for every inhomogeneity, the concentration tensor of the matrix phase is given by the relation

$$\boldsymbol{A}_0 = \frac{1}{c_0}\left[\boldsymbol{I} - \sum_{r=1}^{N} c_r \boldsymbol{A}_r\right]. \tag{5.24}$$

When considering the composite in total, we can expect that the macroscopic constitutive law has a similar form, that is,

$$\bar{\boldsymbol{\sigma}} = \bar{\boldsymbol{L}} : \bar{\boldsymbol{\varepsilon}}. \tag{5.25}$$

BOX 5.4 Macroscopic elasticity tensor in mean–field methods

With the help of the concentration tensors and expression (5.24), the macroscopic tensor $\bar{\boldsymbol{L}}$ can be easily shown to be given by the expression

$$\bar{\boldsymbol{L}} = \sum_{r=0}^{N} c_r \boldsymbol{L}_r : \boldsymbol{A}_r = \boldsymbol{L}_0 + \sum_{r=1}^{N} c_r [\boldsymbol{L}_r - \boldsymbol{L}_0] : \boldsymbol{A}_r. \tag{5.26}$$

The derivation of this expression follows similar steps with Example 1 of Chapter 3: The relation between the stresses and strains in the phase r can be written with the help of the tensor $\boldsymbol{A}_r$ as

$$\boldsymbol{\sigma}_r = \boldsymbol{L}_r : \boldsymbol{\varepsilon}_r = \boldsymbol{L}_r : \boldsymbol{A}_r : \bar{\boldsymbol{\varepsilon}}.$$

Averaging the latter over the whole RVE yields

$$\bar{\boldsymbol{\sigma}} = \sum_{i=0}^{N} c_r \boldsymbol{\sigma}_r = \left[\sum_{i=0}^{N} c_r \boldsymbol{L}_r : \boldsymbol{A}_r\right] : \bar{\boldsymbol{\varepsilon}} \quad \Rightarrow \quad \bar{\boldsymbol{L}} = \sum_{i=0}^{N} c_r \boldsymbol{L}_r : \boldsymbol{A}_r .$$

Moreover, thanks to Eq. (5.24), we have

$$\bar{\boldsymbol{L}} = c_0 \boldsymbol{L}_0 \colon \boldsymbol{A}_0 + \sum_{i=1}^{N} c_r \boldsymbol{L}_r : \boldsymbol{A}_r = \boldsymbol{L}_0 - \boldsymbol{L}_0 \colon \sum_{i=1}^{N} c_r : \boldsymbol{A}_r + \sum_{i=1}^{N} c_r \boldsymbol{L}_r : \boldsymbol{A}_r$$
$$= \boldsymbol{L}_0 + \sum_{r=1}^{N} c_r [\boldsymbol{L}_r - \boldsymbol{L}_0] \colon \boldsymbol{A}_r .$$

As it becomes clear, the knowledge of the tensors $\boldsymbol{A}_r$ for $r = 1, N$ is sufficient to obtain the macroscopic response of the composite. The identification of the concentration tensors is the main task of various Eshelby–based methods, which will be presented in the following subsections.

The consistency of the overall properties when applying displacement or traction boundary conditions has been extensively discussed in the literature [5,6]. In complete analogy with the strain concentration tensors, it is also possible to introduce stress concentration tensors and obtain the macroscopic compliance tensor. A formalism based on stress concentration tensors follows similar steps and is not discussed in the book.

5.2.1 Eshelby dilute

Eshelby has proposed a methodology for identifying the macroscopic properties of fiber or particulate composites with small volume fraction of reinforcement. Considering the composite of Fig. 5.5, the matrix material is characterized by the elasticity tensor $\boldsymbol{L}_0$. The dilute approach assumes that an inhomogeneity with elasticity tensor $\boldsymbol{L}_r$ is not affected by its neighbor inhomogeneities. A single inhomogeneity with its surrounding matrix is seen as an independent system subjected to the macroscopic strain $\bar{\boldsymbol{\varepsilon}}$ (Fig. 5.6). Thus the underlying hypothesis is that in the inhomogeneity problem, we should proceed with the substitutions

$$\boldsymbol{L}' = \boldsymbol{L}_0 \quad \text{and} \quad \boldsymbol{\varepsilon}' = \bar{\boldsymbol{\varepsilon}}. \tag{5.27}$$

As a consequence, the strain in the rth inhomogeneity is directly obtained by the formula

$$\boldsymbol{\varepsilon}_r = \boldsymbol{T}_r : \bar{\boldsymbol{\varepsilon}}, \tag{5.28}$$

where $\boldsymbol{T}_r$ is the interaction tensor

$$\boldsymbol{T}_r = \left[\boldsymbol{I} + \boldsymbol{S}_r : \boldsymbol{L}_0^{-1} \colon [\boldsymbol{L}_r - \boldsymbol{L}_0]\right]^{-1} \tag{5.29}$$

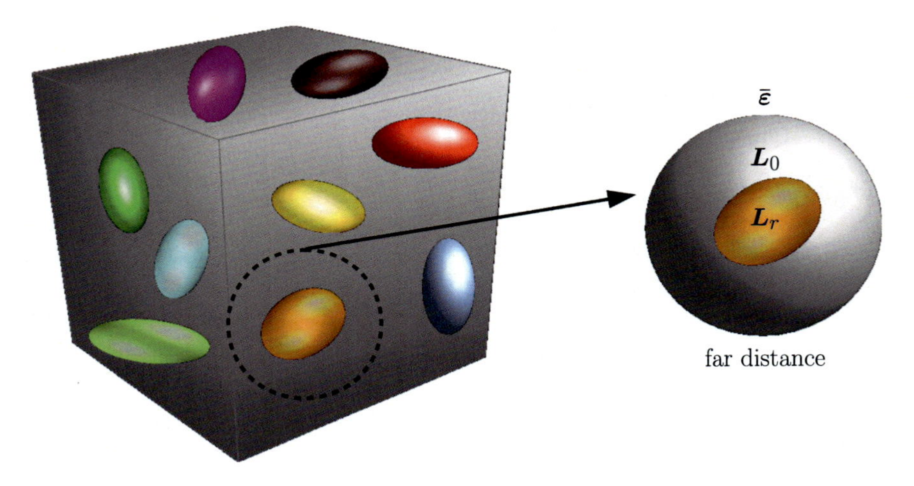

FIGURE 5.6 Hypothesis for the Eshelby dilute method.

corresponding to the analogous inhomogeneity problem, and S_r is the Eshelby tensor per phase, which depends on the matrix properties L_0 and the shape of the rth inhomogeneity. Comparing (5.28) with (5.23), we eventually obtain:

BOX 5.5 Concentration tensors in Eshelby dilute approach

In the Eshelby dilute approach the strain concentration tensor of the rth ellipsoidal inhomogeneity is given by the expression

$$A_r = T_r. \tag{5.30}$$

The advantage of the Eshelby dilute method is that it provides analytical formulas for the composite response. However, since it neglects the interactions between the inhomogeneities, it can be applied only for small volume fractions of reinforcement.

5.2.2 Mori–Tanaka

Consider the composite of Fig. 5.5. For an arbitrary inhomogeneity with elasticity tensor L_r, the influence of the other inhomogeneities on its fields is transferred through the strain and stress in its surrounding matrix material with elasticity tensor L_0. Even though various fields vary spatially from one position to another inside the matrix, the average strain ε_0 and stress σ_0 represent fair approximations of the actual fields of the matrix surrounding each inhomogeneity when a large number of inhomogeneities exist and are randomly distributed in the composite. In addition, it is reasonable to assume that the absence of only one inhomogeneity does not affect the overall elastic behavior of the composite. In other words, when the rth inhomogeneity is removed and replaced by the matrix material, the average fields are not altered substantially. Therefore the rth inhomogeneity can be viewed as an ellipsoidal inhomogeneity, placed within a uniform matrix, which is subjected to the uniform strain field $\boldsymbol{\varepsilon}_0$ (Fig. 5.7). Thus the underlying hypothesis is that in the inhomogeneity problem of Section 5.1, we should proceed with the substitutions [7]

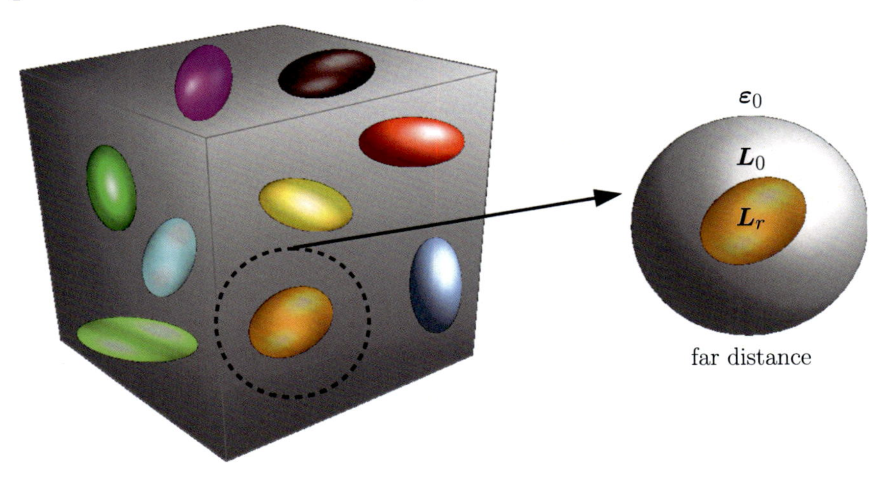

FIGURE 5.7 Hypothesis for the Mori–Tanaka method.

$$\boldsymbol{L}' = \boldsymbol{L}_0 \quad \text{and} \quad \boldsymbol{\varepsilon}' = \boldsymbol{\varepsilon}_0. \tag{5.31}$$

Under this assumption and using the relations presented in the inhomogeneity problem, the strain $\boldsymbol{\varepsilon}_r$ in the rth inhomogeneity is expressed as [8]

$$\boldsymbol{\varepsilon}_r = \boldsymbol{T}_r : \boldsymbol{\varepsilon}_0, \tag{5.32}$$

where

$$\boldsymbol{T}_r = \left[\boldsymbol{I} + \boldsymbol{S}_r : \boldsymbol{L}_0^{-1} : [\boldsymbol{L}_r - \boldsymbol{L}_0] \right]^{-1}, \tag{5.33}$$

and $\boldsymbol{S}_r$ is the Eshelby tensor per phase, which depends on the matrix properties $\boldsymbol{L}_0$ and the shape of the rth inhomogeneity. Averaging Eq. (5.32) over the whole RVE and considering Eq. $(5.22)_1$ yield

$$\bar{\boldsymbol{\varepsilon}} = \left[c_0 \boldsymbol{I} + \sum_{r=1}^{N} c_r \boldsymbol{T}_r \right] : \boldsymbol{\varepsilon}_0 \quad \text{or} \quad \boldsymbol{\varepsilon}_0 = \left[c_0 \boldsymbol{I} + \sum_{n=1}^{N} c_n \boldsymbol{T}_n \right]^{-1} : \bar{\boldsymbol{\varepsilon}}. \tag{5.34}$$

Combining Eqs. (5.32) and (5.34) and considering (5.23), we eventually obtain:

BOX 5.6 Concentration tensors in Mori–Tanaka approach

In the Mori–Tanaka approach the strain concentration tensor of the rth ellipsoidal inhomogeneity is given by the expression

$$\boldsymbol{A}_r = \boldsymbol{T}_r : \left[c_0 \boldsymbol{I} + \sum_{n=1}^{N} c_n \boldsymbol{T}_n \right]^{-1}. \tag{5.35}$$

It should be mentioned that relation (5.34) provides the analytical expression for the matrix strain concentration tensor

$$\boldsymbol{A}_0 = \left[c_0 \boldsymbol{I} + \sum_{n=1}^{N} c_n \boldsymbol{T}_n \right]^{-1}, \tag{5.36}$$

which allows us to re–express the strain concentration tensor of the rth ellipsoidal inhomogeneity as

$$\boldsymbol{A}_r = \boldsymbol{T}_r : \boldsymbol{A}_0. \tag{5.37}$$

The advantage of Mori–Tanaka approach is that it provides analytical formulas for the composite response. It is very efficient method for studying matrix–particle type of composites, where the particles can have arbitrary orientation. Willis [9] has demonstrated that: i) for a soft matrix and stiff reinforcement, the Mori–Tanaka estimate for the composite response coincides with the generalized Hashin–Shtrikman lower bound, ii) for stiff matrix and soft reinforcement, the Mori–Tanaka estimate for the composite response coincides with the generalized Hashin–Shtrikman upper bound.

5.2.3 Self–consistent

Consider the composite of Fig. 5.5. The self–consistent approach proceeds with the hypothesis that the macroscopic tensor $\overline{\boldsymbol{L}}$ is already known. If there are numerous inhomogeneities in the composite, then the macroscopic properties are not altered by the absence of a single inhomogeneity. When the composite is subjected to displacement boundary conditions, we may reasonably assume that the influence of the applied load and the interaction between the inhomogeneities can be accounted for by placing the rth inhomogeneity in a homogeneous body with properties $\overline{\boldsymbol{L}}$ that is subjected to the strain tensor $\overline{\boldsymbol{\varepsilon}}$ (Fig. 5.8). Thus the underlying hypothesis

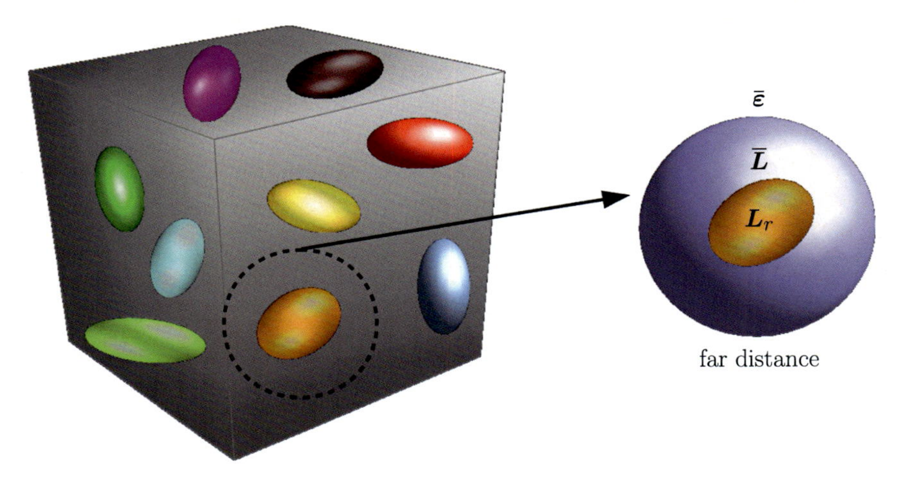

FIGURE 5.8 Hypothesis for the self–consistent method.

is that in the inhomogeneity problem of Section 5.1, we should consider the substitutions

$$L' = \bar{L} \quad \text{and} \quad \varepsilon' = \bar{\varepsilon}. \tag{5.38}$$

Under this assumption and using the relations presented in the inhomogeneity problem, we conclude that:

BOX 5.7 Concentration tensors in self–consistent approach

In the self–consistent approach the strain concentration tensor of the rth ellipsoidal inhomogeneity is given by the expression

$$A_r = \left[I + \bar{S}_r : \bar{L}^{-1} : [L_r - \bar{L}] \right]^{-1}, \tag{5.39}$$

where $\bar{S}_r$ is the Eshelby tensor per phase, which depends on the homogenized medium properties $\bar{L}$ and the shape of the rth inhomogeneity.

The self–consistent approach yields implicit equations for obtaining the macroscopic properties, which can be solved through iterative computational schemes. Its main advantage is that the presence of a matrix phase is not required (the effective medium plays the role of "matrix"), and thus it is an efficient method to study polycrystalline materials.

Concluding, the three presented mean–field homogenization theories consider different hypotheses, which are summarized below:

BOX 5.8 Comparison of mean–field methods

In all Eshelby–based methods the rth inhomogeneity is assumed to be inside a body with elasticity tensor $\boldsymbol{L}'$ and is subjected to far field homogeneous strain $\boldsymbol{\varepsilon}'$ (Fig. 5.3). The three micromechanics approaches, the Eshelby dilute, the Mori–Tanaka, and the self–consistent, assume different body material properties and far field strain field, which are summarized in the following table:

method	body $\boldsymbol{L}'$	far field strain $\boldsymbol{\varepsilon}'$
Eshelby dilute	$\boldsymbol{L}_0$	$\bar{\boldsymbol{\varepsilon}}$
Mori–Tanaka	$\boldsymbol{L}_0$	$\boldsymbol{\varepsilon}_0$
self–consistent	$\bar{\boldsymbol{L}}$	$\bar{\boldsymbol{\varepsilon}}$

5.3 Examples

Example 1: Mori–Tanaka for particulate composite

Let us consider the particulate reinforced composite of Fig. 5.9. The particles are spherical and are distributed randomly inside the matrix material, having a volume fraction equal to c. Both the matrix (phase 0) and the particles (phase 1) are assumed isotropic with bulk moduli K_0 and K_1 and shear moduli μ_0 and μ_1.

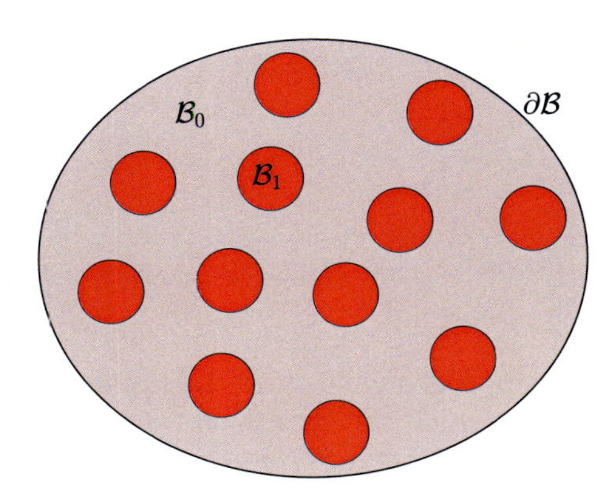

FIGURE 5.9 Schematic representation of particle reinforced composite cross–section.

Due to the geometrical characteristics of this composite, the Eshelby tensor is also isotropic (see Eq. (5.5)). Using the Hill notation for isotropic fourth–order tensors, the elastic properties of the two phases and the Eshelby tensor are expressed as

$$\boldsymbol{L}_0 = (3K_0, 2\mu_0), \quad \boldsymbol{L}_1 = (3K_1, 2\mu_1), \quad \boldsymbol{S}_1 = (3\gamma_0, 2\delta_0),$$

with

$$\gamma_0 = \frac{K_0}{3K_0 + 4\mu_0}, \quad \delta_0 = \frac{3K_0 + 6\mu_0}{15K_0 + 20\mu_0}.$$

As discussed in Example 2 of Chapter 1, Section 1.4, we can perform scalar computations when dealing with isotropic tensors algebra. Thus

$$\begin{aligned}
\boldsymbol{T}_1 &= \left[\boldsymbol{I} + \boldsymbol{S}_1 : \boldsymbol{L}_0^{-1} : [\boldsymbol{L}_1 - \boldsymbol{L}_0]\right]^{-1} \\
&= \left[(1,1) + (3\gamma_0, 2\delta_0) \cdot \left(\frac{1}{3K_0}, \frac{1}{2\mu_0}\right) \cdot (3K_1 - 3K_0, 2\mu_1 - 2\mu_0)\right]^{-1} \\
&= \left[\left(1 + \frac{3\gamma_0[K_1 - K_0]}{K_0}, 1 + \frac{2\delta_0[\mu_1 - \mu_0]}{\mu_0}\right)\right]^{-1} \\
&= \left(\frac{K_0}{K_0 + 3\gamma_0[K_1 - K_0]}, \frac{\mu_0}{\mu_0 + 2\delta_0[\mu_1 - \mu_0]}\right) = \left(3T_1^b, 2T_1^s\right),
\end{aligned}$$

$$\begin{aligned}
\boldsymbol{A}_1 &= \boldsymbol{T}_1 : [[1-c]\boldsymbol{I} + c\boldsymbol{T}_1]^{-1} \\
&= (3T_1^b, 2T_1^s) \cdot \left[(1-c, 1-c) + (3cT_1^b, 2cT_1^s)\right]^{-1} \\
&= \left(\frac{3T_1^b}{1 - c + 3cT_1^b}, \frac{2cT_1^s}{1 - c + 2cT_1^s}\right) = \left(3A_1^b, 2A_1^s\right),
\end{aligned}$$

$$\begin{aligned}
\bar{\boldsymbol{L}} &= \boldsymbol{L}_0 + c[\boldsymbol{L}_1 - \boldsymbol{L}_0] : \boldsymbol{A}_1 \\
&= \left(3K_0 + 9c[K_1 - K_0]A_1^b, 2\mu_0 + 4c[\mu_1 - \mu_0]A_1^s\right) = \left(3\bar{K}, 2\bar{\mu}\right).
\end{aligned}$$

After some algebra, the final expressions for the macroscopic bulk modulus $\bar{K}$ and the macroscopic shear modulus $\bar{\mu}$ are

$$\begin{aligned}
\bar{K} &= K_0 + \frac{c}{\dfrac{1}{K_1 - K_0} + \dfrac{3[1-c]}{3K_0 + 4\mu_0}}, \\
\bar{\mu} &= \mu_0 + \frac{c}{\dfrac{1}{\mu_1 - \mu_0} + \dfrac{6}{5}\dfrac{[1-c][K_0 + 2\mu_0]}{[3K_0 + 4\mu_0]\mu_0}}.
\end{aligned} \tag{5.40}$$

Example 2: three–phase composite

Consider a composite in which the matrix contains two types of spherical particles (one from glass and one from carbon) with the following bulk and shear moduli for every phase:

matrix	glass particles	carbon particles
$K_0 = 2.5$ GPa	$K_1 = 44$ GPa	$K_2 = 160$ GPa
$\mu_0 = 1.15$ GPa	$\mu_1 = 30$ GPa	$\mu_2 = 100$ GPa

The volume fractions of the glass and carbon particles are $c_1 = 20\%$ and $c_2 = 10\%$, respectively. Using the Mori–Tanaka method:

1. calculate the bulk modulus $\bar{K}$ and the shear modulus $\bar{\mu}$ of the composite,
2. for the macroscopic strain $\bar{\boldsymbol{\varepsilon}}$ with

$$\bar{\varepsilon}_{11} = 0.002, \quad \bar{\varepsilon}_{22} = 0.001,$$

and the rest of the terms equal to zero, calculate the average stress in the carbon particles.

Solution:

In the Hill notation the elasticity tensors of the three phases are expressed through the bulk and shear moduli as

$$\begin{aligned}
\boldsymbol{L}_0 &= (3K_0, 2\mu_0) = (7.5, 2.3) \text{ GPa}, \\
\boldsymbol{L}_1 &= (3K_1, 2\mu_1) = (132, 60) \text{ GPa}, \\
\boldsymbol{L}_2 &= (3K_2, 2\mu_2) = (480, 200) \text{ GPa}.
\end{aligned}$$

Since the two particle types are spherical, one Eshelby tensor computation is required for the Mori–Tanaka scheme implementation:

$$\boldsymbol{S}_1 = \boldsymbol{S}_2 = \left(\frac{3K_0}{3K_0 + 4\mu_0}, \frac{6K_0 + 12\mu_0}{15K_0 + 20\mu_0} \right) = (0.6198, 0.476).$$

Consequently,

$$\begin{aligned}
\boldsymbol{T}_1 &= \left[\boldsymbol{I} + \boldsymbol{S}_1 : \boldsymbol{L}_0^{-1} : [\boldsymbol{L}_1 - \boldsymbol{L}_0] \right]^{-1} \\
&= \left[(1, 1) + (3\gamma_0, 2\delta_0) \cdot \left(\frac{1}{3K_0}, \frac{1}{2\mu_0} \right) \cdot (3K_1 - 3K_0, 2\mu_1 - 2\mu_0) \right]^{-1} \\
&= \left[\left(1 + \frac{3\gamma_0[K_1 - K_0]}{K_0}, 1 + \frac{2\delta_0[\mu_1 - \mu_0]}{\mu_0} \right) \right]^{-1} \\
&= \left(\frac{K_0}{K_0 + 3\gamma_0[K_1 - K_0]}, \frac{\mu_0}{\mu_0 + 2\delta_0[\mu_1 - \mu_0]} \right) = (0.08858, 0.07727),
\end{aligned}$$

$$
\begin{aligned}
\boldsymbol{T}_2 &= \left[\boldsymbol{I} + \boldsymbol{S}_2 : \boldsymbol{L}_0^{-1} : [\boldsymbol{L}_2 - \boldsymbol{L}_0]\right]^{-1} \\
&= \left(\frac{K_0}{K_0 + 3\gamma_0[K_2 - K_0]}, \frac{\mu_0}{\mu_0 + 2\delta_0[\mu_2 - \mu_0]}\right) = (0.02497, 0.02386).
\end{aligned}
$$

Expressions (5.36) and (5.37) yield

$$
\begin{aligned}
\boldsymbol{A}_0 &= [c_0\boldsymbol{I} + c_1\boldsymbol{T}_1 + c_2\boldsymbol{T}_2]^{-1} \\
&= [[1 - 0.2 - 0.1] \cdot (1, 1) + 0.2 \cdot (0.08858, 0.07727) \\
&\quad + 0.1 \cdot (0.02497, 0.02386)]^{-1} = (1.3885, 1.3931), \\
\boldsymbol{A}_1 &= \boldsymbol{T}_1 : \boldsymbol{A}_0 = (0.08858, 0.07727) \cdot (1.3885, 1.3931) = (0.123, 0.1076), \\
\boldsymbol{A}_2 &= \boldsymbol{T}_2 : \boldsymbol{A}_0 = (0.02497, 0.02386) \cdot (1.3885, 1.3931) = (0.0347, 0.0332).
\end{aligned}
$$

1. The macroscopic elasticity tensor is given by the Eq. (5.26),

$$
\bar{\boldsymbol{L}} = (3\bar{K}, 2\bar{\mu}) = c_0\boldsymbol{L}_0 : \boldsymbol{A}_0 + c_1\boldsymbol{L}_1 : \boldsymbol{A}_2 + c_1\boldsymbol{L}_2 : \boldsymbol{A}_2.
$$

For the three materials,

$$
\begin{aligned}
\boldsymbol{L}_0 : \boldsymbol{A}_0 &= (7.5, 2.3) \cdot (1.3885, 1.3931) = (10.414, 3.204) \text{ GPa}, \\
\boldsymbol{L}_1 : \boldsymbol{A}_1 &= (132, 60) \cdot (0.123, 0.1076) = (16.236, 6.456) \text{ GPa}, \\
\boldsymbol{L}_2 : \boldsymbol{A}_2 &= (480, 200) \cdot (0.0347, 0.0332) = (16.656, 6.64) \text{ GPa}.
\end{aligned}
$$

Consequently,

$$
\begin{aligned}
\bar{\boldsymbol{L}} &= [1 - 0.2 - 0.1] \cdot (10.414, 3.204) + 0.2 \cdot (16.236, 6.456) \\
&\quad + 0.1 \cdot (16.656, 6.64) = (12.203, 4.198) \text{ GPa},
\end{aligned}
$$

which means that

$$
\bar{K} = 4.068 \text{ GPa}, \qquad \bar{\mu} = 2.099 \text{ GPa}.
$$

2. The average stress in the carbon particles is given by the expressions

$$
\boldsymbol{\sigma}_2 = \boldsymbol{L}_2 : \boldsymbol{\varepsilon}_2 = \boldsymbol{L}_2 : \boldsymbol{A}_2 : \bar{\boldsymbol{\varepsilon}}.
$$

When a fourth–order isotropic tensor $\boldsymbol{C} = (3C^b, 2C^s)$ is multiplied by a second–order symmetric tensor

$$
\boldsymbol{c} = \begin{bmatrix} c_{11} & c_{12} & c_{13} \\ c_{12} & c_{22} & c_{23} \\ c_{13} & c_{23} & c_{33} \end{bmatrix},
$$

the tensor algebra yields

$$\boldsymbol{C}:\boldsymbol{c} = \begin{bmatrix} D_1 & D_4 & D_5 \\ D_4 & D_2 & D_6 \\ D_5 & D_6 & D_3 \end{bmatrix}$$

with

$$\begin{aligned}
D_1 &= \left[3C^b - 2C^s\right] \frac{c_{11} + c_{22} + c_{33}}{3} + 2C^s c_{11}, \\
D_2 &= \left[3C^b - 2C^s\right] \frac{c_{11} + c_{22} + c_{33}}{3} + 2C^s c_{22}, \\
D_3 &= \left[3C^b - 2C^s\right] \frac{c_{11} + c_{22} + c_{33}}{3} + 2C^s c_{33}, \\
D_4 &= 2C^s c_{12}, \quad D_5 = 2C^s c_{13}, \quad D_6 = 2C^s c_{23}.
\end{aligned}$$

So

$$\boldsymbol{\sigma}_2 = \begin{bmatrix} 0.0233 & 0 & 0 \\ 0 & 0.0166 & 0 \\ 0 & 0 & 0.01 \end{bmatrix} \text{GPa}.$$

Example 3: Mori–Tanaka for long fiber composite

Let us consider the fiber reinforced composite of Fig. 5.10. The fibers are cylindrical and are considered long (aspect ratio 1:1:α, $\alpha \to \infty$), and they are distributed randomly, parallel to the x_3 axis, inside the matrix material.

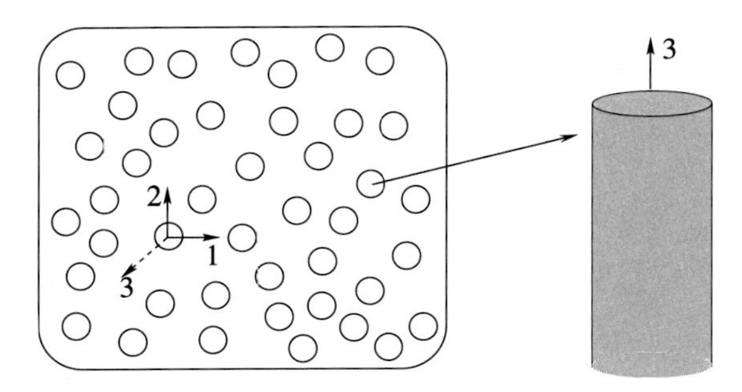

FIGURE 5.10 Schematic representation of unidirectional, long fiber reinforced composite cross–section.

The matrix is assumed to be an elastic isotropic material with Young modulus $E_0 = 2.25$ GPa and Poisson ratio $\nu_0 = 0.19$. The fibers are also considered elastic isotropic with Young modulus $E_1 = 385$ GPa and Poisson ratio $\nu_0 = 0.24$. The fiber volume fraction varies between 20% and 80%.

Due to the geometrical characteristics of the microstructure, the composite is expected to behave like an equivalent transversely isotropic medium. Thus the macroscopic elasticity tensor $\bar{\boldsymbol{L}}$ can be expressed in terms of five elastic constants: the transverse bulk modulus $\bar{K}^{\text{tr}}$, the transverse shear modulus $\bar{\mu}^{\text{tr}}$, the axial shear modulus $\bar{\mu}^{\text{ax}}$, and the stiffness coefficients $\bar{l}$ and $\bar{n}$:

$$\bar{\boldsymbol{L}} = \begin{bmatrix} \bar{K}^{\text{tr}} + \bar{\mu}^{\text{tr}} & \bar{K}^{\text{tr}} - \bar{\mu}^{\text{tr}} & \bar{l} & 0 & 0 & 0 \\ \bar{K}^{\text{tr}} - \bar{\mu}^{\text{tr}} & \bar{K}^{\text{tr}} + \bar{\mu}^{\text{tr}} & \bar{l} & 0 & 0 & 0 \\ \bar{l} & \bar{l} & \bar{n} & 0 & 0 & 0 \\ 0 & 0 & 0 & \bar{\mu}^{\text{tr}} & 0 & 0 \\ 0 & 0 & 0 & 0 & \bar{\mu}^{\text{ax}} & 0 \\ 0 & 0 & 0 & 0 & 0 & \bar{\mu}^{\text{ax}} \end{bmatrix}.$$

In addition, the Eshelby tensor for this unidirectional fiber composite is given by expression (5.8). Fig. 5.11 illustrates the obtained macroscopic shear moduli as a function of the fiber volume fraction according to the Mori–Tanaka theory. As it is observed, the two shear moduli are quite close one to another, with slightly higher values for the axial.

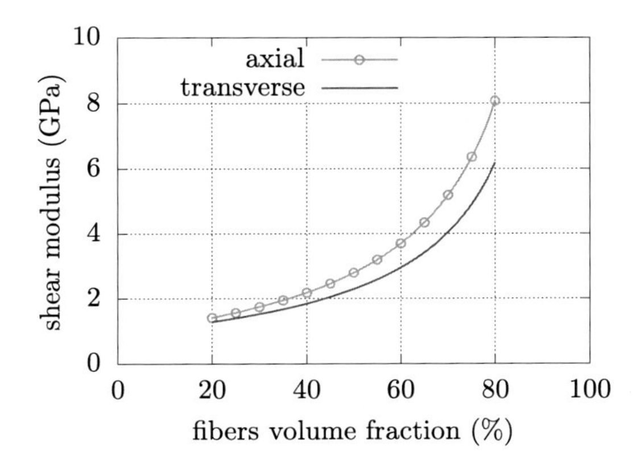

FIGURE 5.11 Macroscopic shear moduli (axial and transverse) vs fiber volume fraction.

The corresponding Python script for obtaining these results is the following:

```
import numpy as np

# Data
E0=2.25
v0=0.19
E1=385.0
```

```
v1=0.24

# Elasticity tensors
K0=E0/(3*(1-2*v0))
mu0=E0/(2*(1+v0))
K1=E1/(3*(1-2*v1))
mu1=E1/(2*(1+v1))
L0=np.matrix([[K0+4*mu0/3, K0-2*mu0/3, K0-2*mu0/3,\
               0., 0., 0.],\
              [K0-2*mu0/3, K0+4*mu0/3, K0-2*mu0/3,\
               0., 0., 0.],\
              [K0-2*mu0/3, K0-2*mu0/3, K0+4*mu0/3,\
               0., 0., 0.],\
              [0., 0., 0., mu0, 0., 0.],\
              [0., 0., 0., 0., mu0, 0.],\
              [0., 0., 0., 0., 0., mu0]])
L1=np.matrix([[K1+4*mu1/3, K1-2*mu1/3, K1-2*mu1/3,\
               0., 0., 0.],\
              [K1-2*mu1/3, K1+4*mu1/3, K1-2*mu1/3,\
               0., 0., 0.],\
              [K1-2*mu1/3, K1-2*mu1/3, K1+4*mu1/3,\
               0., 0., 0.],\
              [0., 0., 0., mu1, 0., 0.],\
              [0., 0., 0., 0., mu1, 0.],\
              [0., 0., 0., 0., 0., mu1]])

# Eshelby tensor
S1=np.matrix([[(5-4*v0)/(8*(1-v0)), (4*v0-1)/(8*(1-v0)),\
                v0/(2*(1-v0)), 0., 0., 0.],\
              [(4*v0-1)/(8*(1-v0)), (5-4*v0)/(8*(1-v0)),\
                v0/(2*(1-v0)), 0., 0., 0.],\
              [0., 0., 0., 0., 0., 0.],\
              [0., 0., 0., (3-4*v0)/(4*(1-v0)), 0., 0.],\
              [0., 0., 0., 0., 0.5, 0.],\
              [0., 0., 0., 0., 0., 0.5]])

# Interaction tensor
I=np.eye(6)
T1=(I+S1*(L0.I)*(L1-L0)).I

tableA=np.zeros((61,3))

for n in range(61):
   c=0.01*n+0.2
```

```
# Mori-Tanaka strain concentration tensor of fiber
    A1=T1*((1-c)*I+c*T1).I

# Mori-Tanaka strain concentration tensor of matrix
    A0=(I-c*A1)/(1-c)

# Effective elastic properties
    Leff=(1-c)*L0*A0+c*L1*A1
    GL=Leff[4,4]
    GT=Leff[3,3]
    tableA[n,:]=[c, GL, GT]

np.savetxt('GL_GT.txt',tableA,fmt='%1.5f')
```

Example 4: self–consistent for particulate composite

Let us again consider the particulate reinforced composite of Fig. 5.9. Given the detailed material properties

matrix	glass particles
$K_0 = 2.5$ GPa	$K_1 = 44$ GPa
$\mu_0 = 1.15$ GPa	$\mu_1 = 30$ GPa

compute the macroscopic Young modulus of the composite using the self–consistent method as a function of the particle volume fraction c. Compare also the results with the prediction of the Mori–Tanaka, Voigt–Reuss, and Hashin–Shtrikman bounds.

Solution:

The main characteristics discussed in Example 1 remain in this problem, that is, the Eshelby tensor is isotropic and is expressed by Eq. (5.5). However, in the self–consistent method the "matrix phase" in the Eshelby problem is considered to be an effective medium. Thus the Eshelby tensor in Hill notation is given by

$$\bar{\boldsymbol{S}}_1 = (3\bar{\gamma}, 2\bar{\delta})$$

with

$$\bar{\gamma} = \frac{\bar{K}}{3\bar{K} + 4\bar{\mu}}, \qquad \bar{\delta} = \frac{3\bar{K} + 6\bar{\mu}}{15\bar{K} + 20\bar{\mu}}.$$

In these expressions, $\bar{K}$ and $\bar{\mu}$ denote the bulk and shear modulus, respectively, of the unknown effective medium. The material properties of the two phases and of the effective medium are written in the Hill notation as

$$\boldsymbol{L}_0 = (3K_0, 2\mu_0), \quad \boldsymbol{L}_1 = (3K_1, 2\mu_1), \quad \bar{\boldsymbol{L}} = (3\bar{K}, 2\bar{\mu}).$$

As mentioned in Example 1, we can perform scalar computations when dealing with isotropic tensors algebra. Thus

$$\begin{aligned}\boldsymbol{A}_1 &= \left[\boldsymbol{I} + \bar{\boldsymbol{S}}_1 : \bar{\boldsymbol{L}}^{-1} : [\boldsymbol{L}_1 - \bar{\boldsymbol{L}}]\right]^{-1} \\ &= \left(\frac{\bar{K}}{\bar{K} + 3\bar{\gamma}[K_1 - \bar{K}]}, \frac{\bar{\mu}}{\bar{\mu} + 2\bar{\delta}[\mu_1 - \bar{\mu}]}\right) = \left(3A_1^b, 2A_1^s\right),\end{aligned}$$

$$\begin{aligned}\bar{\boldsymbol{L}} &= \boldsymbol{L}_0 + c[\boldsymbol{L}_1 - \boldsymbol{L}_0] : \boldsymbol{A}_1 \\ &= \left(3K_0 + 9c[K_1 - K_0]A_1^b, 2\mu_0 + 4c[\mu_1 - \mu_0]A_1^s\right) = (3\bar{K}, 2\bar{\mu}).\end{aligned}$$

After some algebra, we obtain the following nonlinear system of equations for the macroscopic bulk and shear moduli:

$$\begin{aligned}\bar{K} &= K_0 + c[K_1 - K_0]\frac{3\bar{K} + 4\bar{\mu}}{3K_1 + 4\bar{\mu}}, \\ \bar{\mu} &= \mu_0 + c[\mu_1 - \mu_0]\frac{15\bar{K}\bar{\mu} + 20\bar{\mu}^2}{9\bar{K}\bar{\mu} + 8\bar{\mu}^2 + 6\mu_1\bar{K} + 12\mu_1\bar{\mu}}.\end{aligned} \tag{5.41}$$

The solution of system (5.41) can be obtained using an iterative computational scheme. An important factor for such a scheme to work is the choice of a good initial prediction. Here the Mori–Tanaka estimate (5.40) is utilized as a first prediction. The computations per volume fraction are completed, and the moduli are considered correct when the error between two consecutive predictions m and $m+1$,

$$\text{err} = \sqrt{\left[\bar{K}^{m+1} - \bar{K}^m\right]^2 + \left[\bar{\mu}^{m+1} - \bar{\mu}^m\right]^2},$$

is less than the tolerance. Once the macroscopic bulk and shear moduli are determined, we obtain the macroscopic Young modulus through the classical formula

$$\bar{E} = \frac{9\bar{K}\bar{\mu}}{3\bar{K} + \bar{\mu}}.$$

The numerical results for the composite in question are illustrated in Fig. 5.12 for both self–consistent and Mori–Tanaka approaches, as well as the Voigt–Reuss bounds (expressions (3.38) of Chapter 3) and the Hashin–Shtrikman upper bound (expressions (3.40) and (3.41) of Chapter 3). The Hashin–Shtrikman lower bound is omitted since it coincides with the Mori–Tanaka estimate. The corresponding Python script that performs the calculations is the following:

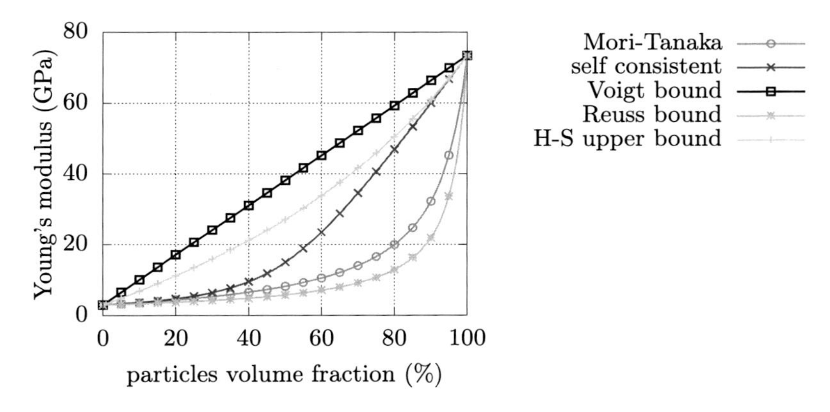

FIGURE 5.12 Macroscopic Young modulus vs particle volume fraction. Self–consistent and Mori–Tanaka estimates, the Voigt–Reuss bounds, and the Hashin–Shtrikman (H-S) upper bound. The H-S lower bound coincides with the Mori–Tanaka estimate.

```
import numpy as np

# Data
K0=2.5
mu0=1.15
K1=44.
mu1=30.
tol=1.0E-4

tableA=np.zeros((101,6))

for n in range(101):
   c=0.01*n

# Voigt-Reuss bounds
   KV=(1-c)*K0+c*K1
   muV=(1-c)*mu0+c*mu1
   EV=9*KV*muV/(3*KV+muV)
   KR=K0*K1/((1-c)*K1+c*K0)
   muR=mu0*mu1/((1-c)*mu1+c*mu0)
   ER=9*KR*muR/(3*KR+muR)

# Hashin-Shtrikman upper bound
   KHSU=K1+(1-c)/(1/(K0-K1)+3*c/(3*K1+4*mu1))
   muHSU=mu1+(1-c)/(1/(mu0-mu1)+\
        6*c*(K1+2*mu1)/(5*(3*K1+4*mu1)*mu1))
   EHSU=9*KHSU*muHSU/(3*KHSU+muHSU)
```

```
# Mori-Tanaka estimate
   Kmt=K0+c/(1/(K1-K0)+3*(1-c)/(3*K0+4*mu0))
   mumt=mu0+c/(1/(mu1-mu0)+\
        6*(1-c)*(K0+2*mu0)/(5*(3*K0+4*mu0)*mu0))
   Emt=9*Kmt*mumt/(3*Kmt+mumt)

# self-consistent iterational scheme
   Ksc0=Kmt
   musc0=mumt
   err=1.
   while err>tol:
     Ksc=K0+c*(K1-K0)*(3*Ksc0+4*musc0)\
           /(3*K1+4*musc0)
     musc=mu0+c*(mu1-mu0)*(15*Ksc0*musc0\
            +20*musc0*musc0)/(9*Ksc0*musc0\
            +8*musc0*musc0+6*mu1*Ksc0+12*mu1*musc0)
     err2=(Ksc-Ksc0)*(Ksc-Ksc0)\
         +(musc-musc0)*(musc-musc0)
     err=np.sqrt(err2)
     Ksc0=Ksc
     musc0=musc

   Esc=9*Ksc*musc/(3*Ksc+musc)
   tableA[n,:]=[c, Emt, Esc, EV, ER, EHSU]

# results
np.savetxt('Emtsc.txt',tableA,fmt='%1.5f')
```

References

[1] J.D. Eshelby, The determination of the elastic field of an ellipsoidal inclusion, and related problems, Proceedings of the Royal Society of London. Series A, Mathematical and Physical Sciences 241 (1226) (1957) 376–396.
[2] T. Mura, Micromechanics of Defects in Solids, second, revised ed., Kluwer Academic Publishers, Dordrecht, 1987.
[3] G. Chatzigeorgiou, N. Charalambakis, Y. Chemisky, F. Meraghni, Thermomechanical Behavior of Dissipative Composite Materials, ISTE Press – Elsevier, London, 2018.
[4] A.C. Gavazzi, D.C. Lagoudas, On the numerical evaluation of Eshelby's tensor and its application to elastoplastic fibrous composites, Computational Mechanics 7 (1990) 13–19.
[5] S. Nemat–Nasser, M. Hori, Micromechanics: Overall Properties of Heterogeneous Materials, 2nd ed., North–Holland, Amsterdam, 1999.
[6] J. Qu, M. Cherkaoui, Fundamentals of Micromechanics of Solids, Wiley, New Jersey, 2006.
[7] T. Mori, K. Tanaka, Average stress in matrix and average elastic energy of materials with misfitting inclusions, Acta Metallurgica 21 (5) (1973) 571–574.
[8] Y. Benveniste, A new approach to the application of Mori–Tanaka's theory in composite materials, Mechanics of Materials 6 (1987) 147–157.
[9] J.R. Willis, Bounds and self–consistent estimates for the overall properties of anisotropic composites, Journal of the Mechanics and Physics of Solids 25 (1977) 185–202.

CHAPTER

6

Periodic homogenization

OUTLINE

6.1 Preliminaries

Highly heterogeneous materials are very complicated, and their behavior is difficult to compute numerically. For composites with periodic microstructure, periodic homogenization is an efficient tool to overcome difficulties posed by the implementation of global finite element methods and bridge the gap between micromechanics and effective behavior of complex structures.

Consider a composite body that occupies the space $\mathcal{B}^\epsilon$ with volume V^ϵ and has a periodic microstructure (Fig. 6.1). The body is bounded by the surface $\partial\mathcal{B}^\epsilon$ with unit normal vector $\boldsymbol{n}^\epsilon$. The position vector at every point in $\mathcal{B}^\epsilon$ is denoted by $\bar{\boldsymbol{x}}$. Moreover, the characteristic length of the periodic microstructure is assigned with scalar parameter ϵ.

The periodic homogenization, similarly to all micromechanics approaches, describes the composite using two scales. The first, microscopic scale takes into account the different material constituents and their geometry inside the microstructure. The second, macroscopic scale considers the overall body as an imaginary homogeneous medium. For the micro-

Multiscale Modeling Approaches for Composites
https://doi.org/10.1016/B978-0-12-823143-2.00016-3

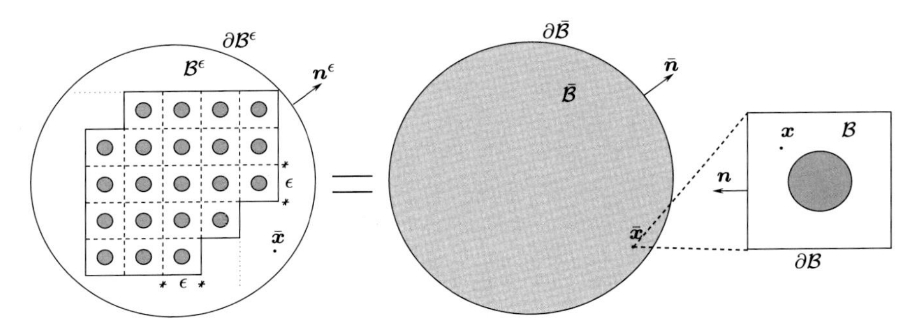

FIGURE 6.1 Composite with periodic microstructure. The homogenization separates the problem into two scales, macroscopic and microscopic.

scopic scale, the term "unit cell" is frequently utilized in periodic media instead of the representative volume element (RVE). Strictly speaking, the RVE is the smallest part of the microstructure where the overall behavior tends to become independent of the applied boundary conditions on it, whereas the unit cell is used for periodic microstructures.

At the macroscale the continuum body occupies the space $\bar{\mathcal{B}}$ with volume $\bar{V}$ bounded by the surface $\partial\bar{\mathcal{B}}$ with unit normal vector $\bar{\boldsymbol{n}}$. Each macroscopic point is assigned with a position vector $\bar{\boldsymbol{x}}$ in $\bar{\mathcal{B}}$. A periodic unit cell that describes the microscopic scale is attached to every position $\bar{\boldsymbol{x}}$. The unit cell occupies the space $\mathcal{B}$ with volume V bounded by the surface $\partial\mathcal{B}$ with unit normal vector $\boldsymbol{n}$. Each microscopic point is assigned with a position vector $\boldsymbol{x}$ in $\mathcal{B}$ (Fig. 6.1). The two scales $\bar{\boldsymbol{x}}$ and $\boldsymbol{x}$ are connected with the characteristic length ϵ through the relation $\boldsymbol{x} = \bar{\boldsymbol{x}}/\epsilon$. The periodic homogenization theory provides accurate results only when the characteristic length tends to zero, that is, when the microstructure is extremely small compared to the actual size of the composite.

6.2 Theoretical background

The composites with periodic microstructure have been extensively studied in the literature. In this section, we discuss the so–called engineering approach for performing periodic homogenization.

For a composite with periodic microstructure, we expect that the main fields (displacements, strains, stresses) should present some sort of periodicity from one unit cell to another. Taking this into account, Suquet [1] has proposed that the displacement field in a unit cell is decomposed in a linear part, dependent on the macroscopic strain, and a periodic part (Fig. 6.2):

$$\boldsymbol{u}(\bar{\boldsymbol{x}},\boldsymbol{x}) = \bar{\boldsymbol{\varepsilon}}(\bar{\boldsymbol{x}})\cdot\boldsymbol{x} + \boldsymbol{u}^{\text{per}}(\bar{\boldsymbol{x}},\boldsymbol{x}), \quad \boldsymbol{u}^{\text{per}} \text{ periodic in } \mathcal{B}. \tag{6.1}$$

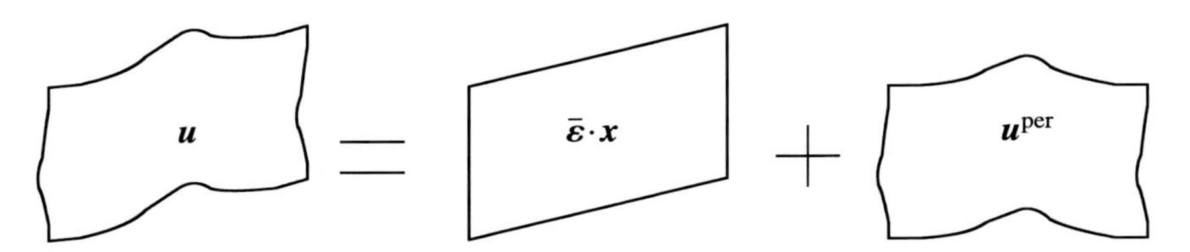

FIGURE 6.2 Decomposition of microscopic displacement into a linear macroscopic and a periodic part [2–4].

In this decomposition an arbitrary rigid body macroscopic displacement $\bar{\boldsymbol{u}}^0$ can be added.

In the engineering approach of the homogenization problem, both macroscopic and microscopic problems are considered to satisfy the classical continuum mechanics principles: the usual correlation between displacements and strains, the equilibrium equation (neglecting inertia and body forces), and a constitutive law that connects stresses with strains. Table 6.1 summarizes the equations in both scales for elastic composites.

TABLE 6.1 Microscopic and macroscopic equations for elastic periodic composites. $\partial\bar{\mathcal{B}}^u$ and $\partial\bar{\mathcal{B}}^t$ are the parts of the macroscopic boundary surface with prescribed displacements and tractions, respectively. It is assumed that $\partial\bar{\mathcal{B}}^u \bigcap \partial\bar{\mathcal{B}}^t = \varnothing$ and $\partial\bar{\mathcal{B}}^u \bigcup \partial\bar{\mathcal{B}}^t = \partial\bar{\mathcal{B}}$.

	Microscale	Macroscale
Equilibrium	$\frac{\partial \sigma_{ij}}{\partial x_j} = 0$	$\frac{\partial \bar{\sigma}_{ij}}{\partial \bar{x}_j} = 0$
Boundary conditions	$u_i = \bar{\varepsilon}_{ij} x_j + u_i^{\text{per}}$ u_i^{per} periodic in $\mathcal{B}$	$\bar{u}_i = 0$ on $\partial\bar{\mathcal{B}}^u$ $\bar{\sigma}_{ij}\bar{n}_j = \bar{t}_i$ on $\partial\bar{\mathcal{B}}^t$
Kinematics	$\varepsilon_{ij} = \frac{1}{2}\left[\frac{\partial u_i}{\partial x_j} + \frac{\partial u_j}{\partial x_i}\right]$	$\bar{\varepsilon}_{ij} = \frac{1}{2}\left[\frac{\partial \bar{u}_i}{\partial \bar{x}_j} + \frac{\partial \bar{u}_j}{\partial \bar{x}_i}\right]$
Hooke's law	$\sigma_{ij} = L_{ijkl}\varepsilon_{kl}$	$\bar{\sigma}_{ij} = \bar{L}_{ijkl}\bar{\varepsilon}_{kl}$

To identify the macroscopic elasticity tensor $\boldsymbol{L}$, the equilibrium equation at the microscale is solved for the periodic boundary conditions (6.1). In indicial notation, this is written as

$$\frac{\partial \sigma_{ij}}{\partial x_j} = 0 \implies \frac{\partial}{\partial x_j}\left(L_{ijkl}\frac{\partial u_k}{\partial x_l}\right) = 0 \implies$$
$$\frac{\partial}{\partial x_j}\left(L_{ijkl}\right)\bar{\varepsilon}_{kl} + \frac{\partial}{\partial x_j}\left(L_{ijkl}\frac{\partial u_k^{\text{per}}}{\partial x_l}\right) = 0. \tag{6.2}$$

Considering an arbitrary known macroscopic strain, the last expression is a linear problem in terms of the periodic displacement $\boldsymbol{u}^{\text{per}}$, and the solution is given in the form [5]

$$u_k^{\text{per}} = N_k^{pq}\overline{\varepsilon}_{pq}, \tag{6.3}$$

where N_k^{pq} is a third–order periodic tensor in $\mathcal{B}$, which is called the corrector tensor. Substituting (6.3) into (6.2) and taking into account the arbitrariness of $\overline{\boldsymbol{\varepsilon}}$ yield

$$\frac{\partial}{\partial x_j}\left(L_{ijkl}A_{klpq}\right) = 0,$$
$$A_{klpq} = \frac{1}{2}\left[\delta_{kp}\delta_{lq} + \delta_{lp}\delta_{kq} + \frac{\partial N_k^{pq}}{\partial x_l} + \frac{\partial N_l^{pq}}{\partial x_k}\right], \tag{6.4}$$

where δ_{ij} denotes the Kronecker delta. This linear system provides the corrector tensor. Once N_k^{pq} is computed, the microscopic stress is expressed as

$$\sigma_{ij} = L_{ijkl}\frac{\partial u_k}{\partial x_l} = L_{ijkl}A_{klpq}\overline{\varepsilon}_{pq}. \tag{6.5}$$

Considering that $\overline{\boldsymbol{\sigma}} = \langle\boldsymbol{\sigma}\rangle = \overline{\boldsymbol{L}}:\overline{\boldsymbol{\varepsilon}}$, we obtain the macroscopic elasticity modulus as

$$\overline{L}_{ijpq} = \langle L_{ijkl}A_{klpq}\rangle. \tag{6.6}$$

6.3 Computation of the overall elasticity tensor

Eqs. (6.4) under periodic conditions for the third–order tensor N_k^{pq} define the unit cell problem, the solution of which provides the overall elasticity tensor of the composite through expression (6.6). Solving (6.4) in one step requires the development of special finite element (or other numerical method) computational technique, which is not possible with the available commercial softwares. Instead, we can rewrite (6.4) in the form of six classical elasticity–type problems using expressions (6.5). Indeed, (6.5) is written in the Voigt notation as

$$\boldsymbol{\sigma}_\alpha = \boldsymbol{L}_{\alpha\gamma}\cdot\widetilde{\boldsymbol{A}}_{\gamma\beta}\cdot\widetilde{\overline{\boldsymbol{\varepsilon}}}_\beta,$$

where $\boldsymbol{L}_{\alpha\gamma}$ and $\widetilde{\boldsymbol{A}}_{\gamma\beta}$ are 6×6 matrices, and

$$\boldsymbol{\sigma}_{\alpha}=\begin{bmatrix}\sigma_{11}\\\sigma_{22}\\\sigma_{33}\\\sigma_{12}\\\sigma_{13}\\\sigma_{23}\end{bmatrix},\qquad \tilde{\bar{\boldsymbol{\varepsilon}}}_{\beta}=\begin{bmatrix}\bar{\varepsilon}_{11}\\\bar{\varepsilon}_{22}\\\bar{\varepsilon}_{33}\\2\bar{\varepsilon}_{12}\\2\bar{\varepsilon}_{13}\\2\bar{\varepsilon}_{23}\end{bmatrix}.$$

Suppose that the unit cell is subjected to the periodicity conditions

$$u_i=\bar{\varepsilon}_{1_{ij}}x_j+u_i^{\text{per}},\qquad \bar{\varepsilon}_{1_{ij}}=\begin{cases}1, & i=j=1,\\ 0 & \text{otherwise.}\end{cases}$$

Technically speaking, the above displacement field can be imposed in a finite element code with the help of an additional ("dummy") degree of freedom, whose displacement is equal to the macroscopic strain (further details can be found in [2,6]). Under this condition, the equilibrium equation reads

$$\operatorname{div}\boldsymbol{\sigma}_{\alpha}^{1}=0\quad\Longrightarrow\quad\operatorname{div}\left(\boldsymbol{L}_{\alpha\gamma}\cdot\widetilde{\boldsymbol{A}}_{\gamma1}\right)=0.$$

The final expression is the same as (6.4) when $p=q=1$. After solving this elasticity problem, the average of the stress tensor over the volume of the unit cell yields

$$\left\langle\boldsymbol{\sigma}_{\alpha}^{1}\right\rangle=\left\langle\boldsymbol{L}_{\alpha\gamma}\cdot\widetilde{\boldsymbol{A}}_{\gamma1}\right\rangle=\bar{\boldsymbol{L}}_{\alpha1}$$

due to (6.6). In other words, the macroscopic boundary condition $\bar{\boldsymbol{\varepsilon}}_1$ provides local stress fields, whose average value gives the first column of the macroscopic elasticity tensor. In a similar manner, we can apply five additional macroscopic strain fields and obtain the rest of the columns of $\bar{\boldsymbol{L}}$.

In summary:

BOX 6.1 Macroscopic elasticity tensor in periodic media

In periodic homogenization the unit cell problem is described by the linear system

$$\sigma_{ij}=L_{ijkl}A_{klpq}\bar{\varepsilon}_{pq},\qquad \frac{\partial\sigma_{ij}}{\partial x_j}=0,$$

$$A_{klpq}=\frac{1}{2}\left[\delta_{kp}\delta_{lq}+\delta_{lp}\delta_{kq}+\frac{\partial N_k^{pq}}{\partial x_l}+\frac{\partial N_l^{pq}}{\partial x_k}\right],$$

N_k^{pq} periodic third-order tensor.

The macroscopic elasticity tensor is computed by the expression

$$\bar{L}_{ijpq} = \langle L_{ijkl} A_{klpq} \rangle .$$

The components of $\bar{\boldsymbol{L}}$ can be obtained through finite elements or other numerical technique by applying six macroscopic strain fields:

1. $\bar{\boldsymbol{\varepsilon}}_1$ with $\bar{\varepsilon}_{11} = 1$, and the rest of the terms are zero.
2. $\bar{\boldsymbol{\varepsilon}}_2$ with $\bar{\varepsilon}_{22} = 1$, and the rest of the terms are zero.
3. $\bar{\boldsymbol{\varepsilon}}_3$ with $\bar{\varepsilon}_{33} = 1$, and the rest of the terms are zero.
4. $\bar{\boldsymbol{\varepsilon}}_4$ with $2\bar{\varepsilon}_{12} = 1$, and the rest of the terms are zero.
5. $\bar{\boldsymbol{\varepsilon}}_5$ with $2\bar{\varepsilon}_{13} = 1$, and the rest of the terms are zero.
6. $\bar{\boldsymbol{\varepsilon}}_6$ with $2\bar{\varepsilon}_{23} = 1$, and the rest of the terms are zero.

For each one of these conditions, the average value of the stress tensor is equal to the corresponding column of the elasticity tensor (in the Voigt notation).

The unit cell problem under periodic conditions requires, for solvability reasons, to add an additional kinematic constraint. Usually, this issue is resolved by imposing zero displacements at an arbitrary node of the unit cell.

It is worth mentioning that for general type of composites, full–field finite element computations can be performed using various types of boundary conditions, as long as they respect the Hill–Mandel principle. By applying uniform traction or linear displacements at the boundary we can obtain an estimation of bounds in the overall behavior. Periodic boundary conditions provide response that lies between these bounds (see Example 3 of the chapter).

6.4 Particular case: multilayered composite

The simplest problem in periodic homogenization with analytical solution is the multilayered composite, that is, a material with unit cell consisting of several stacked constituents. Here we assume a bi–phase composite whose unit cell is demonstrated in Fig. 6.3. Each constituent/layer has volume fraction c_r and elasticity tensor $\boldsymbol{L}^{(r)}$, where r =1 or 2. Obviously, $c_1 + c_2 = 1$.

Due to the form of the microstructure, all fields vary spatially only in the x_1 direction, transforming the unit cell equilibrium into an 1–D problem. In the general case the layers possess full anisotropy, that is, their elasticity tensors have 21 independent components. For the simplicity of the subsequent expressions, the elasticity tensor for each layer is decomposed by

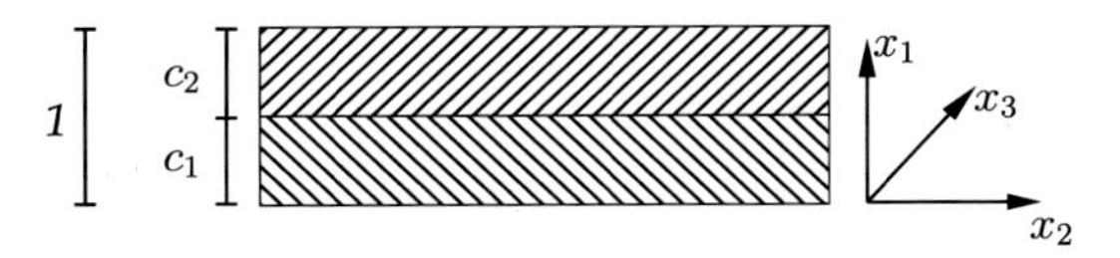

FIGURE 6.3 Unit cell of a multilayered composite with 2 distinct layers in the microstructure.

means of the following three matrices:

$$\boldsymbol{L}_{nn}^{(r)} = \begin{bmatrix} L_{11}^{(r)} & L_{14}^{(r)} & L_{15}^{(r)} \\ L_{14}^{(r)} & L_{44}^{(r)} & L_{45}^{(r)} \\ L_{15}^{(r)} & L_{45}^{(r)} & L_{55}^{(r)} \end{bmatrix}, \quad \boldsymbol{L}_{nt}^{(r)} = \begin{bmatrix} L_{12}^{(r)} & L_{13}^{(r)} & L_{16}^{(r)} \\ L_{24}^{(r)} & L_{34}^{(r)} & L_{46}^{(r)} \\ L_{25}^{(r)} & L_{35}^{(r)} & L_{56}^{(r)} \end{bmatrix},$$
$$\boldsymbol{L}_{tt}^{(r)} = \begin{bmatrix} L_{22}^{(r)} & L_{23}^{(r)} & L_{26}^{(r)} \\ L_{23}^{(r)} & L_{33}^{(r)} & L_{36}^{(r)} \\ L_{26}^{(r)} & L_{36}^{(r)} & L_{66}^{(r)} \end{bmatrix}. \tag{6.7}$$

Considering all fields to depend only on x_1, the equilibrium equations (6.4) are reduced for each layer to

$$\frac{\mathrm{d}}{\mathrm{d}x_1}\left(\boldsymbol{L}_{nn}^{(r)} \cdot \frac{\mathrm{d}\boldsymbol{U}_n^{(r)}}{\mathrm{d}x_1} + \boldsymbol{L}_{nn}^{(r)}\right) = 0, \quad \frac{\mathrm{d}}{\mathrm{d}x_1}\left(\boldsymbol{L}_{nn}^{(r)} \cdot \frac{\mathrm{d}\boldsymbol{U}_t^{(r)}}{\mathrm{d}x_1} + \boldsymbol{L}_{nt}^{(r)}\right) = 0, \tag{6.8}$$

with

$$\boldsymbol{U}_n^{(r)} = \begin{bmatrix} N_1^{11(r)} & N_1^{21(r)} & N_1^{31(r)} \\ N_2^{11(r)} & N_2^{21(r)} & N_2^{31(r)} \\ N_3^{11(r)} & N_3^{21(r)} & N_3^{31(r)} \end{bmatrix}, \quad \boldsymbol{U}_t^{(r)} = \begin{bmatrix} N_1^{22(r)} & N_1^{33(r)} & N_1^{23(r)} \\ N_2^{22(r)} & N_2^{33(r)} & N_2^{23(r)} \\ N_3^{22(r)} & N_3^{33(r)} & N_3^{23(r)} \end{bmatrix}. \tag{6.9}$$

Due to the continuity of tractions along the x_1 direction, integrating once the differential equations (6.8) yields

$$\frac{\mathrm{d}\boldsymbol{U}_n^{(r)}}{\mathrm{d}x_1} = \left[\boldsymbol{L}_{nn}^{(r)}\right]^{-1} \cdot \left[\boldsymbol{m}_n - \boldsymbol{L}_{nn}^{(r)}\right],$$
$$\frac{\mathrm{d}\boldsymbol{U}_t^{(r)}}{\mathrm{d}x_1} = \left[\boldsymbol{L}_{nn}^{(r)}\right]^{-1} \cdot \left[\boldsymbol{m}_t - \boldsymbol{L}_{nt}^{(r)}\right], \tag{6.10}$$

where the second–order tensors $\boldsymbol{m}_n$ and $\boldsymbol{m}_t$ are constant, and the same for both phases. Integrating once more gives

$$\boldsymbol{U}_n^{(r)}(x_1) = \left[\boldsymbol{L}_{nn}^{(r)}\right]^{-1} \cdot \left[\boldsymbol{m}_n - \boldsymbol{L}_{nn}^{(r)}\right] x_1 + \boldsymbol{e}_n^{(r)},$$
$$\boldsymbol{U}_t^{(r)}(x_1) = \left[\boldsymbol{L}_{nn}^{(r)}\right]^{-1} \cdot \left[\boldsymbol{m}_t - \boldsymbol{L}_{nt}^{(r)}\right] x_1 + \boldsymbol{e}_t^{(r)}, \tag{6.11}$$

where the vectors $\boldsymbol{e}_n^{(r)}$ and $\boldsymbol{e}_t^{(r)}$ differ for the two layers. The last expressions define the solution of the unit cell problem. The position x_1 varies from 0 to c_1 for the layer 1 and from c_1 to 1 for the layer 2. The unknown constants $\boldsymbol{m}_n$ and $\boldsymbol{m}_t$ are computed with the help of the periodicity conditions

$$\boldsymbol{U}_n^{(1)}(0) = \boldsymbol{U}_n^{(2)}(1), \quad \boldsymbol{U}_t^{(1)}(0) = \boldsymbol{U}_t^{(2)}(1) \tag{6.12}$$

and the continuity conditions

$$\boldsymbol{U}_n^{(1)}(c_1) = \boldsymbol{U}_n^{(2)}(c_1), \quad \boldsymbol{U}_t^{(1)}(c_1) = \boldsymbol{U}_t^{(2)}(c_1). \tag{6.13}$$

After algebraic calculations, we find that

$$\boldsymbol{m}_n = \left[c_1 \left[\boldsymbol{L}_{nn}^{(1)}\right]^{-1} + c_2 \left[\boldsymbol{L}_{nn}^{(2)}\right]^{-1}\right]^{-1},$$
$$\boldsymbol{m}_t = \boldsymbol{m}_n \cdot \left[c_1 \left[\boldsymbol{L}_{nn}^{(1)}\right]^{-1} \cdot \boldsymbol{L}_{nt}^{(1)} + c_2 \left[\boldsymbol{L}_{nn}^{(2)}\right]^{-1} \cdot \boldsymbol{L}_{nt}^{(2)}\right]. \tag{6.14}$$

Eq. (6.6) and the 1–D representation of the problem yield

$$\overline{\boldsymbol{L}}_{nn} = \left\langle \boldsymbol{L}_{nn} \cdot \frac{\mathrm{d}\boldsymbol{U}_n^{(r)}}{\mathrm{d}x_1} + \boldsymbol{L}_{nn} \right\rangle,$$
$$\overline{\boldsymbol{L}}_{nt} = \left\langle \boldsymbol{L}_{nn} \cdot \frac{\mathrm{d}\boldsymbol{U}_t^{(r)}}{\mathrm{d}x_1} + \boldsymbol{L}_{nt} \right\rangle,$$
$$\overline{\boldsymbol{L}}_{tt} = \left\langle \boldsymbol{L}_{nt}^T \cdot \frac{\mathrm{d}\boldsymbol{U}_t^{(r)}}{\mathrm{d}x_1} + \boldsymbol{L}_{tt} \right\rangle. \tag{6.15}$$

Combining the last relations with (6.10) and (6.14) leads to the conclusion that:

BOX 6.2 Elasticity tensor of a bi–phase multilayered composite

In a bi–phase multilayered composite the components of the macroscopic elasticity tensor $\bar{L}$ are identified from the expressions

$$\bar{L}_{nn} = \left[c_1 \left[L_{nn}^{(1)} \right]^{-1} + c_2 \left[L_{nn}^{(2)} \right]^{-1} \right]^{-1},$$

$$\bar{L}_{nt} = \bar{L}_{nn} \cdot \left[c_1 \left[L_{nn}^{(1)} \right]^{-1} \cdot L_{nt}^{(1)} + c_2 \left[L_{nn}^{(2)} \right]^{-1} \cdot L_{nt}^{(2)} \right],$$

$$\begin{aligned} \bar{L}_{tt} = c_1 & \left[\left[L_{nt}^{(1)} \right]^T \cdot \left[L_{nn}^{(1)} \right]^{-1} \cdot \left[\bar{L}_{nt} - L_{nt}^{(1)} \right] + L_{tt}^{(1)} \right] \\ + c_2 & \left[\left[L_{nt}^{(2)} \right]^T \cdot \left[L_{nn}^{(2)} \right]^{-1} \cdot \left[\bar{L}_{nt} - L_{nt}^{(2)} \right] + L_{tt}^{(2)} \right]. \end{aligned} \tag{6.16}$$

As a final remark, note that the strain concentration tensors for the two phases are expressed in the Voigt notation as

$$\widetilde{A}^{(k)} = \begin{bmatrix} 1+\dfrac{dN_1^{11(k)}}{dx_1} & \dfrac{dN_1^{22(k)}}{dx_1} & \dfrac{dN_1^{33(k)}}{dx_1} & \dfrac{dN_1^{21(k)}}{dx_1} & \dfrac{dN_1^{31(k)}}{dx_1} & \dfrac{dN_1^{23(k)}}{dx_1} \\ 0 & 1 & 0 & 0 & 0 & 0 \\ 0 & 0 & 1 & 0 & 0 & 0 \\ \dfrac{dN_2^{11(k)}}{dx_1} & \dfrac{dN_2^{22(k)}}{dx_1} & \dfrac{dN_2^{33(k)}}{dx_1} & 1+\dfrac{dN_2^{21(k)}}{dx_1} & \dfrac{dN_2^{31(k)}}{dx_1} & \dfrac{dN_2^{23(k)}}{dx_1} \\ \dfrac{dN_3^{11(k)}}{dx_1} & \dfrac{dN_3^{22(k)}}{dx_1} & \dfrac{dN_3^{33(k)}}{dx_1} & \dfrac{dN_3^{21(k)}}{dx_1} & 1+\dfrac{dN_3^{31(k)}}{dx_1} & \dfrac{dN_3^{23(k)}}{dx_1} \\ 0 & 0 & 0 & 0 & 0 & 1 \end{bmatrix}, \tag{6.17}$$

where the necessary derivatives are computed from (6.10) with the help of (6.14).

6.5 Examples

Example 1: multilayered composite

Let us consider the multilayered composite of Fig. 6.4. The RVE of this composite consists of two layers, which have a common surface normal to the x_1 axis.

Both layers are assumed to be elastic isotropic materials. Their Young modulus and Poisson ratio are $E_1 = 3$ GPa and $\nu_1 = 0.3$ for layer 1 and

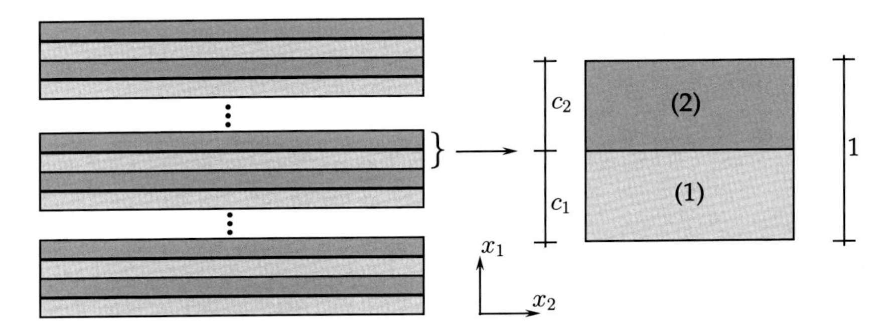

FIGURE 6.4 Schematic representation of multilayered composite and its unit cell.

$E_2 = 55$ GPa and $\nu_2 = 0.25$ for layer 2, respectively. The layer 2 volume fraction varies between 0% and 50%. Identify the macroscopic shear moduli in the normal and parallel directions to the layer direction.

Solution:

Due to the geometrical characteristics of the microstructure, the composite is expected to behave like an equivalent transversely isotropic medium with axis of symmetry parallel to the x_1 direction. Thus the macroscopic elasticity tensor $\bar{\boldsymbol{L}}$ can be expressed in terms of five elastic constants: the transverse bulk modulus $\bar{K}^{\mathrm{tr}}$, the transverse shear modulus $\bar{\mu}^{\mathrm{tr}}$, the axial shear modulus $\bar{\mu}^{\mathrm{ax}}$, and the stiffness coefficients $\bar{l}$ and $\bar{n}$. For the multilayered composite, the term "axial" refers to the plane normal to the layers, whereas the term "transverse" refers to the plane parallel to the layers. The Voigt form of $\bar{\boldsymbol{L}}$ is written as

$$\bar{\boldsymbol{L}} = \begin{bmatrix} \bar{n} & \bar{l} & \bar{l} & 0 & 0 & 0 \\ \bar{l} & \bar{K}^{\mathrm{tr}} + \bar{\mu}^{\mathrm{tr}} & \bar{K}^{\mathrm{tr}} - \bar{\mu}^{\mathrm{tr}} & 0 & 0 & 0 \\ \bar{l} & \bar{K}^{\mathrm{tr}} - \bar{\mu}^{\mathrm{tr}} & \bar{K}^{\mathrm{tr}} + \bar{\mu}^{\mathrm{tr}} & 0 & 0 & 0 \\ 0 & 0 & 0 & \bar{\mu}^{\mathrm{ax}} & 0 & 0 \\ 0 & 0 & 0 & 0 & \bar{\mu}^{\mathrm{ax}} & 0 \\ 0 & 0 & 0 & 0 & 0 & \bar{\mu}^{\mathrm{tr}} \end{bmatrix}.$$

Fig. 6.5 illustrates the obtained macroscopic shear moduli as functions of layer 2 volume fraction according to the periodic homogenization theory. Note that the shear modulus in the parallel plane to the layers is much higher than that in the normal plane to the layers.

The corresponding python script for obtaining these results is the following:

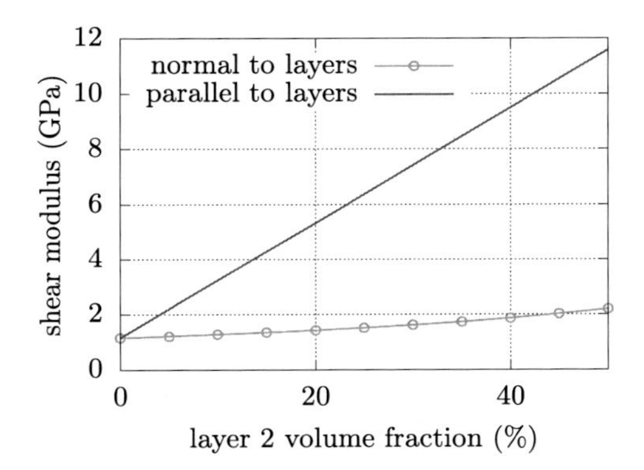

FIGURE 6.5 Macroscopic shear moduli vs layer 2 volume fraction.

```
import numpy as np
# Data
E1=3.0
v1=0.3
E2=55.0
v2=0.25

K1=E1/(3*(1-2*v1))
mu1=E1/(2*(1+v1))
K2=E2/(3*(1-2*v2))
mu2=E2/(2*(1+v2))

# Elasticity tensors
L1=np.matrix([[K1+4*mu1/3, K1-2*mu1/3, K1-2*mu1/3,\
               0., 0., 0.],\
              [K1-2*mu1/3, K1+4*mu1/3, K1-2*mu1/3,\
               0., 0., 0.],\
              [K1-2*mu1/3, K1-2*mu1/3, K1+4*mu1/3,\
               0., 0., 0.],\
              [0., 0., 0., mu1, 0., 0.],\
              [0., 0., 0., 0., mu1, 0.],\
              [0., 0., 0., 0., 0., mu1]])
L2=np.matrix([[K2+4*mu2/3, K2-2*mu2/3, K2-2*mu2/3,\
               0., 0., 0.],\
              [K2-2*mu2/3, K2+4*mu2/3, K2-2*mu2/3,\
               0., 0., 0.],\
              [K2-2*mu2/3, K2-2*mu2/3, K2+4*mu2/3,\
               0., 0., 0.],\
              [0., 0., 0., mu2, 0., 0.],\
```

```
              [0., 0., 0., 0., mu2, 0.],\
              [0., 0., 0., 0., 0., mu2]])

Lnn1=np.matrix([[L1[0,0], L1[0,3], L1[0,4]],\
                [L1[0,3], L1[3,3], L1[3,4]],\
                [L1[0,4], L1[3,4], L1[4,4]]])
Lnt1=np.matrix([[L1[0,1], L1[0,2], L1[0,5]],\
                [L1[1,3], L1[2,3], L1[3,5]],\
                [L1[1,4], L1[2,4], L1[4,5]]])
Ltt1=np.matrix([[L1[1,1], L1[1,2], L1[1,5]],\
                [L1[1,2], L1[2,2], L1[2,5]],\
                [L1[1,5], L1[2,5], L1[5,5]]])
Lnn2=np.matrix([[L2[0,0], L2[0,3], L2[0,4]],\
                [L2[0,3], L2[3,3], L2[3,4]],\
                [L2[0,4], L2[3,4], L2[4,4]]])
Lnt2=np.matrix([[L2[0,1], L2[0,2], L2[0,5]],\
                [L2[1,3], L2[2,3], L2[3,5]],\
                [L2[1,4], L2[2,4], L2[4,5]]])
Ltt2=np.matrix([[L2[1,1], L2[1,2], L2[1,5]],\
                [L2[1,2], L2[2,2], L2[2,5]],\
                [L2[1,5], L2[2,5], L2[5,5]]])

tableA=np.zeros((51,3))

# Periodic homogenization elasticity tensor
for n in range(51):
   c=0.01*n
   Lnn=((1-c)*(Lnn1.I)+c*(Lnn2.I)).I
   Lnt=Lnn*((1-c)*(Lnn1.I)*Lnt1+c*(Lnn2.I)*Lnt2)
   Ltt=(1-c)*((Lnt1.T)*(Lnn1.I)*(Lnt-Lnt1)+Ltt1)\
          +c*((Lnt2.T)*(Lnn2.I)*(Lnt-Lnt2)+Ltt2)
   Leff=np.matrix([[Lnn[0,0], Lnt[0,0], Lnt[0,1],\
                     Lnn[0,1], Lnn[0,2], Lnt[0,2]],\
                    [Lnt[0,0], Ltt[0,0], Ltt[0,1],\
                     Lnt[1,0], Lnt[2,0], Ltt[0,2]],\
                    [Lnt[0,1], Ltt[0,1], Ltt[1,1],\
                     Lnt[1,1], Lnt[2,1], Ltt[1,2]],\
                    [Lnn[0,1], Lnt[1,0], Lnt[1,1],\
                     Lnn[1,1], Lnn[1,2], Lnt[1,2]],\
                    [Lnn[0,2], Lnt[2,0], Lnt[2,1],\
                     Lnn[1,2], Lnn[2,2], Lnt[2,2]],\
                    [Lnt[0,2], Ltt[0,2], Ltt[1,2],\
                     Lnt[1,2], Lnt[2,2], Ltt[2,2]]])
   GL=Leff[4,4]
   GT=Leff[5,5]
```

```
    tableA[n,:]=[c, GL, GT]

np.savetxt('GL_GT.txt',tableA,fmt='%1.5f')
```

Example 2: unidirectional long fiber composite with hexagonal packing

In this example, the macroscopic response of a long fiber composite with hexagonal packing (Fig. 6.6_a) is identified with the help of the periodic homogenization. The corresponding periodic unit cell for such a microstructure is illustrated in Fig. 6.6_b. Recall that these composites behave like transversely isotropic materials (see Chapter 2, Subsection 2.3.2). The matrix (phase 0) and the fibers (phase 1) are considered isotropic with Young moduli $E_0 = 3$ GPa and $E_1 = 72$ GPa and Poisson ratios $\nu_0 = 0.3$ and $\nu_1 = 0.2$, respectively. The fibers have the long axis parallel to the third direction. Their overall volume fraction is equal to 20%.

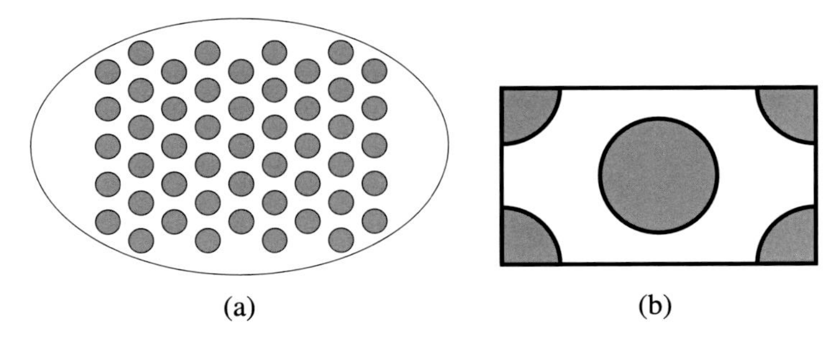

FIGURE 6.6 Schematic representation of (a) unidirectional long fiber composite cross–section and (b) its corresponding unit cell. The fibers are distributed in hexagonal packing inside the matrix.

Solution:

The numerical computations on the unit cell were performed using the commercial finite element software ABAQUS. The discretization of the unit cell is shown in Fig. 6.7. The six virtual macroscopic strain fields were imposed with the help of the concept of the constraint drivers.[1] Fig. 6.8 demonstrates the distribution of the microscopic strains for three imposed macroscopic strain fields. The average stresses for all the cases construct the macroscopic elasticity tensor, as explained in Section 6.3. The obtained

[1]Further details regarding the concept of constraint drivers and the application of periodic boundary conditions in a unit cell are provided by Praud [6].

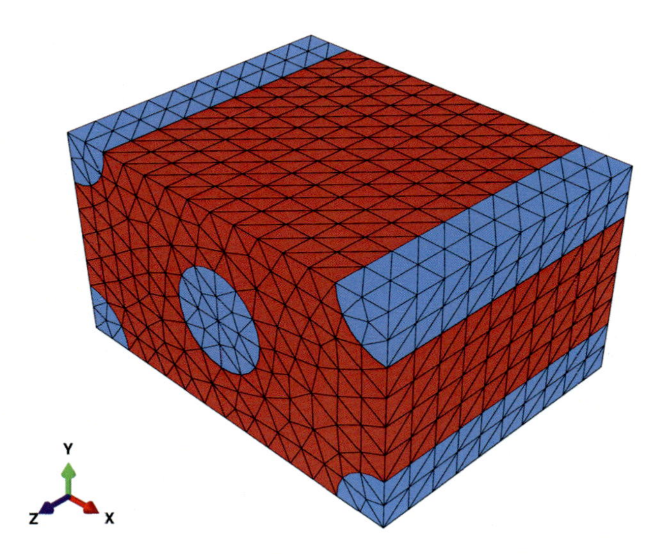

FIGURE 6.7 Finite element discretization of the unit cell using the commercial software ABAQUS.

$\bar{\boldsymbol{L}}$ is

$$\bar{\boldsymbol{L}} = \begin{bmatrix} 5.364 & 2.221 & 2.083 & 0 & 0 & 0 \\ 2.221 & 5.364 & 2.083 & 0 & 0 & 0 \\ 2.083 & 2.083 & 17.886 & 0 & 0 & 0 \\ 0 & 0 & 0 & 1.572 & 0 & 0 \\ 0 & & 00 & 0 & 1.676 & 0 \\ 0 & 0 & 0 & 0 & 0 & 1.676 \end{bmatrix} \text{GPa}.$$

It is worth mentioning that the hexagonal packing of fibers produces a composite with behavior equivalent to that of fibers randomly distributed inside the matrix [8]. Considering the Mori–Tanaka method (computational details are provided in Example 3 of Chapter 5), the macroscopic elasticity tensor is equal to

$$\bar{\boldsymbol{L}}_{\mathrm{MT}} = \begin{bmatrix} 5.367 & 2.226 & 2.084 & 0 & 0 & 0 \\ 2.226 & 5.367 & 2.084 & 0 & 0 & 0 \\ 2.084 & 2.084 & 17.951 & 0 & 0 & 0 \\ 0 & 0 & 0 & 1.570 & 0 & 0 \\ 0 & & 00 & 0 & 1.678 & 0 \\ 0 & 0 & 0 & 0 & 0 & 1.678 \end{bmatrix} \text{GPa}.$$

Note that Mori–Tanaka and periodic homogenization provide quite similar results for these composites.

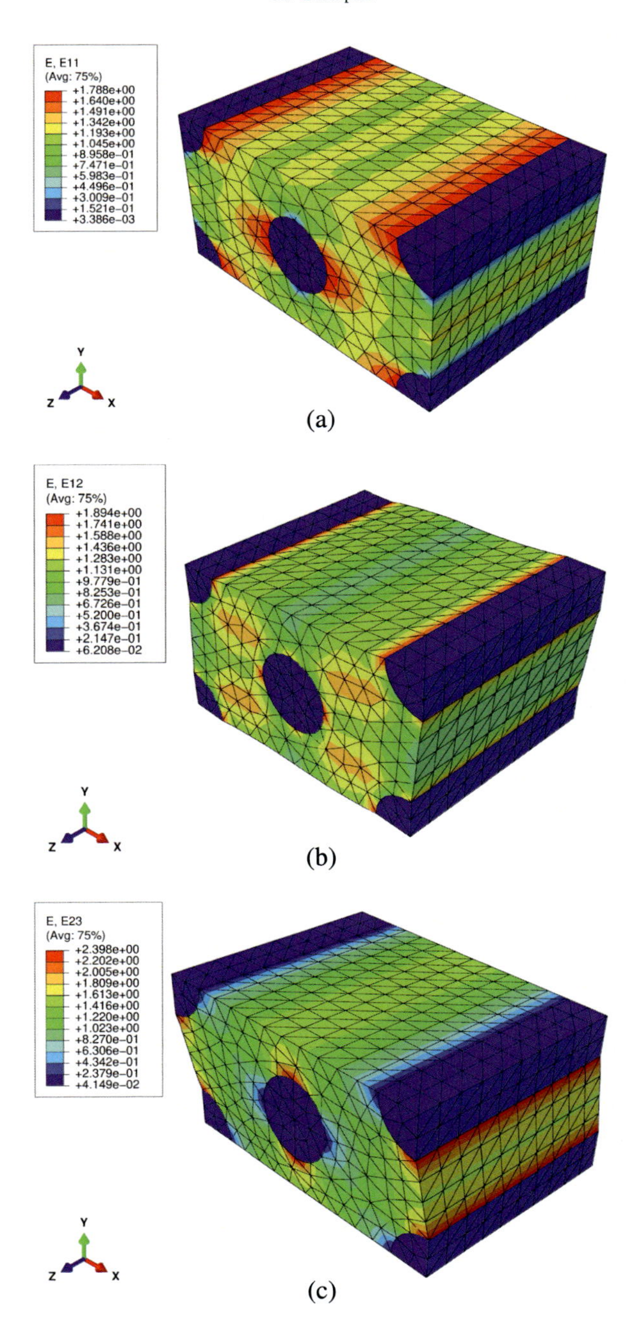

FIGURE 6.8 Spatial distribution of (a) microscopic ε_{11} for $\bar{\varepsilon}_{11} = 1$ and the zero remaining terms, (b) microscopic $2\varepsilon_{12}$ for $2\bar{\varepsilon}_{12} = 1$ and the zero remaining terms, and (c) microscopic $2\varepsilon_{23}$ for $2\bar{\varepsilon}_{23} = 1$ and the zero remaining terms.

Example 3: unidirectional long fiber composite with square packing

Consider a bi–phase composite consisting of a polymer matrix ($E_0 = 3$ GPa, $\nu_0 = 0.3$) reinforced with unidirectional long cylindrical glass fibers ($E_1 = 72$ GPa, $\nu_1 = 0.2$). The fiber arrangement follows square packing. Compute the macroscopic transverse shear modulus of the composite as a function of the fiber volume fraction (from 10% until 50%) using finite element computations for three different types of boundary conditions on the unit cell:

- linear displacements,
- uniform tractions,
- periodicity conditions.

Solution:

The numerical computations were performed on unit cells under the three different types of boundary conditions, using the commercial finite element software ABAQUS.

With regard to the linear displacements and uniform traction boundary conditions, the macroscopic shear stresses and strains are obtained numerically using the stresses and strains at the centroid of each element and its volume. If σ_{12}^i is the microscopic stress at the centroid of element i with volume V^i, for N total finite elements, the macroscopic stress $\bar{\sigma}_{12}$ is computed by the simple formula

$$\bar{\sigma}_{12} = \frac{\sum\limits_{i=1}^{N} \sigma_{12}^i V^i}{\sum\limits_{i=1}^{N} V^i}.$$

With regard to the periodic boundary conditions, the macroscopic transverse shear strain is imposed at a special constraint driver (i.e., a dummy node linked through kinematic constraints with opposite surfaces of the unit cell). Setting the name of this constraint driver as CD12, the macroscopic transverse shear stress is given by

$$\bar{\sigma}_{12} = \frac{\text{reaction force at CD12}}{\text{volume of unit cell}}.$$

Further computational details are provided in [6].

The discretization of the unit cell for 10% volume fraction of the fibers is shown in Fig. 6.9. For the same volume fraction, Fig. 6.10 illustrates the shear stress distribution inside the deformed unit cell for the three loading cases. As we can observe, the deformation pattern due to uniform tractions

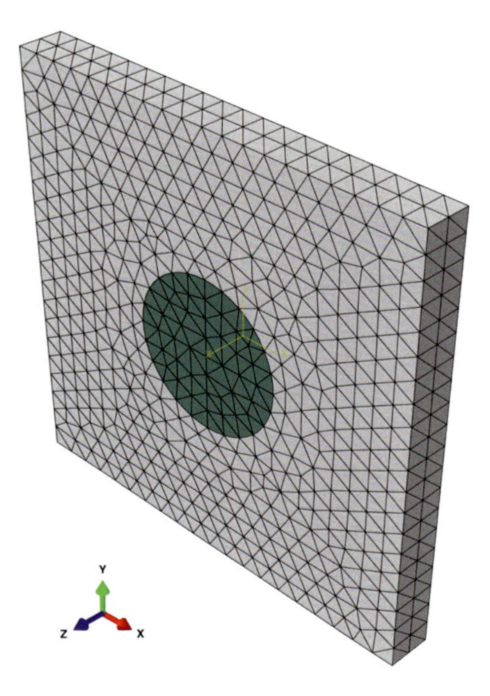

FIGURE 6.9 FE discretization of fiber composite unit cell with square arrangement and 10% fiber volume fraction.

is quite similar to that of the periodicity conditions. Fig. 6.11 demonstrates that the macroscopic transverse shear modulus predicted by periodicity conditions lies between the predictions of the other two loading cases for all volume fractions. This observation also holds for the other macroscopic moduli [7].

References

[1] P.M. Suquet, Elements of Homogenization for Inelastic Solid Mechanics, Lecture Notes in Physics, vol. 272, Springer, Berlin, 1987, pp. 193–278.

[2] J.C. Michel, H. Moulinec, P. Suquet, Effective properties of composite materials with periodic microstructure: a computational approach, Computer Methods in Applied Mechanics and Engineering 172 (1999) 109–143.

[3] M. Bornert, T. Bretheau, P. Gilormini, Homogenization in Mechanics of Materials, ISTE–Wiley, New York, 2008.

[4] N. Charalambakis, G. Chatzigeorgiou, Y. Chemisky, F. Meraghni, Mathematical homogenization of inelastic dissipative materials: a survey and recent progress, Continuum Mechanics and Thermodynamics 30 (1) (2018) 1–51.

[5] E. Sanchez-Palencia, Non–homogeneous Media and Vibration Theory, Lecture Notes in Physics, vol. 127, Springer–Verlag, Berlin, 1978, pp. 1–398.

[6] F. Praud, G. Chatzigeorgiou, F. Meraghni, Fully integrated multi–scale modelling of damage and time–dependency in thermoplastic–based woven composites, International Journal of Damage Mechanics 30 (2) (2021) 163–195.

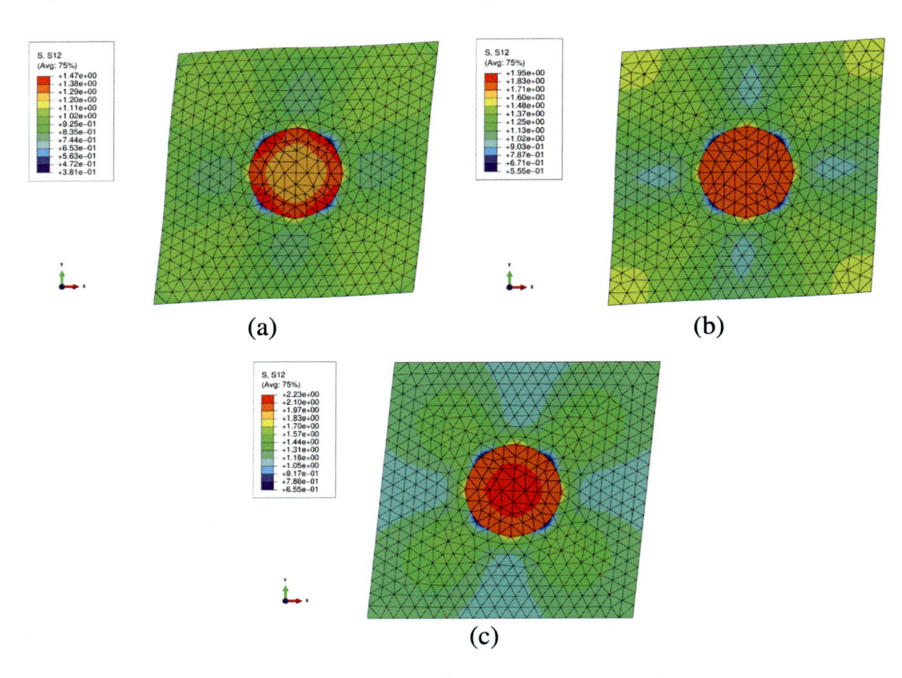

FIGURE 6.10 Distribution of microscopic transverse shear stress inside the unit cell when the latter is subjected to (a) uniform traction (TBC), (b) periodic displacement (PBC), and (c) linear displacement (DBC) boundary conditions. The fiber volume fraction is 10%.

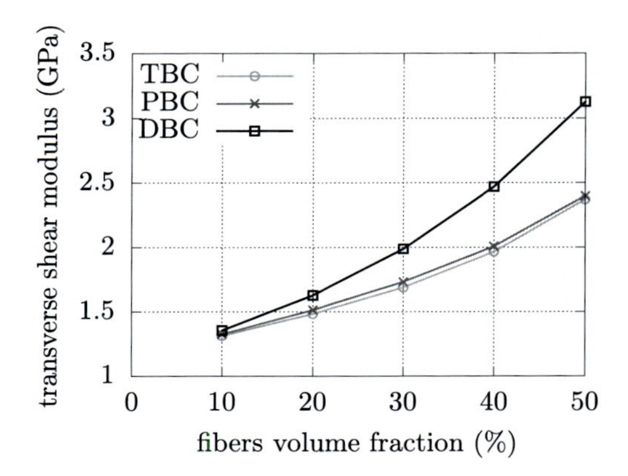

FIGURE 6.11 Macroscopic transverse shear modulus vs fiber volume fraction when the unit cell is subjected to uniform traction (TBC), periodic displacement (PBC), and linear displacement (DBC) boundary conditions.

[7] A. Drago, M.J. Pindera, Micro–macromechanical analysis of heterogeneous materials: macroscopically homogeneous vs periodic microstructures, Composites Science and Technology 67 (2007) 1243–1263.

[8] Z. Hashin, B.W. Rosen, The elastic moduli of fiber–reinforced materials, Journal of Applied Mechanics 31 (1964) 223–232.

CHAPTER

7

Classical laminate theory

OUTLINE

Multiscale Modeling Approaches for Composites
https://doi.org/10.1016/B978-0-12-823143-2.00017-5

7.1 Introduction

Scale transition: from the microstructure to the laminate

A fiber–reinforced composite laminate plate is built as a layup of thin oriented laminae bonded together. A lamina is typically thin and can consist of a matrix reinforced by aligned fibers, short random fibers, woven fabric, and so on. Mechanical, thermomechanical, or even hygrothermal analysis of a laminated composite structure requires a scale transition from the lamina to representative laminate element and coordinate system transformation from the global to local axes, as shown in Fig. 7.1. Using micromechanics methods detailed in the previous chapters, we can estimate the average properties of the lamina from its constitutive phases (fiber, matrix) and their geometrical characteristics such as the fiber aspects ratio (short fiber) or architecture (woven). At this scale, the lamina stiffness and strength can be optimized by varying its microstructure. Then the

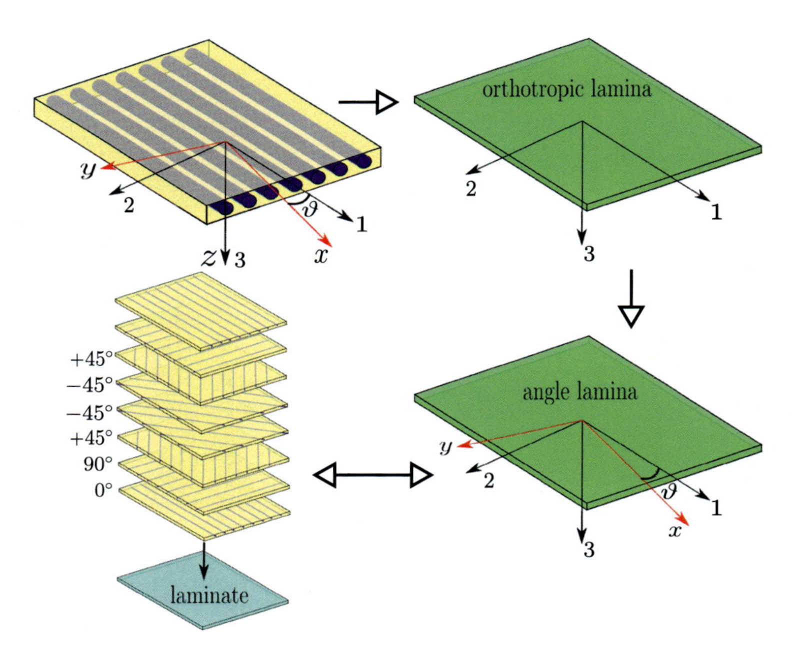

FIGURE 7.1 Schematic of the analysis of a laminate illustrating the scale transition and transformation between local and global axes.

lamina is homogenized and considered as an orthotropic homogeneous material. The properties of the lamina can be also determined experimentally through a series of mechanical tests. Hence the stress–strain relation for orthotropic medium can be expressed in principal lamina directions of symmetry (local axes) and then transformed to an arbitrary coordinate system obtained by a rotation of the local one.

The design of the laminate as a layup of the angle laminae requires the estimation of the strains and stresses in the whole laminate. The failure of the laminate is based on the application of failure criteria for each ply in its local axes.

Plane stress assumption: Kirchhoff–Love plates

The basic assumptions of Kirchhoff plate theory can be summarized as follows: no thickness changes, no transverse shear, linear distributions of the in–plane stresses, and parabolic distribution of the transverse shear stresses. The model can be applied to plates made of classical isotropic materials and for small deflections; it is assumed that any cross–section must be plane and orthogonal to the mid–plane before and after deformation.

7.2 Stress–strain relation for an orthotropic material

7.2.1 From tensor to contracted (Voigt) notation

As discussed in the previous chapters, the generalized Hooke law relating stress to strain tensors of anisotropic media is expressed as

$$\sigma_{ij} = L_{ijkl}\varepsilon_{kl}, \tag{7.1}$$

where σ_{ij} and ε_{kl} are the symmetric stress and strain second–order tensors, respectively, whereas L_{ijkl} is the elastic stiffness symmetric fourth–order tensor.

The strain–stress relation can also be written using the tensor notation as

$$\varepsilon_{ij} = \mathcal{S}_{ijkl}\sigma_{kl}, \tag{7.2}$$

where $\mathcal{S}_{ijkl}$ is the elastic compliance symmetric fourth–order tensor.[1] We can easily note that

$$\mathcal{S}_{ijkl} = L^{-1}_{ijkl} \quad \text{and also} \quad L_{ijkl} = \mathcal{S}^{-1}_{ijkl}. \tag{7.3}$$

[1]In the rest of the book the compliance tensor is denoted as $\boldsymbol{M}$. In this chapter, the notation $\boldsymbol{\mathcal{S}}$ is preferred to avoid confusion with the moments.

The stress and strain tensors as well as elasticity and compliance tensors can be expressed as vectors and matrices, as explained in Chapter 1. Indeed,

$$\begin{aligned}
\sigma_{ij} &\rightarrow \sigma_I, \\
\varepsilon_{ij} &\rightarrow \varepsilon_I, \\
I = i &= j \quad \text{if } i = j, \\
I = 1 + &[i + j] \quad \text{if } i \neq j.
\end{aligned} \tag{7.4}$$

Thus the stress and strain tensors are transformed into a stress and strain vector (6×1) as follows[2]:

$$\begin{bmatrix} \sigma_{11} & \sigma_{12} & \sigma_{13} \\ & \sigma_{22} & \sigma_{23} \\ \text{SYM} & & \sigma_{33} \end{bmatrix} \rightarrow \begin{bmatrix} \sigma_{11} \\ \sigma_{22} \\ \sigma_{33} \\ \sigma_{12} \\ \sigma_{13} \\ \sigma_{23} \end{bmatrix},$$

$$\begin{bmatrix} \varepsilon_{11} & \varepsilon_{12} & \varepsilon_{13} \\ & \varepsilon_{22} & \varepsilon_{23} \\ \text{SYM} & & \varepsilon_{33} \end{bmatrix} \rightarrow \begin{bmatrix} \varepsilon_{11} \\ \varepsilon_{22} \\ \varepsilon_{33} \\ 2\varepsilon_{12} \\ 2\varepsilon_{13} \\ 2\varepsilon_{23} \end{bmatrix}. \tag{7.5}$$

The strain tensor is obtained from the symmetric part of the displacement gradient:

$$\begin{aligned}
\varepsilon_{ij} &= \frac{1}{2}\left[\frac{\partial u_i}{\partial x_j} + \frac{\partial u_j}{\partial x_i}\right], \\
\varepsilon_{11} &= \frac{1}{2}\left[\frac{\partial u_1}{\partial x_1} + \frac{\partial u_1}{\partial x_1}\right] = \frac{\partial u_1}{\partial x_1}, \\
\varepsilon_{22} &= \frac{1}{2}\left[\frac{\partial u_2}{\partial x_2} + \frac{\partial u_2}{\partial x_2}\right] = \frac{\partial u_2}{\partial x_2}, \\
\varepsilon_{12} &= \frac{1}{2}\left[\frac{\partial u_1}{\partial x_2} + \frac{\partial u_2}{\partial x_1}\right] \rightarrow 2\varepsilon_{12} = \frac{\partial u_1}{\partial x_2} + \frac{\partial u_2}{\partial x_1} = \gamma_{12}.
\end{aligned} \tag{7.6}$$

Here ε_{11} and ε_{22} are the axial extension strains in axes 1 and 2, respectively.[3] Moreover, ε_{12} is $(1-2)$ in–plane shear strain.

[2]SYM in a matrix denotes that the rest of the terms are obtained by symmetry.

[3]For simplicity, the axes parallel to x_1, x_2, and x_3 are simply called axes 1, 2, and 3, respectively.

In what follows, we prefer the engineering shear strain γ_{12} for formulations. It corresponds to the angle of shearing under a state of simple shear as illustrated in Fig. 7.2 taken from the reference book by Jones [1].

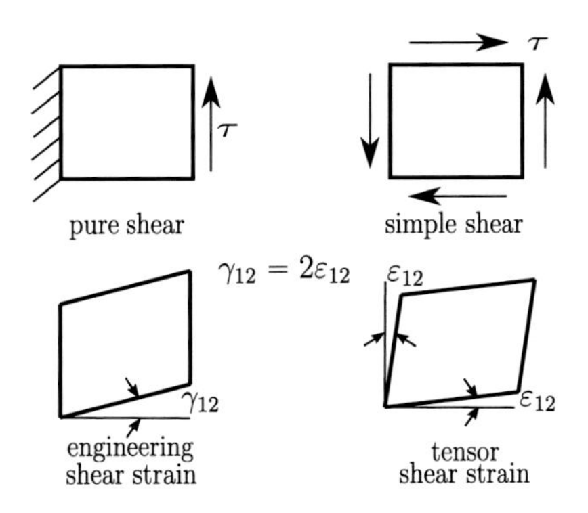

FIGURE 7.2 Simple shear and pure shear [1].

The fourth–order stiffness (elasticity) and compliance tensors can also be expressed as matrices using the contracted notation:

$$\begin{aligned} L_{ijkl} &\rightarrow L_{IJ}, \\ \mathcal{S}_{ijkl} &\rightarrow \mathcal{S}_{IJ}, \end{aligned} \tag{7.7}$$

where the indices i and j are transformed to I according to the rule given in Eq. (7.4). The indices k and l are also transformed to the index J according to the same rule:

$$\begin{aligned} J = k = l \quad &\text{if } k = l, \\ J = 1 + [k + l] \quad &\text{if } k \neq l. \end{aligned} \tag{7.8}$$

Then Eqs. (7.1) and (7.2) can be written using contracted or matrix notation (Table 1.1 of Chapter 1):

$$\begin{aligned} \sigma_I = L_{IJ}\varepsilon_J \quad &\text{or} \quad \boldsymbol{\sigma} = \boldsymbol{L} \cdot \widetilde{\boldsymbol{\varepsilon}}, \\ \varepsilon_I = \mathcal{S}_{IJ}\sigma_J \quad &\text{or} \quad \widetilde{\boldsymbol{\varepsilon}} = \widehat{\boldsymbol{\mathcal{S}}} \cdot \boldsymbol{\sigma}, \end{aligned} \tag{7.9}$$

where L_{IJ} and $\mathcal{S}_{IJ}$ are the stiffness and compliance (6×6) matrices.[4] It is worth mentioning that the compliance tensor is of "e–s" type in the sense that the element $\mathcal{S}_{44}$ corresponds to $4\mathcal{S}_{1212}$, the element $\mathcal{S}_{14}$ corresponds to $2\mathcal{S}_{1112}$, and so on.

[4]For simplicity, the Voigt form of the stiffness and compliance tensors will be further called the stiffness and compliance matrices, respectively.

7.2.2 Hooke's law for orthotropic material in Voigt notation

As example of an orthotropic material, we can consider a composite ply consisting of a flat or curved arrangement of unidirectional fibers embedded in a matrix. It represents the basic element of a laminate composite material. A unidirectional ply exhibits two orthogonal planes of material property symmetry, and a symmetry will exist relative to a third mutually orthogonal plane. For such a material, the stress–strain relations involve nine independent elastic constants and are expressed by Eq. (7.10) in the 3–D coordinate system aligned with the principal material directions (parallel to x_1 and x_2), as shown in Fig. 7.1.

$$\begin{bmatrix} \sigma_{11} \\ \sigma_{22} \\ \sigma_{33} \\ \sigma_{12} \\ \sigma_{13} \\ \sigma_{23} \end{bmatrix} = \begin{bmatrix} L_{11} & L_{12} & L_{13} & 0 & 0 & 0 \\ & L_{22} & L_{23} & 0 & 0 & 0 \\ & & L_{33} & 0 & 0 & 0 \\ & & & L_{44} & 0 & 0 \\ & & & & L_{55} & 0 \\ \text{SYM} & & & & & L_{66} \end{bmatrix} \cdot \begin{bmatrix} \varepsilon_{11} \\ \varepsilon_{22} \\ \varepsilon_{33} \\ 2\varepsilon_{12} \\ 2\varepsilon_{13} \\ 2\varepsilon_{23} \end{bmatrix}. \tag{7.10}$$

The stiffness matrix $\boldsymbol{L}$ can be expressed as a function of the engineering material constants:

$$\boldsymbol{L} = [L_{IJ}] = \begin{bmatrix} \frac{1-\nu_{23}\nu_{32}}{E_2E_3\Delta} & \frac{\nu_{21}+\nu_{23}\nu_{13}}{E_1E_3\Delta} & \frac{\nu_{31}+\nu_{21}\nu_{32}}{E_2E_3\Delta} & 0 & 0 & 0 \\ & \frac{1-\nu_{13}\nu_{31}}{E_1E_3\Delta} & \frac{\nu_{32}+\nu_{12}\nu_{31}}{E_1E_3\Delta} & 0 & 0 & 0 \\ & & \frac{1-\nu_{12}\nu_{21}}{E_1E_2\Delta} & 0 & 0 & 0 \\ & & & G_{12} & 0 & 0 \\ & & & & G_{13} & 0 \\ \text{SYM} & & & & & G_{23} \end{bmatrix}, \tag{7.11}$$

where

$$\Delta = \frac{1-\nu_{12}\nu_{21}-\nu_{32}\nu_{23}-\nu_{13}\nu_{31}-2\nu_{21}\nu_{13}\nu_{32}}{E_1E_2E_3}.$$

The strain–stress equation provides the strain response induced by a specific applied stress:

$$\varepsilon_{ij} = S_{ijkl}\sigma_{kl} \rightarrow \varepsilon_I = S_{IJ}\sigma_J,$$

$$\begin{bmatrix} \varepsilon_1 \\ \varepsilon_2 \\ \varepsilon_3 \\ \varepsilon_4 = 2\varepsilon_{12} \\ \varepsilon_5 = 2\varepsilon_{13} \\ \varepsilon_6 = 2\varepsilon_{23} \end{bmatrix} = \begin{bmatrix} \mathcal{S}_{11} & \mathcal{S}_{12} & \mathcal{S}_{13} & 0 & 0 & 0 \\ & \mathcal{S}_{22} & \mathcal{S}_{23} & 0 & 0 & 0 \\ & & \mathcal{S}_{33} & 0 & 0 & 0 \\ & & & \mathcal{S}_{44} & 0 & 0 \\ & & & & \mathcal{S}_{55} & 0 \\ \text{SYM} & & & & & \mathcal{S}_{66} \end{bmatrix} \cdot \begin{bmatrix} \sigma_1 \\ \sigma_2 \\ \sigma_3 \\ \sigma_4 = \sigma_{12} \\ \sigma_5 = \sigma_{32} \\ \sigma_6 = \sigma_{23} \end{bmatrix}. \tag{7.12}$$

The components of the compliance matrix $\widehat{\mathcal{S}}$ have more obvious physical meaning than those of the stiffness matrix $\boldsymbol{L}$. In fact, the components $\mathcal{S}_{IJ}$ are simply expressed in terms of the material engineering constants. For an orthotropic material, the compliance matrix is expressed as

$$\widehat{\mathcal{S}} = [\mathcal{S}_{IJ}] = \begin{bmatrix} \frac{1}{E_1} & -\frac{\nu_{12}}{E_1} & -\frac{\nu_{13}}{E_1} & 0 & 0 & 0 \\ & \frac{1}{E_2} & -\frac{\nu_{23}}{E_2} & 0 & 0 & 0 \\ & & \frac{1}{E_3} & 0 & 0 & 0 \\ & & & \frac{1}{G_{12}} & 0 & 0 \\ & & & & \frac{1}{G_{13}} & 0 \\ \text{SYM} & & & & & \frac{1}{G_{23}} \end{bmatrix}. \tag{7.13}$$

Moreover,

$$\begin{aligned} \widehat{\mathcal{S}} = \boldsymbol{L}^{-1} \quad &\text{or} \quad [\mathcal{S}_{IJ}] = [L_{IJ}]^{-1}, \\ \boldsymbol{L} = \widehat{\mathcal{S}}^{-1} \quad &\text{or} \quad [L_{IJ}] = [\mathcal{S}_{IJ}]^{-1}. \end{aligned} \tag{7.14}$$

Furthermore:

- E_1, E_2, E_3 are the Young moduli in the principal directions 1, 2, and 3, respectively.
- ν_{12}, ν_{13}, ν_{23} are the Poisson ratios. Note that

$$\nu_{ij} = -\frac{\varepsilon_j}{\varepsilon_i} \quad i, j = 1, 2, 3, \quad i \neq j,$$

  in the case where the applied stress is σ_i and all other stresses are zero.
- G_{12}, G_{13}, and G_{23} are the shear moduli in the planes $(1-2)$, $(1-3)$, and $(2-3)$, respectively.

Using Eqs. (7.12) and (7.13), we can express the strain in the orthotropic ply induced by the applied stress in the principal material directions (local coordinate system) as follows:

$$
\begin{aligned}
\varepsilon_1 &= \frac{\sigma_1}{E_1} - \frac{\nu_{12}}{E_1}\sigma_2 - \frac{\nu_{13}}{E_1}\sigma_3,\\
\varepsilon_2 &= \frac{\sigma_2}{E_2} - \frac{\nu_{21}}{E_2}\sigma_1 - \frac{\nu_{23}}{E_2}\sigma_3,\\
\varepsilon_3 &= \frac{\sigma_3}{E_3} - \frac{\nu_{31}}{E_3}\sigma_1 - \frac{\nu_{32}}{E_3}\sigma_2,\\
2\varepsilon_{12} &= \frac{\sigma_{12}}{G_{12}},\\
2\varepsilon_{13} &= \frac{\sigma_{13}}{G_{13}},\\
2\varepsilon_{23} &= \frac{\sigma_{23}}{G_{23}}.
\end{aligned}
\tag{7.15}
$$

Properties of L_{IJ} and $\mathcal{S}_{IJ}$

Based on the first law of thermodynamics, the stiffness and compliance matrices must be positive definite (see Chapter 10 for a detailed definition of positive definite matrices), which means that they have positive eigenvalues. This condition induces that the diagonal terms of $\boldsymbol{L}$ and $\widehat{\boldsymbol{\mathcal{S}}}$ are necessarily positive:

$$
\begin{aligned}
L_{II} > 0 \quad &\longmapsto \quad L_{11}, L_{22}, L_{33}, L_{44}, L_{55}, L_{66} > 0,\\
\mathcal{S}_{II} > 0 \quad &\longmapsto \quad \mathcal{S}_{11}, \mathcal{S}_{22}, \mathcal{S}_{33}, \mathcal{S}_{44}, \mathcal{S}_{55}, \mathcal{S}_{66} > 0.
\end{aligned}
$$

The double indices in the above relations do not imply summation. Knowing that E_1, E_2, E_3 are positive, we have:

$$
\begin{aligned}
1 - \nu_{12}\nu_{21} > 0, \quad 1 - \nu_{13}\nu_{31} > 0, \quad 1 - \nu_{23}\nu_{32} > 0,\\
E_1E_2E_3\Delta = 1 - \nu_{12}\nu_{21} - \nu_{32}\nu_{23} - \nu_{13}\nu_{31} - 2\nu_{21}\nu_{13}\nu_{32} > 0.
\end{aligned}
\tag{7.16}
$$

Considering the symmetry of the compliance matrix ($\mathcal{S}_{IJ} = \mathcal{S}_{JI}$), we can easily deduce the following reciprocal equations:

$$
\frac{\nu_{12}}{E_1} = \frac{\nu_{21}}{E_2}, \quad \frac{\nu_{13}}{E_1} = \frac{\nu_{31}}{E_3}, \quad \frac{\nu_{23}}{E_2} = \frac{\nu_{32}}{E_3},
$$

or, in compact form (double indices do not imply summation),

$$
\frac{\nu_{ij}}{E_i} = \frac{\nu_{ji}}{E_j}, \quad i, j = 1, 2, 3, \quad i \neq j. \tag{7.17}
$$

The positive definite nature of $\widehat{\boldsymbol{\mathcal{S}}}$ or $\boldsymbol{L}$ implies the positiveness of their determinants, which leads to the following inequalities (double indices do not imply summation):

$$
\mathcal{S}_{II}\mathcal{S}_{JJ} - [\mathcal{S}_{IJ}]^2 > 0 \Rightarrow |\mathcal{S}_{IJ}| < \sqrt{\mathcal{S}_{II}\mathcal{S}_{JJ}} \quad I, J = 1, 2, 3, \quad I \neq J. \tag{7.18}
$$

Consequently,

$$|\mathcal{S}_{12}| < \sqrt{\mathcal{S}_{11}\mathcal{S}_{22}}, \quad |\mathcal{S}_{13}| < \sqrt{\mathcal{S}_{11}\mathcal{S}_{33}}, \quad |\mathcal{S}_{23}| < \sqrt{\mathcal{S}_{22}\mathcal{S}_{33}}. \tag{7.19}$$

From Eqs. (7.19) and (7.13) we obtain that

$$\frac{1}{E_1}\frac{1}{E_2} - \left[\frac{\nu_{12}}{E_1}\right]^2 > 0 \quad \Rightarrow \quad \frac{E_1}{E_2} - [\nu_{12}]^2 > 0. \tag{7.20}$$

Finally, we deduce that

$$|\nu_{12}| < \sqrt{\frac{E_1}{E_2}} \quad \text{and also} \quad |\nu_{21}| < \sqrt{\frac{E_2}{E_1}}, \tag{7.21}$$

which leads to

$$|\nu_{12}\nu_{21}| < 1.$$

Similarly, we can express four other relationships linking the material Poisson ratios and the Young moduli:

$$\begin{aligned} |\nu_{13}| &< \sqrt{\frac{E_1}{E_3}}, \qquad & |\nu_{31}| &< \sqrt{\frac{E_3}{E_1}}, \\ |\nu_{23}| &< \sqrt{\frac{E_2}{E_3}}, \qquad & |\nu_{32}| &< \sqrt{\frac{E_3}{E_2}}. \end{aligned} \tag{7.22}$$

The previous inequalities (7.21) and (7.22) between the elastic moduli in the three principal directions and the Poisson ratios of an orthotropic composite can be helpful to assess the experimental data in terms of their physical consistency. They show also that optimizing an independent constant may necessarily affect the limits of the others.

Exercise:

Using the reciprocal relations (7.17) and the previous inequalities, demonstrate the following inequality bounding the Poisson ratio of an orthotropic material:

$$2\nu_{21}\nu_{13}\nu_{32} < \left[1 - \nu_{21}^2\frac{E_1}{E_2} - \nu_{32}^2\frac{E_2}{E_3} - \nu_{13}^2\frac{E_3}{E_1}\right],$$

which leads to

$$\nu_{21}\nu_{13}\nu_{32} < \frac{1}{2}.$$

7.3 Hooke's law for an orthotropic lamina under the assumption of plane stress

In the rest of this chapter, the tilde and hat symbols of the Voigt notation are omitted for simplicity of the expressions. Whenever necessary, we provide clarification and connection with the notation of Chapter 1.

Let us consider a thin unidirectional reinforced lamina that falls under the orthotropic material. In the case that this lamina has a low thickness t compared to its other dimensions in the plane $(1-2)$ $(t \ll \min(l_1, l_2))$ and is subjected to loads that do not induce any out–of–plane stresses, we can assume that the lamina satisfies the plane stress assumption. The latter consists in reducing the 3–D problem to a 2–D one by setting the following terms in Eqs. (7.10) or (7.12):

$$\sigma_{33} = \sigma_3 = 0, \quad \sigma_{13} = \sigma_5 = 0, \quad \sigma_{23} = \sigma_6 = 0, \tag{7.23}$$

whereas

$$\sigma_{11} = \sigma_1 \neq 0, \quad \sigma_{22} = \sigma_2 \neq 0, \quad \sigma_{12} = \sigma_4 \neq 0. \tag{7.24}$$

However, based on Eq. $(7.15)_3$, we can notice that the plane stress assumption results in an axial out–of–plane strain component $\varepsilon_3 \neq 0$. Indeed, in this case, ε_3 represents the out–of–plane Poisson effect, which is expressed as a function of σ_1 and σ_2:

$$\varepsilon_3 = \mathcal{S}_{13}\,\sigma_1 + \mathcal{S}_{23}\,\sigma_2 = -\left[\frac{\nu_{31}}{E_3}\sigma_1 + \frac{\nu_{32}}{E_3}\sigma_2\right] = -\left[\frac{\nu_{13}}{E_1}\sigma_1 + \frac{\nu_{23}}{E_2}\sigma_2\right]. \tag{7.25}$$

Thus ε_3 is not independent since it can be related to the strain components ε_1 and ε_2. Therefore it can be omitted from Eqs. (7.12) expressing the 3–D Hooke law.

Strain–stress relation and reduced compliance matrix

Given the low thickness of the lamina, we can assume the transverse shear stress components ($\sigma_{13} = \sigma_5 = 0$ and $\sigma_{23} = \sigma_6 = 0$) to be zero within the lamina. Consequently, we also omit the transverse shearing strains ε_5 and ε_6 in the strain–stress relationship (7.12). Then the latter is reduced to

$$\begin{bmatrix} \varepsilon_1 \\ \varepsilon_2 \\ \varepsilon_4 \end{bmatrix} = \begin{bmatrix} \dfrac{1}{E_1} & -\dfrac{\nu_{12}}{E_1} & 0 \\ & \dfrac{1}{E_2} & 0 \\ \text{SYM} & & \dfrac{1}{G_{12}} \end{bmatrix} \cdot \begin{bmatrix} \sigma_1 \\ \sigma_2 \\ \sigma_4 \end{bmatrix}. \tag{7.26}$$

The previous strain–stress relationship (7.26) is 2–D plane stress in the (1 − 2) plane. It requires four independent constants E_1, E_2, ν_{12}, and G_{12}. Nevertheless, the estimation of ε_3 further requires the Poisson ratios ν_{13} and ν_{23}.

Eq. (7.26) can be written using the compliance matrix $[\mathcal{S}_{IJ}]$ expressed in 2–D:

$$\varepsilon_I = \mathcal{S}_{IJ}\sigma_J \quad \Rightarrow \quad \begin{bmatrix} \varepsilon_1 \\ \varepsilon_2 \\ \varepsilon_4 \end{bmatrix} = \begin{bmatrix} \mathcal{S}_{11} & \mathcal{S}_{12} & 0 \\ & \mathcal{S}_{22} & 0 \\ \text{SYM} & & \mathcal{S}_{44} \end{bmatrix} \cdot \begin{bmatrix} \sigma_1 \\ \sigma_2 \\ \sigma_4 \end{bmatrix}. \tag{7.27}$$

Stress–strain relation and reduced stiffness matrix

The stress–strain relationship can easily by obtained by inverting Eq. (7.26), which leads to the reduced stiffness matrix $\boldsymbol{Q}$:

$$\sigma_I = Q_{IJ}\varepsilon_J \quad \Rightarrow \quad \begin{bmatrix} \sigma_1 \\ \sigma_2 \\ \sigma_4 \end{bmatrix} = \begin{bmatrix} Q_{11} & Q_{12} & 0 \\ & Q_{22} & 0 \\ \text{SYM} & & Q_{44} \end{bmatrix} \cdot \begin{bmatrix} \varepsilon_1 \\ \varepsilon_2 \\ \varepsilon_4 \end{bmatrix}, \tag{7.28}$$

where

$$[\mathcal{S}_{IJ}]^{-1} = [Q_{IJ}] \text{ and } [Q_{IJ}]^{-1} = [\mathcal{S}_{IJ}]. \tag{7.29}$$

The components of the reduced stiffness matrix can be expressed in terms of engineering constants by inverting the reduced compliance matrix:

$$\begin{aligned} Q_{11} &= \frac{E_1}{1 - \nu_{12}\nu_{21}}, \\ Q_{22} &= \frac{E_2}{1 - \nu_{12}\nu_{21}}, \\ Q_{12} &= \frac{\nu_{21}E_1}{1 - \nu_{12}\nu_{21}} = \frac{\nu_{12}E_2}{1 - \nu_{12}\nu_{21}}, \\ Q_{44} &= G_{12}. \end{aligned} \tag{7.30}$$

It is worth noticing that only four independent material constants E_1, E_2, ν_{12}, and G_{12} are required for determining the in–plane elastic response of an orthotropic lamina. Indeed, ν_{21} is obtained using the reciprocal relation (7.17):

$$\nu_{21} = \nu_{12}\frac{E_2}{E_1}.$$

Besides the unidirectional reinforced lamina, the plane stress assumption reasonably stands also for a thin woven composite ply having its weaves perpendicular to each other. We can also consider short fiber composites whose fibers are randomly oriented in the plane or even aligned in one direction of the plane.

7.4 Stress–strain relations for a lamina of arbitrary orientation: off–axis loading

The previous stress–strain relationships (7.12) and (7.10) have been developed for an orthotropic material in the principal material directions defined by the coordinate system $(1-2-3)$. Direction 1 is generally parallel to the fibers, whereas the axis 2 is perpendicular to the fiber direction. However, for a general orthotropic lamina (angle lamina), the orthotropic material axes $(1-2)$, also called local axes, often do not coincide with the global coordinates axes $(x-y)$, which are the axes of the loading directions. Then the global axes $(x-y)$ are obtained by a rotation of an angle ϑ around the out–of–plane axis of the orthotropic material directions. Fig. 7.3 shows the rotation of the material orthotropy axes $(1-2)$ from the arbitrary global axes $(x-y)$.

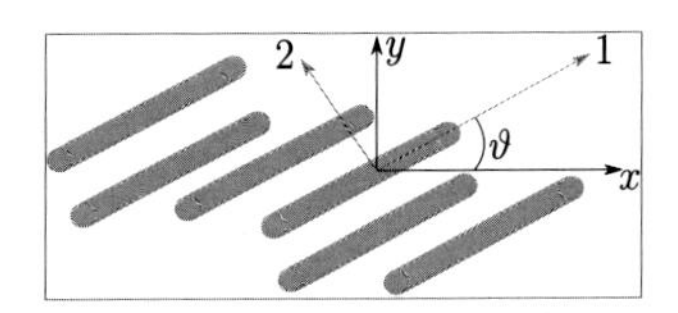

FIGURE 7.3 Local material orthotropy axes $(1-2)$ rotated from the arbitrary global axes $(x-y)$ of angle lamina. Direction 1 is parallel to the fibers.

Such a situation happens in most laminated plates or shells consisting of unidirectional laminae at different orientations. The stresses and strains should be expressed in the global coordinate system as well as in the local one. Then a method is needed for transforming the lamina stress–strain relationship from the local axes to the arbitrary global coordinate system and vice versa. In some literature, direction 1 is referred to as longitudinal direction L, and direction 2 is called the transverse direction (T). Considering the rotation from the local to the global axes, denoted ϑ, we can deduce the stress tensor in the global system by using the in–plane rotation second–order tensor $\boldsymbol{R}$ discussed in Chapter 1:

$$\boldsymbol{R} = \begin{bmatrix} m & -n & 0 \\ n & m & 0 \\ 0 & 0 & 1 \end{bmatrix}, \qquad m = \cos\vartheta, \;\; n = \sin\vartheta. \tag{7.31}$$

7.4.1 Stress and strain in global axes $(x - y)$

In an arbitrary coordinate system an orthotropic lamina exhibits an anisotropic response due to the coupling terms. Thus in this case the lamina is called a *generally orthotropic lamina* (Fig. 7.4). The stresses and strains have to be expressed in the global coordinate axes by using the in–plane rotation matrix. The stress–strain relations can be readily obtained accordingly:

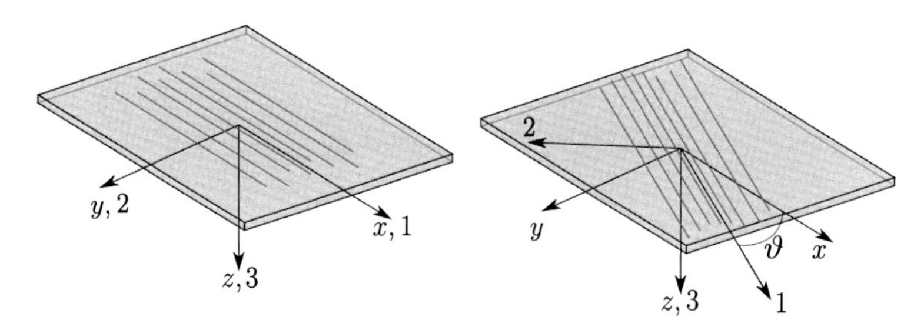

FIGURE 7.4 Orthotropic lamina axes and generally orthotropic lamina with arbitrary global axes $(x - y)$.

$$
\begin{aligned}
\sigma_{ij}^{(x-y)} &= R_{ik} R_{jl} \sigma_{kl}^{(1-2)}, \\
\sigma_{ij}^{(1-2)} &= R_{ki} R_{lj} \sigma_{kl}^{(x-y)},
\end{aligned} \tag{7.32}
$$

where $\sigma_{ij}^{(x-y)}$ and $\sigma_{ij}^{(1-2)}$ are the stress tensor expressed in the global axes $(x - y)$ and the local axes $(1 - 2)$, respectively. In matrix notation the previous equations can be rewritten as follows:

$$
\begin{aligned}
\boldsymbol{\sigma}^{(x-y)} &= \boldsymbol{R} \cdot \boldsymbol{\sigma}^{(1-2)} \cdot \boldsymbol{R}^T, \\
\boldsymbol{\sigma}^{(1-2)} &= \boldsymbol{R}^T \cdot \boldsymbol{\sigma}^{(x-y)} \cdot \boldsymbol{R}.
\end{aligned} \tag{7.33}
$$

As shown in Chapter 1, in the Voigt notation for 2–D cases, assuming the in–plane stress, after rearranging the terms and keeping only the in–plane stress components, we can rewrite Eq. $(7.33)_1$ as follows:

$$
\begin{bmatrix} \sigma_x \\ \sigma_y \\ \sigma_s = \sigma_{xy} \end{bmatrix} = \underbrace{\begin{bmatrix} m^2 & n^2 & -2mn \\ n^2 & m^2 & 2mn \\ mn & -mn & m^2 - n^2 \end{bmatrix}}_{\boldsymbol{T}_\sigma} \cdot \begin{bmatrix} \sigma_1 \\ \sigma_2 \\ \sigma_4 = \sigma_{12} \end{bmatrix}. \tag{7.34}
$$

In compact form, it is written as

$$
\boldsymbol{\sigma}^{(x-y)} = \boldsymbol{T}_\sigma \cdot \boldsymbol{\sigma}^{(1-2)}, \tag{7.35}
$$

where $\boldsymbol{\sigma}^{(x-y)}$ and $\boldsymbol{\sigma}^{(1-2)}$ correspond to the Voigt forms of the global and local stress tensors, respectively. Similarly, we can express the Voigt form of the strain $\boldsymbol{\varepsilon}^{(x-y)}$ as a function of the Voigt form of the strain $\boldsymbol{\varepsilon}^{(1-2)}$:

$$\begin{bmatrix} \varepsilon_x \\ \varepsilon_y \\ \varepsilon_s = 2\varepsilon_{xy} \end{bmatrix} = \underbrace{\begin{bmatrix} m^2 & n^2 & -mn \\ n^2 & m^2 & mn \\ 2mn & -2mn & m^2 - n^2 \end{bmatrix}}_{\boldsymbol{T}_{\varepsilon}} \cdot \begin{bmatrix} \varepsilon_1 \\ \varepsilon_2 \\ \varepsilon_4 = 2\varepsilon_{12} \end{bmatrix}, \tag{7.36}$$

or, in compact form,

$$\boldsymbol{\varepsilon}^{(x-y)} = \boldsymbol{T}_{\varepsilon} \cdot \boldsymbol{\varepsilon}^{(1-2)}. \tag{7.37}$$

The matrices $\boldsymbol{T}_{\sigma}$ and $\boldsymbol{T}_{\varepsilon}$ are called 2–D stress– and strain–transformation matrices, respectively.[5] They are orthogonal matrices since it is easy to demonstrate that

$$\boldsymbol{T}_{\sigma}^{-1} = \boldsymbol{T}_{\varepsilon}^{T}, \quad \boldsymbol{T}_{\varepsilon}^{-1} = \boldsymbol{T}_{\sigma}^{T}, \tag{7.38}$$

where the superscripts -1 and T denote the matrix inverse and matrix transpose, respectively.

7.4.2 Off–axis stress–strain relations

In this subsection, we express the relationship between the stress $\boldsymbol{\sigma}^{(x-y)}$ and the strain $\boldsymbol{\varepsilon}^{(x-y)}$ for an angle lamina in the global coordinate system. By substituting Eqs. (7.35) and (7.37) into the 2–D Hooke law defined in local coordinates system (7.28) we can write:

$$\begin{aligned} \boldsymbol{\sigma}^{(1-2)} &= \boldsymbol{Q} \cdot \boldsymbol{\varepsilon}^{(1-2)} \quad \Rightarrow \\ \boldsymbol{T}_{\sigma}^{-1} \cdot \boldsymbol{\sigma}^{(x-y)} &= \boldsymbol{Q} \cdot \boldsymbol{T}_{\varepsilon}^{-1} \cdot \boldsymbol{\varepsilon}^{(x-y)}. \end{aligned} \tag{7.39}$$

Using Eq. $(7.38)_2$, we obtain the relation

$$\boldsymbol{\sigma}^{(x-y)} = \boldsymbol{T}_{\sigma} \cdot \boldsymbol{Q} \cdot \boldsymbol{T}_{\sigma}^{T} \cdot \boldsymbol{\varepsilon}^{(x-y)}. \tag{7.40}$$

Finally, we can write

$$\boldsymbol{\sigma}^{(x-y)} = \widehat{\boldsymbol{Q}} \cdot \boldsymbol{\varepsilon}^{(x-y)}, \tag{7.41}$$

where

$$\widehat{\boldsymbol{Q}} = \boldsymbol{T}_{\sigma} \cdot \boldsymbol{Q} \cdot \boldsymbol{T}_{\sigma}^{T}, \tag{7.42}$$

[5]The symbols $\boldsymbol{T}_{\sigma}$ and $\boldsymbol{T}_{\varepsilon}$ are used for simplicity in this chapter, and they correspond to the 2–D forms of the rotator tensors $\widetilde{\boldsymbol{Q}}^{T}$ and $\breve{\boldsymbol{Q}}^{T}$, respectively, of Table 1.2 in Chapter 1.

or, in expanded form,

$$\begin{bmatrix} \widehat{Q}_{xx} & \widehat{Q}_{xy} & \widehat{Q}_{xs} \\ & \widehat{Q}_{yy} & \widehat{Q}_{ys} \\ \text{SYM} & & \widehat{Q}_{ss} \end{bmatrix} = \begin{bmatrix} m^2 & n^2 & -2mn \\ n^2 & m^2 & 2mn \\ mn & -mn & m^2-n^2 \end{bmatrix} \cdot$$

$$\begin{bmatrix} Q_{11} & Q_{12} & 0 \\ & Q_{22} & 0 \\ \text{SYM} & & Q_{44} \end{bmatrix} \cdot \begin{bmatrix} m^2 & n^2 & mn \\ n^2 & m^2 & -mn \\ -2mn & 2mn & m^2-n^2 \end{bmatrix}. \tag{7.43}$$

The matrix $\widehat{Q}(\vartheta)$ is expressed as a function of the rotation angle and is called the 2–D transformed reduced stiffness matrix. It relates the strain to the stress in the global axes as follows:

$$\begin{bmatrix} \sigma_x \\ \sigma_y \\ \sigma_s \end{bmatrix} = \begin{bmatrix} \widehat{Q}_{xx} & \widehat{Q}_{xy} & \widehat{Q}_{xs} \\ & \widehat{Q}_{yy} & \widehat{Q}_{ys} \\ \text{SYM} & & \widehat{Q}_{ss} \end{bmatrix} \cdot \begin{bmatrix} \varepsilon_x \\ \varepsilon_y \\ \varepsilon_s \end{bmatrix}. \tag{7.44}$$

The six components of $\widehat{\boldsymbol{Q}}$ are

$$\begin{aligned} \widehat{Q}_{xx} &= Q_{11}m^4 + Q_{22}n^4 + 2\left[Q_{12} + 2Q_{44}\right]m^2n^2, \\ \widehat{Q}_{yy} &= Q_{11}n^4 + Q_{22}m^4 + 2\left[Q_{12} + 2Q_{44}\right]m^2n^2, \\ \widehat{Q}_{xy} &= \left[Q_{11} + Q_{22} - 4Q_{44}\right]m^2n^2 + Q_{12}\left[m^4 + n^4\right], \\ \widehat{Q}_{ss} &= \left[Q_{11} + Q_{22} - 2Q_{12}\right]m^2n^2 + Q_{44}\left[m^2 - n^2\right]^2, \\ \widehat{Q}_{xs} &= \left[Q_{11} - Q_{12} - 2Q_{44}\right]m^3n - \left[Q_{22} - Q_{12} - 2Q_{44}\right]mn^3, \\ \widehat{Q}_{ys} &= \left[Q_{11} - Q_{12} - 2Q_{44}\right]mn^3 - \left[Q_{22} - Q_{12} - 2Q_{44}\right]m^3n. \end{aligned} \tag{7.45}$$

It is worth noticing that the determination of $\widehat{\boldsymbol{Q}}$ still requires only four independent constants, namely Q_{11}, Q_{22}, Q_{12}, and Q_{44} or in–plane engineering constants E_1, E_2, ν_{12}, and G_{12}.

Nevertheless, the transformed reduced stiffness matrix has two other extra–diagonal components $\widehat{Q}_{xs}$ and $\widehat{Q}_{ys}$ expressing the shear/extension (or compression) coupling in the lamina. To illustrate the shear–extension coupling, let us consider an angle lamina subjected to a tension σ_x (Fig. 7.5). Besides the expected axial strains ε_x and ε_y, the strain response implies a shearing ε_s induced by the fact that $\widehat{Q}_{xs} \neq 0$. Using Eq. (7.44) yields

$$\sigma_x = \widehat{Q}_{xx}\varepsilon_x + \widehat{Q}_{xy}\varepsilon_y + \widehat{Q}_{xs}\varepsilon_s.$$

Schematically, the lamina's response is shown in Fig. 7.5.

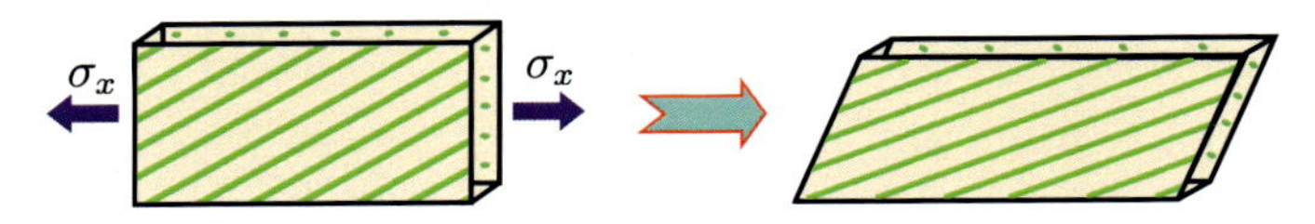

FIGURE 7.5 Illustration of the extension–shear coupling term ($\widehat{Q}_{xy}$) effect in an angle lamina.

It is easy to demonstrate that no coupling occurs between the normal and shearing terms of strains and stresses for a unidirectional reinforced lamina ($\vartheta = 0°$) as well as for a ply oriented at $\vartheta = 90°$.

7.4.3 Off–axis strain–stress relations

Strains in an angle lamina (global axes) can be obtained by inverting directly Eq. (7.44) or by transforming the strain–stress relation (7.27), leading to an analytical expression of the transformed reduced compliance $\widehat{S}$. Following similar developments as above, the strain–stress relation expressed in the orthotropy directions (7.26) is transformed by using the $\boldsymbol{T}_\varepsilon$ matrix:

$$\begin{aligned} \boldsymbol{\varepsilon}^{(1-2)} &= \boldsymbol{S}\cdot\boldsymbol{\sigma}^{(1-2)} \quad \Rightarrow \\ \boldsymbol{T}_\varepsilon^{-1}\cdot\boldsymbol{\varepsilon}^{(x-y)} &= \boldsymbol{S}\cdot\boldsymbol{T}_\sigma^{-1}\cdot\boldsymbol{\sigma}^{(x-y)} \quad \Rightarrow \\ \boldsymbol{\varepsilon}^{(x-y)} &= \boldsymbol{T}_\varepsilon\cdot\boldsymbol{S}\cdot\boldsymbol{T}_\varepsilon^T\cdot\boldsymbol{\sigma}^{(x-y)}. \end{aligned}$$

Finally, we obtain the expression

$$\boldsymbol{\varepsilon}^{(x-y)} = \widehat{\boldsymbol{S}}\cdot\boldsymbol{\sigma}^{(x-y)}, \tag{7.46}$$

where

$$\widehat{\boldsymbol{S}} = \boldsymbol{T}_\varepsilon\cdot\boldsymbol{S}\cdot\boldsymbol{T}_\varepsilon^T. \tag{7.47}$$

Then the off–axis strain–stress relation in the 2–D plane $(x-y)$ is

$$\begin{bmatrix} \varepsilon_x \\ \varepsilon_y \\ \varepsilon_s \end{bmatrix} = \begin{bmatrix} \widehat{S}_{xx} & \widehat{S}_{xy} & \widehat{S}_{xs} \\ & \widehat{S}_{yy} & \widehat{S}_{ys} \\ \text{SYM} & & \widehat{S}_{ss} \end{bmatrix} \cdot \begin{bmatrix} \sigma_x \\ \sigma_y \\ \sigma_s \end{bmatrix}. \tag{7.48}$$

The matrix $\widehat{\mathcal{S}}$ is expressed in expanded form as

$$\begin{bmatrix} \widehat{\mathcal{S}}_{xx} & \widehat{\mathcal{S}}_{xy} & \widehat{\mathcal{S}}_{xs} \\ & \widehat{\mathcal{S}}_{yy} & \widehat{\mathcal{S}}_{ys} \\ \text{SYM} & & \widehat{\mathcal{S}}_{ss} \end{bmatrix} = \begin{bmatrix} m^2 & n^2 & -mn \\ n^2 & m^2 & mn \\ 2mn & -2mn & m^2-n^2 \end{bmatrix} \cdot$$

$$\begin{bmatrix} \mathcal{S}_{11} & \mathcal{S}_{12} & 0 \\ & \mathcal{S}_{22} & 0 \\ \text{SYM} & & \mathcal{S}_{44} \end{bmatrix} \cdot \begin{bmatrix} m^2 & n^2 & 2mn \\ n^2 & m^2 & -2mn \\ -mn & mn & m^2-n^2 \end{bmatrix}. \tag{7.49}$$

The matrix $\widehat{\mathcal{S}}(\vartheta)$ is the 2–D transformed reduced compliance matrix of a generally orthotropic ply with components

$$\begin{aligned} \widehat{\mathcal{S}}_{xx} &= \mathcal{S}_{11}m^4 + \mathcal{S}_{22}n^4 + [2\mathcal{S}_{12} + \mathcal{S}_{44}]\, m^2n^2, \\ \widehat{\mathcal{S}}_{yy} &= \mathcal{S}_{11}n^4 + \mathcal{S}_{22}m^4 + [2\mathcal{S}_{12} + \mathcal{S}_{44}]\, m^2n^2, \\ \widehat{\mathcal{S}}_{xy} &= [\mathcal{S}_{11} + \mathcal{S}_{22} - \mathcal{S}_{44}]\, m^2n^2 + \mathcal{S}_{12}\left[m^4 + n^4\right], \\ \widehat{\mathcal{S}}_{ss} &= 4\,[\mathcal{S}_{11} + \mathcal{S}_{22} - 2\mathcal{S}_{12}]\, m^2n^2 + \mathcal{S}_{44}\left[m^2 - n^2\right]^2, \\ \widehat{\mathcal{S}}_{xs} &= [2\mathcal{S}_{11} - 2\mathcal{S}_{12} - \mathcal{S}_{44}]\, m^3n - [2\mathcal{S}_{22} - 2\mathcal{S}_{12} - \mathcal{S}_{44}]\, mn^3, \\ \widehat{\mathcal{S}}_{ys} &= [2\mathcal{S}_{11} - 2\mathcal{S}_{12} - \mathcal{S}_{44}]\, mn^3 - [2\mathcal{S}_{22} - 2\mathcal{S}_{12} - \mathcal{S}_{44}]\, m^3n. \end{aligned} \tag{7.50}$$

Exercise:

Demonstrate the following invariant relations:

$$\begin{aligned} \widehat{\mathcal{Q}}_{xx} + \widehat{\mathcal{Q}}_{yy} + 2\,\widehat{\mathcal{Q}}_{xy} &= \mathcal{Q}_{11} + \mathcal{Q}_{22} + 2\mathcal{Q}_{12}, \\ \widehat{\mathcal{Q}}_{xx} + \widehat{\mathcal{Q}}_{yy} + \widehat{\mathcal{Q}}_{xy} + \widehat{\mathcal{Q}}_{ss} &= \mathcal{Q}_{11} + \mathcal{Q}_{22} + \mathcal{Q}_{12} + \mathcal{Q}_{44}, \\ \widehat{\mathcal{Q}}_{ss} - \widehat{\mathcal{Q}}_{xy} &= \mathcal{Q}_{44} - \mathcal{Q}_{12}. \end{aligned} \tag{7.51}$$

These relations lead to the conclusion that for an angle lamina, there are some stiffness invariant terms [2] with respect to a rotation ϑ.

7.4.4 Engineering constants and induced coefficients of shear–axial strain mutual influence in an angle lamina

In Eq. (7.26) the four compliance components of a lamina in its orthotropy directions are expressed as functions of the engineering constants. In this section, we relate the components of the transformed compliance matrix to engineering constants of an angle lamina (generally orthotropic material), which can be easily obtained experimentally through uniaxial or shear load configurations. Indeed, let us consider, for instance, an angle lamina under a tension on the x direction ($\sigma_x \neq 0, \sigma_y = 0, \sigma_s = 0$). The strain components induced by this tension loading configuration are an axial strain ε_x, a transverse strain due to Poisson effect ε_y, and, finally, a shear strain induced by the presence of the axial–shear coupling term $\widehat{\mathcal{S}}_{xs}$ as given in Eq. (7.48):

$$\varepsilon_x = \widehat{\mathcal{S}}_{xx}\sigma_x, \quad \varepsilon_y = \widehat{\mathcal{S}}_{xy}\sigma_x, \quad \varepsilon_s = \widehat{\mathcal{S}}_{xs}\sigma_x.$$

Then the elastic engineering constants of the angle lamina are defined as

$$E_x = E_x(\vartheta) = \frac{\sigma_x}{\varepsilon_x} = \frac{1}{\widehat{\mathcal{S}}_{xx}} \quad \Rightarrow \quad \widehat{\mathcal{S}}_{xx} = \frac{1}{E_x},$$

$$\nu_{xy} = -\frac{\varepsilon_y}{\varepsilon_x} = -\frac{\widehat{\mathcal{S}}_{xy}}{\widehat{\mathcal{S}}_{xx}} \quad \Rightarrow \quad \widehat{\mathcal{S}}_{xy} = -\frac{\nu_{xy}}{E_x}.$$

Furthermore, in this axial tension configuration the axial–shear coupling induces an additional non–dimensional term, denoted $\eta_{xy,x}$. The latter characterizes the in–plane shear strain caused by the axial strain. It is expressed as

$$\eta_{xy,x} = \frac{\varepsilon_s}{\varepsilon_x} = \frac{\widehat{\mathcal{S}}_{xs}}{\widehat{\mathcal{S}}_{xx}} \quad \Rightarrow \quad \widehat{\mathcal{S}}_{xs} = \frac{\eta_{xy,x}}{E_x}.$$

Similarly, by considering the configuration of a transverse tension ($\sigma_x = 0$, $\sigma_y \neq 0, \sigma_s = 0$) we can find a pure shear loading configuration ($\sigma_x = 0, \sigma_y = 0, \sigma_s \neq 0$) relations expressing the other terms of the transformed reduced compliance. The six components of $\widehat{\mathcal{S}}$ are:

$$\widehat{\mathcal{S}}_{xx} = \frac{1}{E_x}, \quad \widehat{\mathcal{S}}_{yy} = \frac{1}{E_y}, \quad \widehat{\mathcal{S}}_{xy} = -\frac{\nu_{xy}}{E_x} = -\frac{\nu_{yx}}{E_y},$$
$$\widehat{\mathcal{S}}_{ss} = \frac{1}{G_{xy}}, \quad \widehat{\mathcal{S}}_{xs} = \frac{\eta_{xy,x}}{E_x} = \frac{\eta_{x,xy}}{G_{xy}}, \quad \widehat{\mathcal{S}}_{ys} = \frac{\eta_{xy,y}}{E_y} = \frac{\eta_{y,xy}}{G_{xy}}. \tag{7.52}$$

The off–axis strain–stress relation in the 2–D plane $(x - y)$ can be expressed using the engineering constants of the angle lamina:

$$\begin{bmatrix} \varepsilon_x \\ \varepsilon_y \\ \varepsilon_s \end{bmatrix} = \begin{bmatrix} \frac{1}{E_x} & -\frac{\nu_{xy}}{E_x} & \frac{\eta_{xy,x}}{E_x} \\ -\frac{\nu_{yx}}{E_y} & \frac{1}{E_y} & \frac{\eta_{xy,y}}{E_y} \\ \frac{\eta_{x,xy}}{G_{xy}} & \frac{\eta_{y,xy}}{G_{xy}} & \frac{1}{G_{xy}} \end{bmatrix} \cdot \begin{bmatrix} \sigma_x \\ \sigma_y \\ \sigma_s \end{bmatrix}. \tag{7.53}$$

As explained above, besides the Poisson effect, an off–axis lamina exhibits two other interactions unlike in a unidirectional lamina. These mutual influence terms are proposed by Lekhnitskii [3], and they fold into two types:

- The first interaction deals with the axial strain (ε_x or ε_y) generated by applying pure shear stress σ_s (shear/axial strain coupling). It involves the parameters $\eta_{x,xy}$ and $\eta_{y,xy}$ called the *coefficients of mutual influence of the first type*:

$$\begin{aligned}\eta_{x,xy} &= \frac{\varepsilon_x}{2\varepsilon_{xy}} = \frac{\varepsilon_x}{\varepsilon_s},\\ \eta_{y,xy} &= \frac{\varepsilon_y}{2\varepsilon_{xy}} = \frac{\varepsilon_y}{\varepsilon_s},\end{aligned} \tag{7.54}$$

  for $\sigma_s \neq 0$, and all other stresses are equal to zero.
- The second interaction occurs through the generation of in–plane shearing strain ε_s induced by an axial stress (σ_x or σ_y). It involves the parameters $\eta_{xy,x}$ and $\eta_{xy,y}$ called the *coefficients of mutual influence of the second type*:

$$\begin{aligned}\eta_{xy,x} &= \frac{2\varepsilon_{xy}}{\varepsilon_x} = \frac{\varepsilon_s}{\varepsilon_x} \quad \text{for } \sigma_x \neq 0, \text{ and all other stresses are equal to zero},\\ \eta_{xy,y} &= \frac{2\varepsilon_{xy}}{\varepsilon_y} = \frac{\varepsilon_s}{\varepsilon_y} \quad \text{for } \sigma_y \neq 0, \text{ and all other stresses are equal to zero.}\end{aligned} \tag{7.55}$$

The presence of the shear–extension coupling in an angle ply can be evidenced by the off–axis tensile test, which may result in pure in–plane strain ε_4 besides the axial strain ε_1 in the material orthotropy directions. With appropriate experimental cautions, the off–axis test can be performed for the experimental determination of the in–plane shear modulus G_{12} and the shear strength. To this end, some considerations should be taken in terms of sample geometry (Saint–Venant effects) to prevent instabilities and to avoid end–grip difficulties, which can perturb the pure shear stress/strain state.

The reader can refer to the excellent book by Jones [1] presenting several test configurations to determine the shear modulus and shear strength of thin composites. However, because each has drawbacks and interests, there is no universal agreement on a suitable experimental configuration leading to reliable shear properties.

Given the symmetry of $\widehat{\mathcal{S}}$, we can also note that the coefficients of mutual influence obey reciprocal relations like those defined for the Poisson ratio (7.17):

$$\frac{\eta_{xy,x}}{E_x} = \frac{\eta_{x,xy}}{G_{xy}},$$
$$\frac{\eta_{xy,y}}{E_y} = \frac{\eta_{y,xy}}{G_{xy}}. \tag{7.56}$$

By combining Eqs. (7.48), (7.50), and (7.53) we easily obtain expressions of the engineering constants (apparent moduli) of an angle lamina as functions of its orientation angle ϑ:

$$\frac{1}{E_x} = \frac{1}{E_1}m^4 + \frac{1}{E_2}n^4 + \left[\frac{1}{G_{12}} - \frac{2\nu_{12}}{E_1}\right]m^2n^2,$$
$$\frac{1}{E_y} = \frac{1}{E_1}n^4 + \frac{1}{E_2}m^4 + \left[\frac{1}{G_{12}} - \frac{2\nu_{12}}{E_1}\right]m^2n^2,$$
$$\nu_{xy} = E_x\left[\frac{\nu_{12}}{E_1}[m^4 + n^4] - \left[\frac{1}{E_1} + \frac{1}{E_2} - \frac{1}{G_{12}}\right]m^2n^2\right],$$
$$\frac{1}{G_{xy}} = 4\left[\frac{1}{E_1} + \frac{1}{E_2} + \frac{2\nu_{12}}{E_1}\right]m^2n^2 + \frac{1}{G_{12}}[m^2 - n^2]^2,$$
$$\eta_{xy,x} = E_x\left[\left[\frac{2}{E_1} + \frac{2\nu_{12}}{E_1} - \frac{1}{G_{12}}\right]m^3n - \left[\frac{2}{E_2} + \frac{2\nu_{12}}{E_1} - \frac{1}{G_{12}}\right]mn^3\right],$$
$$\eta_{xy,y} = E_y\left[\left[\frac{2}{E_1} + \frac{2\nu_{12}}{E_1} - \frac{1}{G_{12}}\right]mn^3 - \left[\frac{2}{E_2} + \frac{2\nu_{12}}{E_1} - \frac{1}{G_{12}}\right]m^3n\right]. \tag{7.57}$$

The previous equations of the apparent moduli highlight the anisotropic response of an orthotropic lamina when subjected to off–axis loading. Figs. 7.6, 7.7, 7.8, and 7.9 show the variations with respect to the fiber orientation angle ϑ of the engineering constants for a composite lamina

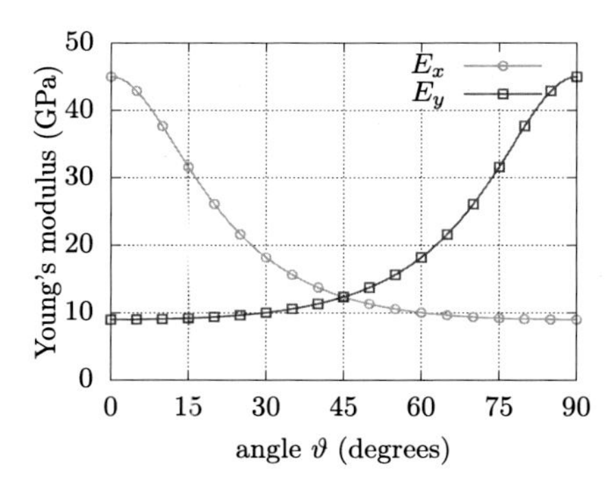

FIGURE 7.6 Evolution of E_x and E_y with respect to the fiber orientation angle of a glass/epoxy lamina.

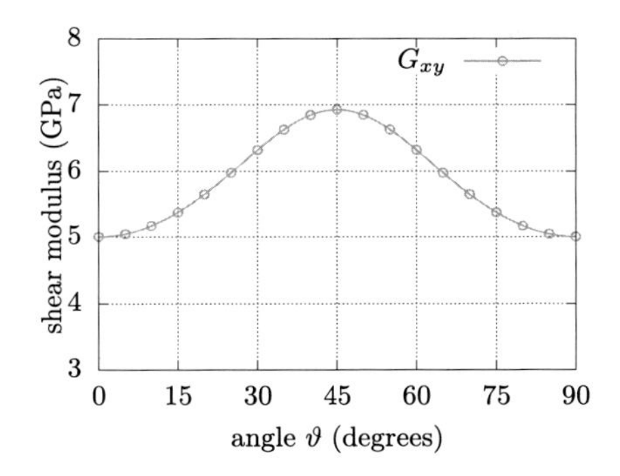

FIGURE 7.7 In–plane shear modulus in the $(x - y)$ plane G_{xy} as a function of angle of lamina for a glass/epoxy lamina.

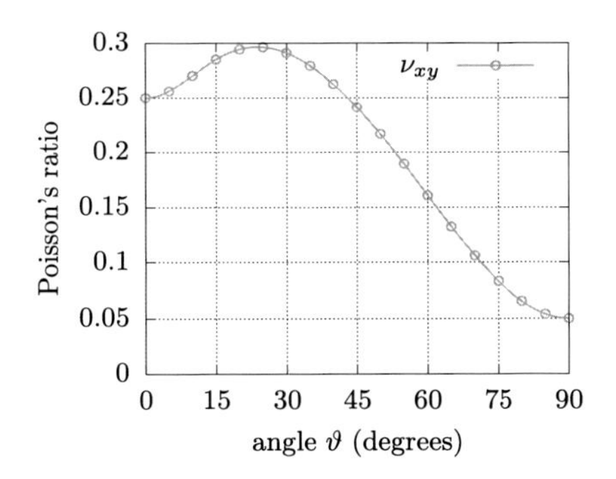

FIGURE 7.8 Poisson ratio ν_{xy} as a function of angle of lamina for a glass/epoxy lamina.

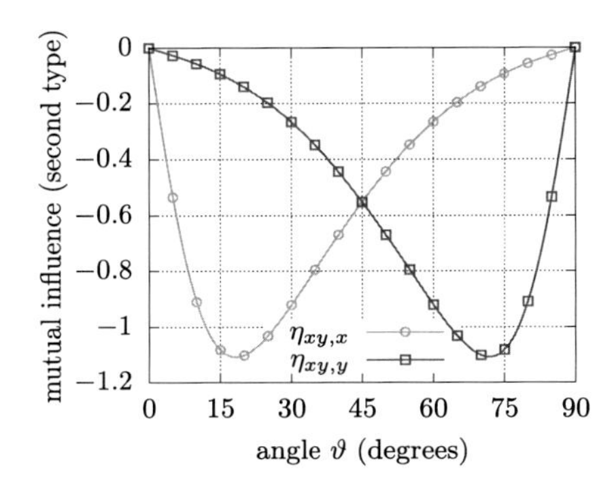

FIGURE 7.9 Evolution of the mutual axial–shear strain coupling $\eta_{xy,x}$ and $\eta_{xy,y}$ as a function of angle of lamina for a glass/epoxy lamina.

consisting of 60% glass fiber reinforced epoxy matrix, where

$$E_1 = 45\ \text{GPa},\quad E_2 = 9\ \text{GPa},\quad G_{12} = 5\ \text{GPa},\quad \text{and}\quad \nu_{12} = 0.25.$$

7.4.5 Example

Consider ϑ angle lamina subjected only to an axial tension stress ($\sigma_x = p$) in the global axes $(x - y)$. A strain gage rosette is utilized for measuring the strains in the direction (0°, 45°, 90°) at the center of the lamina as shown in Fig. 7.10.

1. Find the expression of ε_s as a function of the measured strains (ε_{0°, ε_{90°, ε_{45°).
2. Find the stress tensor $\boldsymbol{\sigma}^{(1-2)}$ in the local axes $(1 - 2)$ and then deduce the local strain tensor $\boldsymbol{\varepsilon}^{(1-2)}$ as a function of the measured strains.
3. Give an analytical expression for the shear modulus of the lamina G_{12}, which can be a function of ε_x, ε_y, ε_{45°, ϑ, and p.
4. What would be this expression for the case $\vartheta = 45^\circ$?

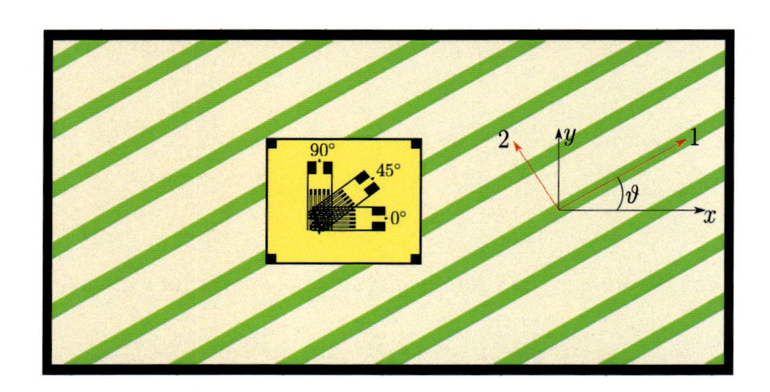

FIGURE 7.10 Off–axis configuration.

Solution:

1. Considering Fig. 7.10, let us assume that the local axis (1) is oriented at $\vartheta = 45^\circ$, which means that the local axial strain ε_1 is directly provided by the strain gage placed at $\vartheta = 45^\circ$. Hence using the transformation matrix $\boldsymbol{T}_\varepsilon$ of Eq. (7.37) yields

$$\begin{bmatrix} \varepsilon_x \\ \varepsilon_y \\ \varepsilon_s \end{bmatrix} = \boldsymbol{T}_\varepsilon(45^\circ) \cdot \begin{bmatrix} \varepsilon_{45^\circ} \\ \varepsilon_? \\ \varepsilon_? \end{bmatrix}.$$

It is known that $\varepsilon_x = \varepsilon_{0^\circ}$ and $\varepsilon_y = \varepsilon_{90^\circ}$. Consequently,

$$\begin{bmatrix} \varepsilon_1 = \varepsilon_{45^\circ} \\ \varepsilon_2 \\ \varepsilon_4 \end{bmatrix} = [\boldsymbol{T}_\varepsilon(45^\circ)]^{-1} \cdot \begin{bmatrix} \varepsilon_x \\ \varepsilon_y \\ \varepsilon_s \end{bmatrix} \Rightarrow$$
$$\begin{bmatrix} \varepsilon_{45^\circ} \\ \varepsilon_2 \\ \varepsilon_4 \end{bmatrix} = [\boldsymbol{T}_\sigma(45^\circ)]^T \begin{bmatrix} \varepsilon_{0^\circ} \\ \varepsilon_{90^\circ} \\ \varepsilon_s \end{bmatrix}.$$

By expressing the first component ε_{45° the above expression gives

$$\varepsilon_{45^\circ} = m^2\varepsilon_{0^\circ} + n^2\varepsilon_{90^\circ} + mn\varepsilon_s.$$

For $\vartheta = 45^\circ$, $m = n = \dfrac{\sqrt{2}}{2}$. Thus

$$\varepsilon_{45^\circ} = \frac{\varepsilon_{0^\circ}}{2} + \frac{\varepsilon_{90^\circ}}{2} + \frac{\varepsilon_s}{2},$$

finally deducing

$$\varepsilon_s = 2\varepsilon_{45^\circ} - [\varepsilon_{0^\circ} + \varepsilon_{90^\circ}].$$

Then we obtain the strain components in the global axes $(x - y)$ as a function of the measured strains:

$$\begin{bmatrix} \varepsilon_x \\ \varepsilon_y \\ \varepsilon_s \end{bmatrix} = \begin{bmatrix} \varepsilon_{0^\circ} \\ \varepsilon_{90^\circ} \\ 2\varepsilon_{45^\circ} - [\varepsilon_{0^\circ} + \varepsilon_{90^\circ}] \end{bmatrix}.$$

2. Knowing the stresses in the global axes, we can obtain the stresses in the local axes by inverting the stress transformation equation (7.35).

$$\boldsymbol{\sigma}^{(1-2)} = \boldsymbol{T}_\sigma^{-1} \cdot \boldsymbol{\sigma}^{(x-y)} \quad \Rightarrow$$
$$\begin{bmatrix} \sigma_1 \\ \sigma_2 \\ \sigma_4 \end{bmatrix} = \boldsymbol{T}_\varepsilon^T \cdot \begin{bmatrix} \sigma_x = p \\ \sigma_y = 0 \\ \sigma_s = 0 \end{bmatrix} \Rightarrow$$
$$\begin{bmatrix} \sigma_1 \\ \sigma_2 \\ \sigma_4 \end{bmatrix} = \begin{bmatrix} m^2 p \\ n^2 p \\ -mnp \end{bmatrix}.$$

3. Similarly, we easily obtain the local strain components:

$$\begin{bmatrix} \varepsilon_1 \\ \varepsilon_2 \\ \varepsilon_4 \end{bmatrix} = \boldsymbol{T}_\varepsilon^{-1} \cdot \begin{bmatrix} \varepsilon_x \\ \varepsilon_y \\ \varepsilon_s \end{bmatrix} \Rightarrow$$

$$\begin{bmatrix} \varepsilon_1 \\ \varepsilon_2 \\ \varepsilon_4 \end{bmatrix} = \boldsymbol{T}_\sigma^T \cdot \begin{bmatrix} \varepsilon_{0^\circ} \\ \varepsilon_{90^\circ} \\ 2\varepsilon_{45^\circ} - [\varepsilon_{0^\circ} + \varepsilon_{90^\circ}] \end{bmatrix} \quad \Rightarrow$$

$$\begin{bmatrix} \varepsilon_1 \\ \varepsilon_2 \\ \varepsilon_4 \end{bmatrix} = \begin{bmatrix} m^2 & n^2 & mn \\ n^2 & m^2 & -mn \\ -2mn & 2nm & m^2 - n^2 \end{bmatrix} \cdot \begin{bmatrix} \varepsilon_{0^\circ} \\ \varepsilon_{90^\circ} \\ 2\varepsilon_{45^\circ} - [\varepsilon_{0^\circ} + \varepsilon_{90^\circ}] \end{bmatrix},$$

where $m = \cos\vartheta$ and $n = \sin\vartheta$.
By expressing the component ε_4 this yields

$$\varepsilon_4 = -2[\varepsilon_{0^\circ} - \varepsilon_{90^\circ}]\cos\vartheta\sin\vartheta + [2\varepsilon_{45^\circ} - \varepsilon_{0^\circ} - \varepsilon_{90^\circ}][\cos^2\vartheta - \sin^2\vartheta]$$
$$\varepsilon_4 = -[\varepsilon_{0^\circ} - \varepsilon_{90^\circ}]\sin 2\vartheta + [2\varepsilon_{45^\circ} - \varepsilon_{0^\circ} - \varepsilon_{90^\circ}]\cos 2\vartheta.$$

Moreover,

$$\sigma_4 = -\frac{p}{2}\sin 2\vartheta.$$

Having expressed σ_4 and ε_4, we can determine G_{12}:

$$G_{12} = \frac{\sigma_4}{\varepsilon_4} \quad \Rightarrow$$
$$G_{12} = \frac{p\sin 2\vartheta}{2[\varepsilon_{0^\circ} - \varepsilon_{90^\circ}]\sin 2\vartheta - 2[2\varepsilon_{45^\circ} - \varepsilon_{0^\circ} - \varepsilon_{90^\circ}]\cos 2\vartheta}.$$

4. For a lamina whose fibers are oriented at angle $\vartheta = 45^\circ$, the last expression of the previous question gives

$$G_{12} = \frac{p}{2[\varepsilon_{0^\circ} - \varepsilon_{90^\circ}]}.$$

Note that for this specific 45° lamina, the strain ε_{45° measurement is not needed, and hence only a biaxial strain gage should be utilized for the strain measurement at directions 0° and 90°.

7.5 Macromechanical response of a laminate composite thin plate

A fiber–reinforced composite laminate consists of a stacking of n thin layers (laminae) bonded together. Each lamina is typically thin (0.2 mm) and can be a matrix reinforced by aligned fibers, short random fibers, woven fabric, and so on. In this section, we consider the lamina as an orthotropic ply with constant thickness and reinforced by unidirectionally aligned fibers. It is identified by its position z in the total laminate thickness, its constituents (in terms of fibers and matrix), and its arbitrary orientation angle with respect to the global axes, as illustrated in Figs. 7.11 and 7.12.

Using the previous sections providing the mechanical responses of an orthotropic lamina in its principal directions and in an arbitrary coordinate system (off–axis), in the next sections, we outline the mechanics of reinforced laminated plates using the classical laminate theory (CLT). This theory is treated in several standard references dealing with mechanics of composites [1,4,5] and will be briefly recalled here. The CLT is considered as a simplified theory for laminate thin plates (flat) and shells subjected to bending and stretching loads. The Kirchhoff–Love hypothesis is invoked when deriving the kinematic equations of such laminated thin plates under the plane stress conditions in the framework of small perturbations. The objective is the analysis of the stress and strain distributions in the laminate in relation with the fiber orientation of each lamina, the layup sequences, and the stiffness of each ply.

7.5.1 Laminate code and convention

In this section, we clarify the convention of the $(x - y - z)$ global Cartesian coordinate system adopted for the laminate analysis, as well as the laminae numbering and the related codes used for a rapid description of a laminate. Fig. 7.11 shows an example of a general laminate consisting of n plies. The laminate has a total thickness h, which is the sum of the individual lamina thicknesses, noted hereafter t_k, where k is the layer's number,

$$h = \sum_{k=1}^{n} t_k.$$

Not all layers necessarily have the same thickness. The position of the mid–plane is $h/2$, located at the origin of the through–thickness coordinate z. The out–of–plane z–axis is oriented toward the bottom surface of the laminate, which means that $+z$ axis should be downward. The geomet-

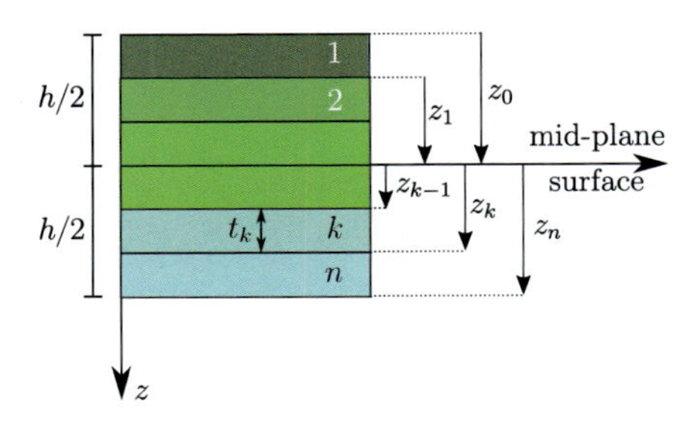

FIGURE 7.11 Laminate layers and adopted convention.

ric middle surface (reference surface) may be within a particular lamina or at an interface between two layers. The through–thickness coordinate z hence extends:

$$-\frac{h}{2} \leqslant z \leqslant \frac{h}{2}.$$

Considering that the $+z$ axis is downward, the laminae are hence numbered from the top to the bottom of the laminate. The first ply is the top ply of the laminate. So the layer at the most negative location is referred to as layer 1, whereas the layer at an arbitrary location z_k is layer k, and hence the layer at the most positive position is layer n.

Each lamina k has a through–thickness coordinate z_k referencing the bottom of the lamina k, that is, the interface between layers k and $k+1$. Thus the coordinate z_{k-1} corresponds to the top of the lamina k:

$$t_k = z_k - z_{k-1}.$$

Laminate composite structures are generally designed by a code consisting of laminae angles and the related possible repetitions separated by slashes. Laminae angles are expressed in degrees and are positive in the counter-clockwise direction. The overall code is written between a pair of square brackets. Let us consider the example of the laminate given in Fig. 7.12. This laminate is denoted a:

$$[60/(90_2)/(0_4)/(45_3)].$$

It consists of 10 plies: a single 60° ply is on the top, then two cross–plies 90°, four plies oriented at the 0°, and finally the ply oriented at 45° repeated three times on the bottom of the laminate. The previous laminate code implies that each ply is made of the same material and is of the same thickness. More details can also be provided such as the constitutive lamina material or the thicknesses. In that case, superscripts are added to the code referring to the laminate.

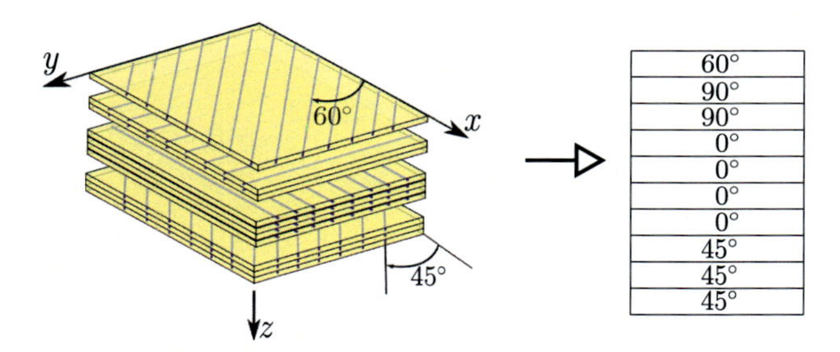

FIGURE 7.12 An example of angle ply laminate and the corresponding schematic representation.

A laminate can be *symmetric about the mid–plane* if for every ply to one side of the mid–plane there is another identical ply (in terms of thickness, material, fibers orientation) from the same distance but on the opposite side of the middle surface. Subscript S is then added to the laminate code. A symmetric laminate can have either even or odd number of plies. For an odd number of plies, the mid–surface will be at a mid–thickness of the central ply. The latter is denoted with a bar on the top in the laminate code.

An *antisymmetric about the mid–plane* laminate consists of even total number of plies, which are symmetric in terms of material and ply thickness, but each couple of plies at the same distance above and below the mid–plane should have angles negative of each other.

A laminate with even total number and having for each ply oriented at $+\vartheta$ a corresponding one oriented at $-\vartheta$ is called a *balanced laminate*. The pair of plies can be located anywhere within the thickness.

A *cross–ply laminate* is composed of plies oriented only at $0°$ and $90°$. Cross–ply laminates are balanced.

7.5.2 Laminated thin plates and Kirchhoff–Love hypothesis

A thin plate, like that of Fig. 7.13, is a flat medium that the out–of–plane dimension (thickness) can be neglected compared to its other two dimensions ($t_{3=z} \ll \min(l_x, l_y)$), length and width. For a laminated plate, we can consider that $10t_{3=z} < \min(l_x, l_y)$, and the 3–D problem can be transformed into a 2–D planar problem. In addition, the equilibrium is expressed by integrating through the thickness of the laminate.

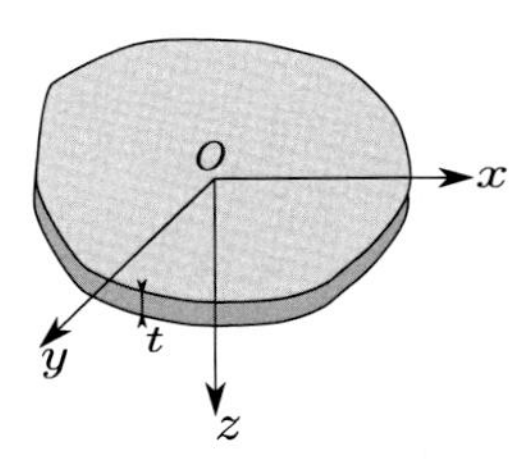

FIGURE 7.13 Thin plate: the thickness is much smaller than the length and width of the plate.

For a thin laminate plate, the Kirchhoff–Love hypothesis implies that a normal (originally straight and perpendicular) to the middle surface of the laminate remains straight and perpendicular to this middle surface during deformation of the laminate (Fig. 7.14). The latter can exhibit extension–compression, shearing, bending, and twisting. A consequence of the Kirchhoff–Love hypothesis is that the transverse shearing strains can be reasonably considered vanished in planes perpendicular to the middle surface: $\varepsilon_{xz} = 0$ and $\varepsilon_{yz} = 0$. Furthermore, the out–of–plane strain ε_z is negligible due to the low thickness.

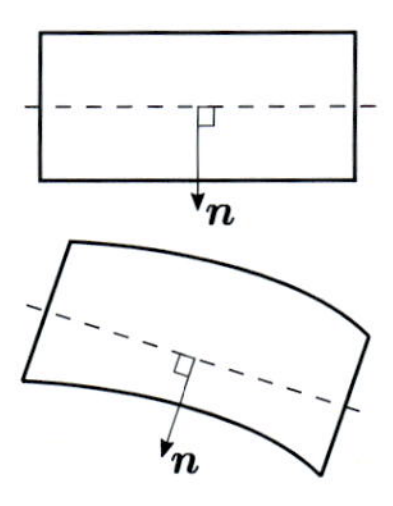

FIGURE 7.14 The normal of thin plate in Kirchhoff–Love assumption.

7.5.3 Kinematics of thin laminated plates and strain–displacement relation

Fig. 7.15 shows the side view of the plate in the $(x-z)$ and $(y-z)$ planes. By assuming that Kirchhoff–Love hypothesis holds for a bent plate the initially straight vertical cross–section remains straight but rotates around the mid–surface ($z=0$) with respect to the y– and x–axes. Consequently, besides the in–plane displacements u_0 and v_0 in the x– and y–directions,

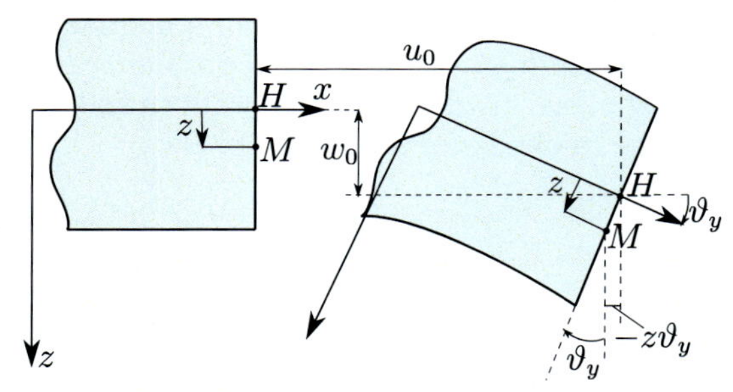

(a) Plate cross-section $(x-z)$ exhibiting displacement $u_0(x,y)$ and rotation ϑ_y

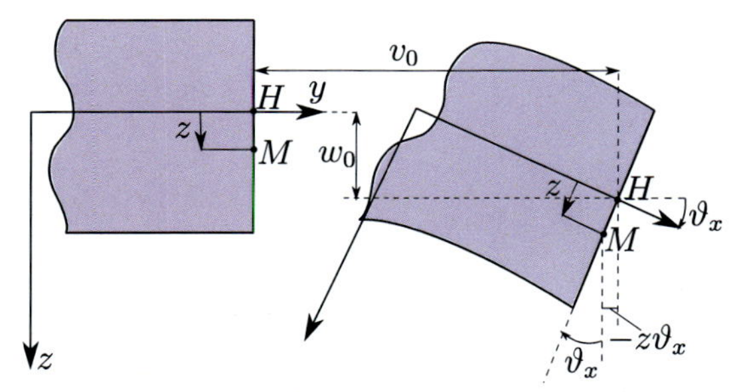

(b) Plate cross-section $(y-z)$ exhibiting displacement $v_0(x,y)$ and rotation ϑ_x

FIGURE 7.15 Thin plate kinematics showing the through–thickness displacement and rotations.

respectively, additional displacements in both directions are induced by these rotations of the cross–sections $(y - z)$ and $(x - z)$, respectively. Under the assumption of small deflections and rotations, these two additional displacements can be approximated as functions of the tangent of the rotation angle (the slope) multiplied by the distance from mid–plane (axial vertical location). For the displacement in the x–direction, this quantity is $-z \tan \vartheta_y \approx -z\vartheta_y$ for small rotation angle.

The two additional rotational displacements are added to the mid–plane displacements u_0 and v_0. Considering the first–order plate theory, we write the displacement fields at any point z through the laminate thickness as follows:

$$\begin{aligned} u(x, y, z) &= u_0(x, y) - z\vartheta_y = u_0(x, y) - z\frac{\partial w_0}{\partial x}, \\ v(x, y, z) &= v_0(x, y) - z\vartheta_x = v_0(x, y) - z\frac{\partial w_0}{\partial y}, \\ w(x, y, z) &= w_0(x, y), \\ \vartheta_y &= \frac{\partial w_0}{\partial x}, \\ \vartheta_x &= \frac{\partial w_0}{\partial y}. \end{aligned} \tag{7.58}$$

Note that $u_0(x, y)$, $v_0(x, y)$, and $w_0(x, y)$ are the displacements of the mid–plane of the laminate plate (reference surface).

We derive the strains from the displacement fields from their gradients:

$$\varepsilon_{ij} = \frac{1}{2}\left[\frac{\partial u_i}{\partial x_j} + \frac{\partial u_j}{\partial x_i}\right], \tag{7.59}$$

where $i = x, y$ and $j = x, y$. In the case of the thin laminated plate the Kirchhoff–Love hypothesis is assumed to hold. Thus the strains are reduced to ε_x, ε_y, and ε_{xy}:

$$\begin{aligned} \varepsilon_x &= \frac{1}{2}\left[\frac{\partial u}{\partial x} + \frac{\partial u}{\partial x}\right] = \frac{\partial u}{\partial x} = \frac{\partial u_0}{\partial x} - z\frac{\partial^2 w_0}{\partial x^2}, \\ \varepsilon_y &= \frac{1}{2}\left[\frac{\partial v}{\partial y} + \frac{\partial v}{\partial y}\right] = \frac{\partial v}{\partial y} = \frac{\partial v_0}{\partial y} - z\frac{\partial^2 w_0}{\partial y^2}, \\ \varepsilon_{xy} &= \frac{1}{2}\left[\frac{\partial u}{\partial y} + \frac{\partial v}{\partial x}\right] = \frac{1}{2}\left[\frac{\partial u_0}{\partial y} + \frac{\partial v_0}{\partial x}\right] - z\frac{\partial^2 w_0}{\partial x \partial y}. \end{aligned} \tag{7.60}$$

As explained in Subsection 7.2.1, we consider the engineering shear strain:

$$\varepsilon_s = 2\varepsilon_{xy} = \frac{\partial u_0}{\partial y} + \frac{\partial v_0}{\partial x} - 2z\frac{\partial^2 w_0}{\partial x \partial y}. \tag{7.61}$$

So the strain components of the laminate are expressed as

$$\begin{bmatrix} \varepsilon_x \\ \varepsilon_y \\ \varepsilon_s \end{bmatrix} = \begin{bmatrix} \dfrac{\partial u_0}{\partial x} \\ \dfrac{\partial v_0}{\partial y} \\ \dfrac{\partial u_0}{\partial y} + \dfrac{\partial v_0}{\partial x} \end{bmatrix} + z \begin{bmatrix} -\dfrac{\partial^2 w_0}{\partial x^2} \\ -\dfrac{\partial^2 w_0}{\partial y^2} \\ -2\dfrac{\partial^2 w_0}{\partial x \partial y} \end{bmatrix}. \tag{7.62}$$

Thus we can write

$$\begin{bmatrix} \varepsilon_x \\ \varepsilon_y \\ \varepsilon_s \end{bmatrix} = \begin{bmatrix} \varepsilon_x^0 \\ \varepsilon_y^0 \\ \varepsilon_s^0 \end{bmatrix} + z \begin{bmatrix} \kappa_x \\ \kappa_y \\ \kappa_{xy} \end{bmatrix} \tag{7.63}$$

or, in compact form,

$$\boldsymbol{\varepsilon}^{(x-y)} = \boldsymbol{\varepsilon}^0 + z\,\boldsymbol{\kappa}, \tag{7.64}$$

where $\boldsymbol{\varepsilon}^0$ is the mid–plane strain vector, which can be easily expressed through the partial derivatives of the mid–plane displacements $u_0(x, y)$ and $v_0(x, y)$. The mid–plane strain components are constant through the laminate thickness. The vector $\boldsymbol{\kappa}$ is called the curvature of the deformed middle surface of the laminate and is expressed in terms of the second partial derivatives of the deflection $w_0(x, y)$:

$$\begin{aligned} \kappa_x &= -\frac{\partial^2 w_0}{\partial x^2} && \text{bending curvature on } x, \\ \kappa_y &= -\frac{\partial^2 w_0}{\partial y^2} && \text{bending curvature on } y, \\ \kappa_{xy} &= -2\frac{\partial^2 w_0}{\partial x \partial y} && \text{twisting curvature of the mid–plane.} \end{aligned} \tag{7.65}$$

The term κ_{xy} quantifies the changes of the mid–plane slope in the x–direction with respect to y. Reciprocally, it provides how the y–direction slope changes with x. It is worth noticing that the components of the curvature vector $\boldsymbol{\kappa}$ are also constant through the laminate thickness.

Eq. (7.63) (or (7.64)) implies a linear variation of the total strain through the laminate thickness. Indeed, since the mid–plane strain $\boldsymbol{\varepsilon}^0$ and the curvatures $\boldsymbol{\kappa}$ are constant through the thickness, the total strain varies linearly as a function of the position z of the considered point in the laminate thickness, as illustrated in Fig. 7.16.

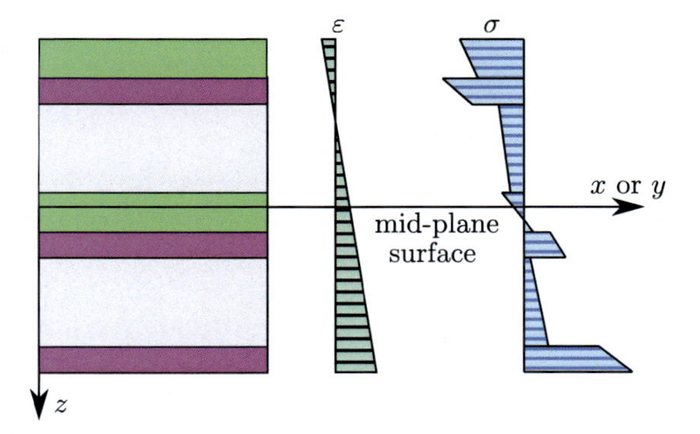

FIGURE 7.16 An illustration of the through–thickness linear variation of the strain (Eq. (7.64)) and the piecewise linear variation of the stress (Eq. (7.67)).

7.5.4 Stress variation in a laminate

After expressing the strain variation through the thickness of a laminate, we can easily obtain the stress in the ply k by substituting Eq. (7.64) into the stress–strain relation (7.41) expressed in the $(x-y)$ axes. Then we write the stresses in terms of mid–surface strains and curvatures. The classical laminate theory assumes a plane stress state in each ply of the laminate. For a lamina k having a thickness $t_k = z_k - z_{k-1}$, we can write the expression

$$\boldsymbol{\sigma}_k^{(x-y)} = \widehat{\boldsymbol{Q}}_k \cdot \boldsymbol{\varepsilon}_k^{(x-y)}. \tag{7.66}$$

The last relation is rewritten as

$$\boldsymbol{\sigma}_k^{(x-y)} = \widehat{\boldsymbol{Q}}_k \cdot \left[\boldsymbol{\varepsilon}^0 + z\,\boldsymbol{\kappa}\right] \qquad \text{for } z_{k-1} \leqslant z \leqslant z_k, \tag{7.67}$$

which, in expanded form, gives

$$\begin{bmatrix} \sigma_x \\ \sigma_y \\ \sigma_s \end{bmatrix}_{(k)} = \begin{bmatrix} \widehat{Q}_{xx} & \widehat{Q}_{xy} & \widehat{Q}_{xs} \\ & \widehat{Q}_{yy} & \widehat{Q}_{ys} \\ \text{SYM} & & \widehat{Q}_{ss} \end{bmatrix}_{(k)} \cdot \left[\begin{bmatrix} \varepsilon_x^0 \\ \varepsilon_y^0 \\ \varepsilon_s^0 \end{bmatrix} + z \begin{bmatrix} \kappa_x \\ \kappa_y \\ \kappa_{xy} \end{bmatrix}\right] \tag{7.68}$$

for $z_{k-1} \leqslant z \leqslant z_k$. Eq. (7.67) gives the global stress relative to the $(x-y)$ axes in a lamina k as a function of the strain and the transformed reduced stiffness of the considered lamina k. We can easily notice that the stress components vary linearly through thickness of the lamina with vertical position z ($z_{k-1} \leqslant z \leqslant z_k$). However, the transformed reduced stiffness matrix can be different from ply to ply within the same laminate (orientation, material). Consequently, a stress jump (a discontinuity) from a ply to ply appears. The stress components exhibit then step linear variation as a function of z (piecewise linear), as shown in Fig. 7.16.

The next remaining step in laminates analysis consists of expressing the mid–plane strains and curvatures in terms of the applied loads. Indeed, once the global strains are determined, we can compute the global stresses in the laminate from Eq. (7.67). However, each lamina should be designed by introducing local stresses and strains that satisfy specific failure criteria. Thus, for each lamina, global stresses should be transformed to local stresses by inverting the stress transformation equation (7.35). Local strains are calculated either by the strain–stress relation in the principal axes $(1-2)$ per Eq. (7.48) or preferably by transforming the global strains using Eq. (7.37).

7.5.5 Force and moment resultants related to midplane strains and curvatures

As explained above, in this section, we relate the force and moment resultants (per unit width of the plate) to the global strains in the laminate. To this end, middle surface in–plane strains and curvatures should be expressed as a function of in–plane tractions $\boldsymbol{N}$ and moments $\boldsymbol{M}$, which are known and applied to the laminate plate.

Force resultants

In the global axes, we obtain the equilibrium of an element of the laminate by integrating through the thickness each lamina stress $\boldsymbol{\sigma}^{(x-y)}$ to balance the traction per unit width $\boldsymbol{N}$, as shown in Fig. 7.17. For a laminate of total thickness h consisting of n plies, we can express the traction balance as

$$\boldsymbol{N} = \int_{-h/2}^{h/2} \boldsymbol{\sigma}^{(x-y)} \mathrm{d}z \tag{7.69}$$

or, in expanded form,

$$\boldsymbol{N} = \begin{bmatrix} N_x \\ N_y \\ N_{xy} \end{bmatrix} = \int_{-h/2}^{h/2} \begin{bmatrix} \sigma_x \\ \sigma_y \\ \sigma_s \end{bmatrix} \mathrm{d}z, \tag{7.70}$$

where N_x and N_y are the normal forces (membrane forces) per unit length in the x– and y–directions, respectively, whereas N_{xy} is the shear force per unit length.

As mentioned before, due to the material and orientation of each ply, the global stresses vary linearly only through the thickness of each lamina $t_k = z_k - z_{k-1}$. Nevertheless, they exhibit a jump at the interface between two adjacent plies. So we can rewrite the integral in Eq. (7.70) as a discrete

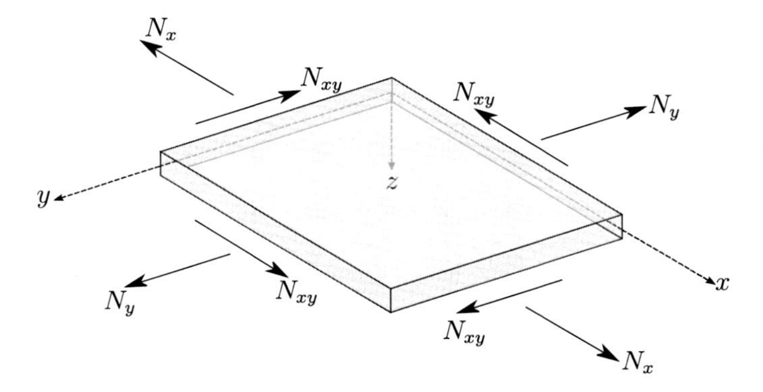

FIGURE 7.17 Schematic illustration of the force resultants applied to an element of composite laminate plate.

summation on the plies:

$$\boldsymbol{N} = \sum_{k=1}^{n} \int_{z_{k-1}}^{z_k} \begin{bmatrix} \sigma_x \\ \sigma_y \\ \sigma_s \end{bmatrix}_{(k)} \mathrm{d}z. \tag{7.71}$$

Using Eq. (7.67) yields

$$\boldsymbol{N} = \sum_{k=1}^{n} \int_{z_{k-1}}^{z_k} \widehat{\boldsymbol{Q}}_k \cdot \left[\boldsymbol{\varepsilon}^0 + z\,\boldsymbol{\kappa}\right] \mathrm{d}z.$$

In the above equations the middle surface in–plane strains and curvatures are independent of the through–thickness coordinate z. Furthermore, the reduced transformed stiffness matrix is constant for each lamina if not subjected to a temperature gradient or moisture gradient. Consequently, $\boldsymbol{\varepsilon}^0$, $\boldsymbol{\kappa}$, and $\widehat{\boldsymbol{Q}}_k$ should go outside the integration but remain within the summation on each lamina, leading to

$$\begin{aligned} \boldsymbol{N} &= \sum_{k=1}^{n} \widehat{\boldsymbol{Q}}_k \cdot \boldsymbol{\varepsilon}^0 \int_{z_{k-1}}^{z_k} \mathrm{d}z + \sum_{k=1}^{n} \widehat{\boldsymbol{Q}}_k \cdot \boldsymbol{\kappa} \int_{z_{k-1}}^{z_k} z\mathrm{d}z \\ &= \underbrace{\left[\sum_{k=1}^{n} \widehat{\boldsymbol{Q}}_k \left[z_k - z_{k-1}\right]\right]}_{\boldsymbol{A}} \cdot \boldsymbol{\varepsilon}^0 + \underbrace{\frac{1}{2}\left[\sum_{k=1}^{n} \widehat{\boldsymbol{Q}}_k \left[z_k^2 - z_{k-1}^2\right]\right]}_{\boldsymbol{B}} \cdot \boldsymbol{\kappa}. \end{aligned} \tag{7.72}$$

Consequently, the force resultants are related to the mid–plane strain and curvature according to the equation

$$\boldsymbol{N} = \boldsymbol{A} \cdot \boldsymbol{\varepsilon}^0 + \boldsymbol{B} \cdot \boldsymbol{\kappa}, \tag{7.73}$$

where $\boldsymbol{A}$ is the extensional stiffness matrix of the laminate, and $\boldsymbol{B}$ stands for its bending–extension coupling matrix:

$$\boldsymbol{A} = \sum_{k=1}^{n} \widehat{\boldsymbol{Q}}_k \left[z_k - z_{k-1} \right],$$

$$\boldsymbol{B} = \frac{1}{2} \sum_{k=1}^{n} \widehat{\boldsymbol{Q}}_k \left[z_k^2 - z_{k-1}^2 \right]. \tag{7.74}$$

By the symmetry of the reduced stiffness matrix $\widehat{\boldsymbol{Q}}_k$ the matrices $\boldsymbol{A}$ and $\boldsymbol{B}$ are also symmetric. Furthermore, we can easily notice that the components of the laminate extensional stiffness matrix $\boldsymbol{A}$ have units $\mathrm{N \cdot m^{-1}}$ and those of the coupling matrix $\boldsymbol{B}$ have units N.
We can also write Eq. (7.73)

$$\begin{bmatrix} N_x \\ N_y \\ N_{xy} \end{bmatrix} = \begin{bmatrix} A_{11} & A_{12} & A_{14} \\ A_{12} & A_{22} & A_{24} \\ A_{14} & A_{24} & A_{44} \end{bmatrix} \cdot \begin{bmatrix} \varepsilon_x^0 \\ \varepsilon_y^0 \\ \varepsilon_s^0 \end{bmatrix} + \begin{bmatrix} B_{11} & B_{12} & B_{14} \\ B_{12} & B_{22} & B_{24} \\ B_{14} & B_{24} & B_{44} \end{bmatrix} \cdot \begin{bmatrix} \kappa_x \\ \kappa_y \\ \kappa_{xy} \end{bmatrix}. \tag{7.75}$$

From the last equation we can show that the general expression of N_x, for instance, is given by

$$N_x = A_{11}\varepsilon_x^0 + A_{12}\varepsilon_y^0 + A_{14}\varepsilon_s^0 + B_{11}\kappa_x + B_{12}\kappa_y + B_{14}\kappa_{xy}.$$

This expression implies that due to the coupling terms, when pulling such general laminate with an extensional force on the x–axis, the response will be anisotropic. Indeed, besides the expected extensional strains ε_x^0 and ε_y^0 (Poisson effect), N_x will bring about a shearing strain ε_s^0, two bending curvatures κ_x, κ_y, and finally a twisting curvature κ_{xy}.

It is obvious to conclude that the terms A_{14} and A_{24} are the shear–extension coupling terms, whereas the terms B_{ij} are the bending/twisting–extension coupling terms. It is worth mentioning that the coupling effects are reciprocal. Indeed, assuming that $B_{ij} = 0$ but $A_{14} \neq 0$ and $A_{24} \neq 0$, we can show that the shear force N_{xy} induces elongations in the x– and y–directions besides the expected shear strain.

In the forthcoming Subsection 7.5.6, we present the significance of a laminate coupling stiffness matrices. We will demonstrate that some coupling terms vanish from the overall responses for certain ply arrangements (e.g., laminates with symmetric, antisymmetric, balanced layups, cross–ply, etc.).

Moment resultants

Similarly to the developments done for expressing the force resultants, we obtain the equilibrium of a plate element in terms of moment resultants per unit width by integrating through the laminate thickness the global stress in each lamina:

$$\boldsymbol{M} = \int_{-h/2}^{h/2} \boldsymbol{\sigma}^{(x-y)} z\mathrm{d}z, \tag{7.76}$$

or, in expanded form,

$$\boldsymbol{M} = \begin{bmatrix} M_x \\ M_y \\ M_{xy} \end{bmatrix} = \int_{-h/2}^{h/2} \begin{bmatrix} \sigma_x \\ \sigma_y \\ \sigma_s \end{bmatrix} z\mathrm{d}z. \tag{7.77}$$

From Fig. 7.18 it becomes clear that:

M_x is the bending moment with respect to the y–axis per unit width,
M_y is the bending moment with respect to the x–axis per unit width,
M_{xy} is twisting (torsion) moment per unit width.

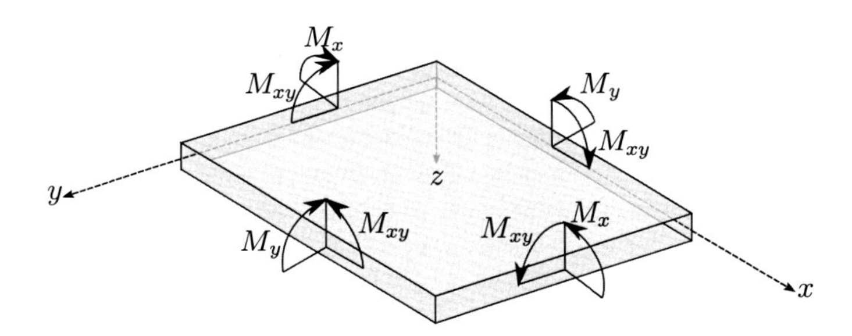

FIGURE 7.18 Schematic illustration of the moment resultants applied to an element of composite laminate plate.

Using Eq. (7.67) yields

$$\boldsymbol{M} = \sum_{k=1}^{n} \int_{z_{k-1}}^{z_k} \widehat{\boldsymbol{Q}}_k \cdot \left[\boldsymbol{\varepsilon}^0 + z\,\boldsymbol{\kappa}\right] z\mathrm{d}z.$$

As in the case of the force resultants, $\boldsymbol{\varepsilon}^0$, $\boldsymbol{\kappa}$, and $\widehat{\boldsymbol{Q}}_k$ can be brought outside the integral but remain within the summation on each lamina:

$$\begin{aligned} \boldsymbol{M} &= \sum_{k=1}^{n} \widehat{\boldsymbol{Q}}_k \cdot \boldsymbol{\varepsilon}^0 \int_{z_{k-1}}^{z_k} z\mathrm{d}z + \sum_{k=1}^{n} \widehat{\boldsymbol{Q}}_k \cdot \boldsymbol{\kappa} \int_{z_{k-1}}^{z_k} z^2\mathrm{d}z \\ &= \underbrace{\frac{1}{2}\left[\sum_{k=1}^{n} \widehat{\boldsymbol{Q}}_k \left[z_k^2 - z_{k-1}^2\right]\right]}_{\boldsymbol{B}} \cdot \boldsymbol{\varepsilon}^0 + \underbrace{\frac{1}{3}\left[\sum_{k=1}^{n} \widehat{\boldsymbol{Q}}_k \left[z_k^3 - z_{k-1}^3\right]\right]}_{\boldsymbol{D}} \cdot \boldsymbol{\kappa}. \end{aligned} \tag{7.78}$$

Then we express the moment resultants vector as a function of the mid–plane strain and curvature:

$$\boldsymbol{M} = \boldsymbol{B} \cdot \boldsymbol{\varepsilon}^0 + \boldsymbol{D} \cdot \boldsymbol{\kappa}, \tag{7.79}$$

where $\boldsymbol{B}$ is the bending–extension coupling stiffness matrix given in $(7.74)_2$, and $\boldsymbol{D}$ is the bending/twisting stiffness matrix whose components have units N · m. Similarly to $\boldsymbol{A}$ and $\boldsymbol{B}$, it is obvious that the matrix $\boldsymbol{D}$ is also symmetric. It reads

$$\boldsymbol{D} = \frac{1}{3} \sum_{k=1}^{n} \widehat{\boldsymbol{Q}}_k \left[z_k^3 - z_{k-1}^3 \right]. \tag{7.80}$$

Similarly to $\boldsymbol{N}$, we can also write Eq. (7.79) as

$$\begin{bmatrix} M_x \\ M_y \\ M_{xy} \end{bmatrix} = \begin{bmatrix} B_{11} & B_{12} & B_{14} \\ B_{12} & B_{22} & B_{24} \\ B_{14} & B_{24} & B_{44} \end{bmatrix} \cdot \begin{bmatrix} \varepsilon_x^0 \\ \varepsilon_y^0 \\ \varepsilon_s^0 \end{bmatrix} + \begin{bmatrix} D_{11} & D_{12} & D_{14} \\ D_{12} & D_{22} & D_{24} \\ D_{14} & D_{24} & D_{44} \end{bmatrix} \cdot \begin{bmatrix} \kappa_x \\ \kappa_y \\ \kappa_{xy} \end{bmatrix}. \tag{7.81}$$

According to the last relation, if the terms D_{14} and D_{24} are not zero, then the response of laminate will exhibit a bend–twist coupling. Eqs. (7.75) and (7.81) express the global laminate force and moment resultants as functions of the mid–plane surface strain and curvature. They can be combined in one system of linear equations:

$$\begin{bmatrix} N_x \\ N_y \\ N_{xy} \\ M_x \\ M_y \\ M_{xy} \end{bmatrix} = \begin{bmatrix} A_{11} & A_{12} & A_{14} & B_{11} & B_{12} & B_{14} \\ A_{12} & A_{22} & A_{24} & B_{12} & B_{22} & B_{24} \\ A_{14} & A_{24} & A_{44} & B_{14} & B_{24} & B_{44} \\ B_{11} & B_{12} & B_{14} & D_{11} & D_{12} & D_{14} \\ B_{12} & B_{22} & B_{24} & D_{12} & D_{22} & D_{24} \\ B_{14} & B_{24} & B_{44} & D_{14} & D_{24} & D_{44} \end{bmatrix} \cdot \begin{bmatrix} \varepsilon_x^0 \\ \varepsilon_y^0 \\ \varepsilon_s^0 \\ \kappa_x \\ \kappa_y \\ \kappa_{xy} \end{bmatrix}, \tag{7.82}$$

or, in compact form,

$$\begin{bmatrix} \boldsymbol{N} \\ \boldsymbol{M} \end{bmatrix} = \begin{bmatrix} \boldsymbol{A} & \boldsymbol{B} \\ \boldsymbol{B} & \boldsymbol{D} \end{bmatrix} \cdot \begin{bmatrix} \boldsymbol{\varepsilon}^0 \\ \boldsymbol{\kappa} \end{bmatrix}. \tag{7.83}$$

In mechanics of composite laminates, generally, the force and moment resultants applied to a plate element are known. It is then required to determine the induced strains in the laminate. To this end, Eq. (7.83) should be inverted to express the strain and curvature of the mid–plane surface:

$$\begin{bmatrix} \boldsymbol{\varepsilon}^0 \\ \boldsymbol{\kappa} \end{bmatrix} = \begin{bmatrix} \boldsymbol{A} & \boldsymbol{B} \\ \boldsymbol{B} & \boldsymbol{D} \end{bmatrix}^{-1} \cdot \begin{bmatrix} \boldsymbol{N} \\ \boldsymbol{M} \end{bmatrix}. \tag{7.84}$$

Nevertheless, even though the inverse of a 6×6 matrix can be easily performed numerically, we can find a formulation [1] analytically expressing the laminate global compliance matrices $\boldsymbol{\alpha}$, $\boldsymbol{\beta}$, and $\boldsymbol{\delta}$ as functions of the global stiffness matrices $\boldsymbol{A}$, $\boldsymbol{B}$, and $\boldsymbol{D}$. Eq. (7.84) is expressed in terms of the global compliance matrices as

$$\begin{bmatrix} \boldsymbol{\varepsilon}^0 \\ \boldsymbol{\kappa} \end{bmatrix} = \begin{bmatrix} \boldsymbol{\alpha} & \boldsymbol{\beta} \\ \boldsymbol{\beta}^T & \boldsymbol{\delta} \end{bmatrix} \cdot \begin{bmatrix} \boldsymbol{N} \\ \boldsymbol{M} \end{bmatrix}, \tag{7.85}$$

where $\boldsymbol{\alpha}$, $\boldsymbol{\beta}$, and $\boldsymbol{\delta}$ are the extensional, extension–bending coupling, and bending compliance matrices of the laminate. They are given by the following relations [1]:

$$\begin{aligned} \boldsymbol{\alpha} &= \left[\boldsymbol{A} - \boldsymbol{B}\cdot\boldsymbol{D}^{-1}\cdot\boldsymbol{B}\right]^{-1} && \text{expressed in } \mathrm{m}\cdot\mathrm{N}^{-1}, \\ \boldsymbol{\delta} &= \left[\boldsymbol{D} - \boldsymbol{B}\cdot\boldsymbol{A}^{-1}\cdot\boldsymbol{B}\right]^{-1} && \text{expressed in } \mathrm{m}^{-1}\cdot\mathrm{N}^{-1}, \\ \boldsymbol{\beta} &= -\boldsymbol{A}^{-1}\cdot\boldsymbol{B}\cdot\boldsymbol{\delta} && \text{expressed in } \mathrm{N}^{-1}. \end{aligned} \tag{7.86}$$

Like the combined Eq. (7.82), we can write a combined equation expressing the laminate in–plane strains and curvatures as functions of the applied force and moment resultants:

$$\begin{bmatrix} \varepsilon_x^0 \\ \varepsilon_y^0 \\ \varepsilon_s^0 \\ \kappa_x \\ \kappa_y \\ \kappa_{xy} \end{bmatrix} = \begin{bmatrix} \alpha_{11} & \alpha_{12} & \alpha_{14} & \beta_{11} & \beta_{12} & \beta_{14} \\ \alpha_{12} & \alpha_{22} & \alpha_{24} & \beta_{21} & \beta_{22} & \beta_{24} \\ \alpha_{14} & \alpha_{24} & \alpha_{44} & \beta_{41} & \beta_{42} & \beta_{44} \\ \beta_{11} & \beta_{21} & \beta_{41} & \delta_{11} & \delta_{12} & \delta_{14} \\ \beta_{12} & \beta_{22} & \beta_{42} & \delta_{12} & \delta_{22} & \delta_{24} \\ \beta_{14} & \beta_{24} & \beta_{44} & \delta_{14} & \delta_{24} & \delta_{44} \end{bmatrix} \cdot \begin{bmatrix} N_x \\ N_y \\ N_{xy} \\ M_x \\ M_y \\ M_{xy} \end{bmatrix}. \tag{7.87}$$

Since the combined 6×6 matrix of Eq. (7.82) is symmetric, its inverse matrix relating the strains to the applied loads given in Eq. (7.85), or (7.87), is also symmetric. However, the matrix $\boldsymbol{\beta}$ is not necessarily symmetric, whereas the matrices $\boldsymbol{\alpha}$ and $\boldsymbol{\delta}$ are necessarily symmetric and positive definite.

7.5.6 Physical meaning of some coupling components of the laminates stiffness matrices

From Eq. (7.82) let us consider some particular cases to illustrate the physical significance of certain coupling terms in an anisotropic laminate and to highlight their effects on the laminate global response. It should be pointed out that only *two loading configurations* are considered for all the following studied cases and the related layups:

1. laminate submitted to axial force N_x only;
2. laminate submitted to a bending moment M_x only.

- **Case 1: no coupling** ($A_{14} = 0$, $D_{14} = 0$, and $B_{ij} = 0$).
  For this case, the laminate behaves like an orthotropic medium. Indeed, when applying a force per unit width N_x, the strain response will imply the expected extensional strains ε_x^0 and ε_y^0 (Fig. 7.19$_a$). In addition, under an applied moment per unit width M_x, only a bending curvature will occur as shown in Fig. 7.19$_b$. Thus

$$N_x = A_{11}\varepsilon_x^0 + A_{12}\varepsilon_y^0,$$
$$M_x = D_{11}\kappa_x + D_{12}\kappa_y.$$

This uncoupled response can be typically obtained for a laminate with *symmetric cross–ply* layup consisting of plies with 0° and 90°, for example, the four plies laminate [0/90/90/0] = $[0/90]_S$. It should be mentioned that, in general, for a symmetric laminate, we can easily demonstrate that the components of the coupling matrix $\boldsymbol{B}$ vanish (see Eq. $(7.74)_2$).

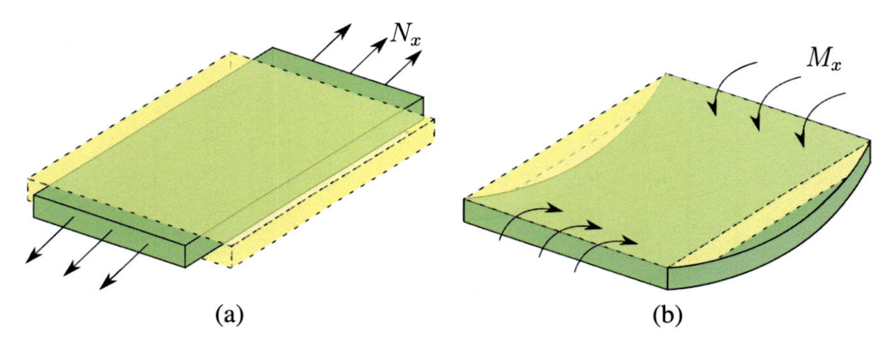

FIGURE 7.19 Classical response of an orthotropic laminate without coupling (case 1) exhibiting axial (a) stretching induced by N_x and (b) bending curvature caused by M_x.

- **Case 2: Extension–bending coupling and bending–stretching coupling** ($A_{14} = 0$ and $D_{14} = 0$, whereas $B_{11} \neq 0$).
  This type of response happens for an *antisymmetric cross–ply* laminate, for instance, the four plies laminate: [90/0/90/0]. Besides the response expected for an orthotropic laminate (case 1), the fact that $B_{11} \neq 0$ will generate a bending curvature if loaded by an axial force N_x. If this laminate is bent by a moment M_x, then an extensional strain ε_x^0 appears besides the bending curvatures κ_x, κ_y, as shown in Fig. 7.20. Thus

$$N_x = A_{11}\varepsilon_x^0 + A_{12}\varepsilon_y^0 + B_{11}\kappa_x,$$
$$M_x = B_{11}\varepsilon_x^0 + D_{11}\kappa_x + D_{12}\kappa_y.$$

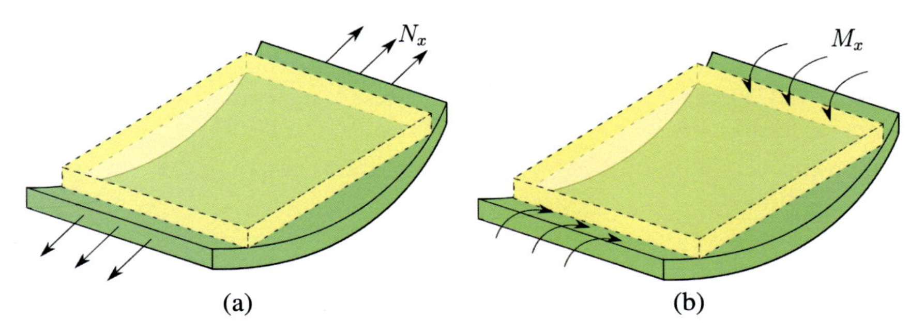

FIGURE 7.20 Coupled response of a laminate (case 2) exhibiting (a) elongation and bending induced by N_x and (b) bending curvatures and elongation caused by M_x.

- **Case 3: Extension–shear coupling or bending–twist coupling** ($A_{14} \neq 0$, $D_{14} \neq 0$, but $B_{ij} = 0$).
  Such a coupled response is obtained for *unbalanced symmetric* laminates with positive angles, for instance, $[+\vartheta_1/+\vartheta_2/+\vartheta_3,...]_S$. When loaded with N_x, the global response will imply the expected extensional strains ε_x^0 and ε_y^0 but also an in–plane shearing strain ε_s^0. A bending moment M_x will cause a twisting κ_{xy} besides the expected curvatures κ_x and κ_y. It can thus be written as

$$N_x = A_{11}\varepsilon_x^0 + A_{12}\varepsilon_y^0 + A_{14}\varepsilon_s^0,$$
$$M_x = D_{11}\kappa_x + D_{12}\kappa_y + D_{14}\kappa_{xy}.$$

  Fig. 7.21$_a$ exhibits the axial–shear coupling induced by N_x, and Fig. 7.21$_b$ shows the bend–twist coupling if M_x is applied to such a laminate.

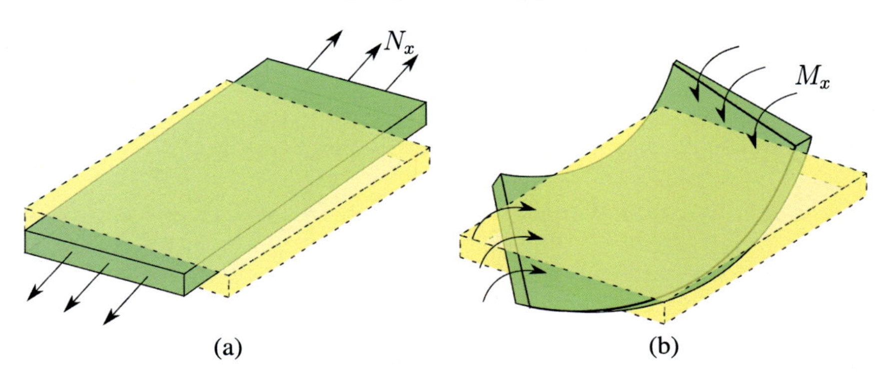

FIGURE 7.21 Coupled response of a laminate (case 3) exhibiting (a) stretching and shearing induced by N_x and (b) a bending and twisting curvature caused by M_x.

- **Case 4: Extension–twist coupling or bending–shear coupling** ($A_{14} = 0$, $D_{14} = 0$ but $B_{14} \neq 0$).

This type of response can be obtained for *balanced antisymmetric about the mid–plane* laminates, for instance, the four plies laminate: $[+\vartheta_1/-\vartheta_2/-\vartheta_1/+\vartheta_2]$. Using Eqs. (7.45) expressing the transformed reduced stiffness components, we easily notice that only the stiffness components $\widehat{Q}_{xs}$ and $\widehat{Q}_{ys}$ with odd powers in m or n are sensitive to the sign of the orientation angle:

$$\widehat{Q}_{xs}(-\vartheta) = -\widehat{Q}_{xs}(+\vartheta), \widehat{Q}_{ys}(-\vartheta) = -\widehat{Q}_{ys}(+\vartheta).$$

The other components are insensitive to the sign of ϑ.
An axial force N_x will bring about extensional strains ε_x^0 and ε_y^0 and a twisting curvature κ_{xy}. A bending moment M_x will result in a in–plane shearing strain ε_s^0 besides the curvatures κ_x and κ_y. The global responses are shown in Fig. 7.22. Thus we can write:

$$N_x = A_{11}\varepsilon_x^0 + A_{12}\varepsilon_y^0 + B_{14}\kappa_{xy},$$
$$M_x = B_{14}\varepsilon_s^0 + D_{11}\kappa_x + D_{12}\kappa_y.$$

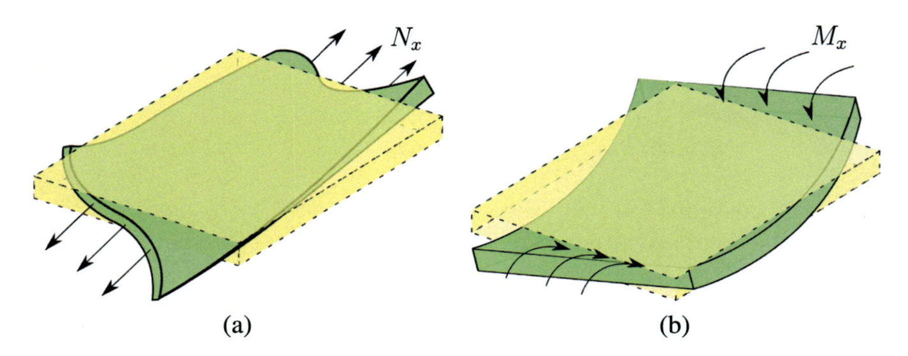

FIGURE 7.22 Coupled response of a laminate (case 4) exhibiting (a) elongation and twisting induced by N_x and (b) a bending curvature and shearing caused by M_x.

- **Case 5: Fully coupled anisotropic laminate** ($A_{ij} \neq 0$, $B_{ij} \neq 0$ and $D_{ij} \neq 0$).
  This type of response occurs in a general case laminate when no symmetries nor balance can be found in the layup. Such an anisotropic laminate consists in a layup of arbitrary angle laminae, for instance, the six–ply laminate [0/45/90/60/0/45]. An axial force N_x will bring about all the in–plane strains including shearing and curvatures, as well as twisting κ_{xy}. The bending moment M_x will also result in a similar behavior. The global responses are shown in Fig. 7.23. We can write:

$$N_x = A_{11}\varepsilon_x^0 + A_{12}\varepsilon_y^0 + A_{14}\varepsilon_s^0 + B_{11}\kappa_x + B_{12}\kappa_y + B_{14}\kappa_{xy},$$

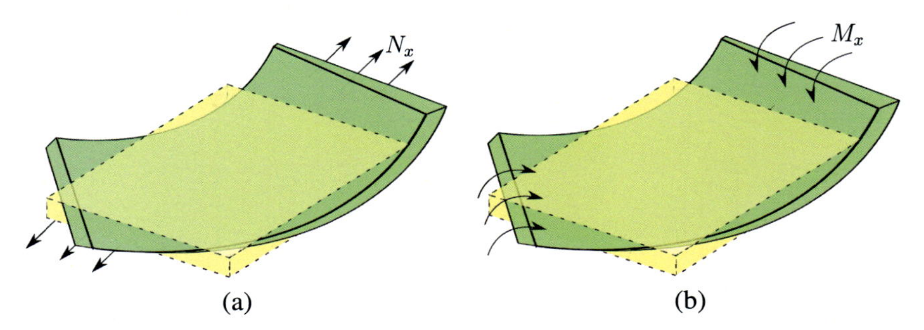

FIGURE 7.23 Coupled response of a laminate (case 5) exhibiting (a) elongation, shearing, bending, and twisting induced by N_x and (b) a bending curvature, twisting, elongation, and shearing caused by M_x.

$$M_x = B_{11}\varepsilon_x^0 + B_{12}\varepsilon_y^0 + B_{14}\varepsilon_s^0 + D_{11}\kappa_x + D_{12}\kappa_y + D_{14}\kappa_{xy}.$$

The previous five cases lead to the obvious conclusion that the response of a laminate depends not only on the material behavior of the single layers, but also on the stacking sequence (symmetric, antisymmetric, cross–ply, unsymmetric sequences) [6].

Fig. 7.24 schematically summarizes the previously analyzed cases and the related effects of the coupling terms on the deformation responses of the laminate. It illustrates also the classical effect of the axial A_{11} and the flexural D_{11} components of the global laminate stiffness matrices.

7.5.7 Workflow and summary

The composite laminate analysis may look complex and tedious since it requires fastidious matrix calculations, notably for laminate with high number of plies. Indeed, the stresses and strains should be determined for each ply, eventually even at its bottom and top, and then rotated from local to global axes for each ply toward applying the lamina failure criteria. Generally, the calculation procedure applied for a laminate plate relies on a systematic approach, which can be summarized in the following workflow and easily written as a software code [8]:

BOX 7.1 Workflow for laminate composite

- Knowing the applied loads and the dimensions of the laminate plate, the force $\boldsymbol{N}$ and moment $\boldsymbol{M}$ resultants are determined.
- From the material properties of each lamina E_1, E_2, G_{12}, and ν_{12}, the reduced local stiffness matrix $\boldsymbol{Q}$ is calculated for each ply using Eqs. (7.30).

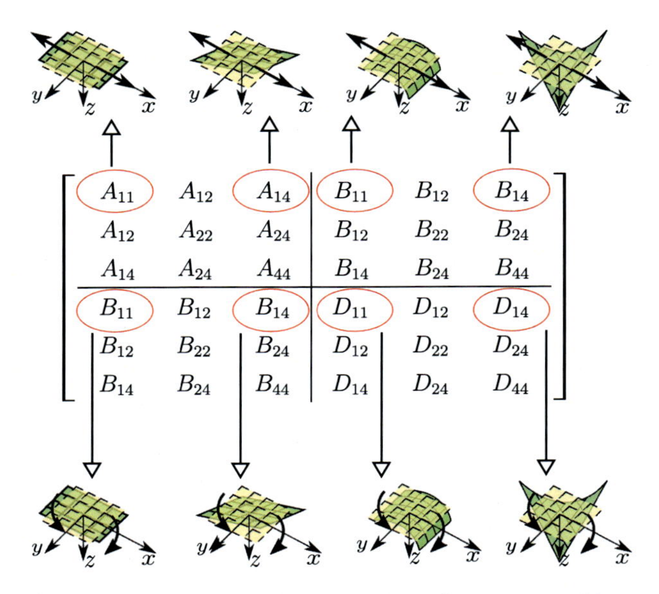

FIGURE 7.24 Schematic illustration of main coupling effect in a general laminate composite.

- Using Eq. (7.42) and having the orientation angle of each ply, the reduced stiffness matrices $\widehat{\boldsymbol{Q}}$ are determined by rotating the local stiffness matrices from local to global axes.
- Determine the through–thickness locations z_k and z_{k-1} for each ply k.
- Then the global stiffness laminate matrices $\boldsymbol{A}$, $\boldsymbol{B}$, and $\boldsymbol{D}$ are calculated using Eqs. (7.74) and (7.80).
- Determine the global compliance matrices $\boldsymbol{\alpha}$, $\boldsymbol{\beta}$, and $\boldsymbol{\delta}$ by inverting numerically the combined matrix of system (7.82) or by using Eqs. (7.86).
- Then the in–plane strains $\boldsymbol{\varepsilon}^0$ and curvatures $\boldsymbol{\kappa}$ of laminate middle surface are computed using Eq. (7.87).
- The global strain can hence be calculated as a linear variation with respect to the coordinate z using Eq. (7.64).
- If required, the global stresses can be eventually determined for each ply using Eq. (7.67) or (7.68).
- The local strains $\boldsymbol{\varepsilon}^{(1-2)}$ are determined for each ply by transforming the global strains using the rotation matrix $\boldsymbol{T}_{\sigma}^{T}$ (inversion of Eq. (7.37)). If the curvatures are not zero, then the local strains have to be calculated at the coordinates z_{k-1} and z_k corresponding the top and bottom, respectively, of a ply k.
- Determine the local stresses $\boldsymbol{\sigma}^{(1-2)}$ in each ply (each position) using the stress–strain relation (7.28).

- Apply a chosen failure criterion (Tsai–Hill, Tsai–Wu) [4,7] for each lamina toward its design in terms of strength.

7.5.8 Example

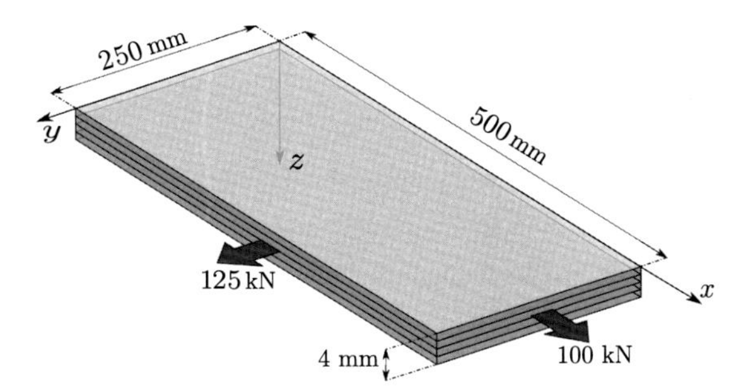

FIGURE 7.25 Rectangular laminate composite plate submitted to biaxial loading condition.

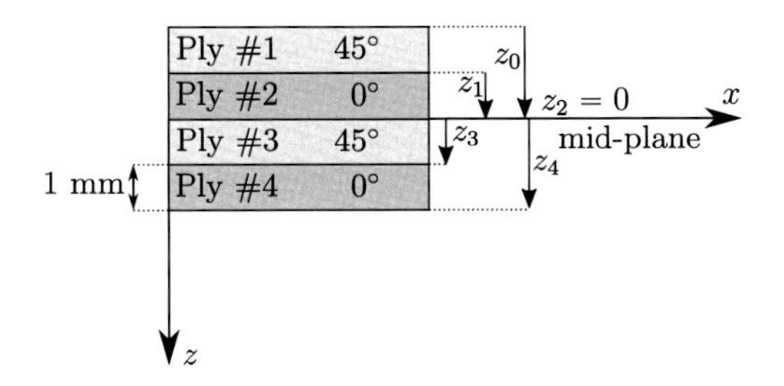

FIGURE 7.26 Antisymmetric four–ply laminate [45/0/45/0].

Let us consider a rectangular laminate composite plate submitted to biaxial loading conditions. The plate has the dimensions given in Fig. 7.25; it is made of a balanced antisymmetrical layup of glass/epoxy unidirectional laminae [45/0/45/0] (Fig. 7.26). The engineering constants of the unidirectional (orthotropic) lamina are:

$$E_1 = 45 \text{ GPa}, \quad E_2 = 9 \text{ GPa}, \quad G_{12} = 5 \text{ GPa}, \text{ and } \nu_{12} = 0.25.$$

- Determine the global strain $\boldsymbol{\varepsilon}^{(x-y)}$ in the laminate and plot the distribution through the thickness of its components ε_x, ε_y, and ε_s.
- Determine the local strain $\boldsymbol{\varepsilon}^{(1-2)}$ and stress $\boldsymbol{\sigma}^{(1-2)}$ in each ply and plot the local strain components for each ply ε_1, ε_2, and ε_4.

Solution:

- Knowing the applied biaxial loads and the dimensions of the laminate plate, we determine the force $\boldsymbol{N}$ and moment $\boldsymbol{M}$ resultants as follows:

$$N_x = \int_{-2}^{2} \left[\frac{100 \cdot 10^3}{4 \cdot 250} \right] dz = 400 \ \text{N} \cdot \text{mm}^{-1},$$
$$N_y = \int_{-2}^{2} \left[\frac{125 \cdot 10^3}{4 \cdot 500} \right] dz = 250 \ \text{N} \cdot \text{mm}^{-1},$$
$$N_{xy} = 0, \quad \text{and}$$
$$M_x = 0, \quad M_y = 0, \quad M_{xy} = 0.$$

- Calculation of the reduced local stiffness matrix $\boldsymbol{Q}$ for the 0° lamina using Eqs. (7.30). We obtain

$$\boldsymbol{Q} = \begin{bmatrix} 45569.62 & 2278.48 & 0 \\ 2278.48 & 9113.92 & 0 \\ 0 & 0 & 5000 \end{bmatrix} \text{MPa}.$$

- Calculation of the reduced global stiffness matrix $\widehat{\boldsymbol{Q}}$ of each oriented angle lamina in the global axes $(x - y)$:

$$\text{Ply \#2 and Ply \#4}: \quad \widehat{\boldsymbol{Q}}(0^\circ) = \boldsymbol{Q} = \begin{bmatrix} 45569.62 & 2278.48 & 0 \\ 2278.48 & 9113.92 & 0 \\ 0 & 0 & 5000 \end{bmatrix} \text{MPa},$$

$$\text{Ply \#1 and Ply \#3}: \quad \widehat{\boldsymbol{Q}}(45^\circ) = \begin{bmatrix} 19810.13 & 9810.13 & 9113.92 \\ 9810.13 & 19810.13 & 9113.92 \\ 9113.92 & 9113.92 & 12531.65 \end{bmatrix} \text{MPa}.$$

- Through–thickness locations z_{k-1} and z_k of each ply k are:

$$z_0 = -2 \text{ mm}, \quad z_1 = -1 \text{ mm}, \quad z_2 = 0 \text{ mm}, \quad z_3 = 1 \text{ mm}, \quad \text{and} \quad z_4 = 2 \text{ mm}.$$

- Calculation of the global stiffness laminate matrices $\boldsymbol{A}$, $\boldsymbol{B}$, and $\boldsymbol{D}$ using Eqs. (7.74) and (7.80):

$$\boldsymbol{A} = \begin{bmatrix} 130759.5 & 24177.2 & 18227.8 \\ 24177.2 & 57848.1 & 18227.8 \\ 18227.8 & 18227.8 & 35063.3 \end{bmatrix} \text{MPa} \cdot \text{mm or N} \cdot \text{mm}^{-1},$$

$$\boldsymbol{B} = \begin{bmatrix} 25759.5 & -7531.6 & -9113.9 \\ -7531.6 & -10696.2 & -9113.9 \\ -9113.9 & -9113.9 & -7531.6 \end{bmatrix} \text{MPa} \cdot \text{mm}^2 \text{ or N},$$

$$\boldsymbol{D} = \begin{bmatrix} 174346.0 & 32236.3 & 24303.8 \\ 32236.3 & 77130.8 & 24303.8 \\ 24303.8 & 24303.8 & 46751.1 \end{bmatrix} \text{ MPa} \cdot \text{mm}^3 \text{ or N} \cdot \text{mm}.$$

- Calculation of the laminate global compliance matrices $\boldsymbol{\alpha}$, $\boldsymbol{\beta}$, and $\boldsymbol{\delta}$ by inverting numerically the combined matrix of system (7.82) or by using Eqs. (7.86):

$$\boldsymbol{\alpha} = \begin{bmatrix} 9.13 & -2.67 & -3.42 \\ -2.67 & 21.79 & -9.24 \\ -3.42 & -9.24 & 36.27 \end{bmatrix} \cdot 10^{-6} \, [\text{MPa} \cdot \text{mm}]^{-1} \text{ or mm} \cdot \text{N}^{-1},$$

$$\boldsymbol{\beta} = \begin{bmatrix} & -1.93 & 0.46 & 1.47 \\ c & 0.46 & 1.01 & 1.47 \\ & 1.47 & 1.47 & 1.84 \end{bmatrix} \cdot 10^{-6} \, [\text{MPa} \cdot \text{mm}^2]^{-1} \text{ or N}^{-1},$$

$$\boldsymbol{\delta} = \begin{bmatrix} 6.85 & -2.00 & -2.57 \\ -2.00 & 16.35 & -6.93 \\ -2.57 & -6.93 & 27.20 \end{bmatrix} \cdot 10^{-6} \, [\text{MPa} \cdot \text{mm}^3]^{-1} \text{ or mm}^{-1} \cdot \text{N}^{-1}.$$

- Determination of the in–plane strain $\boldsymbol{\varepsilon}^0$ and curvature $\boldsymbol{\kappa}$ of the laminate middle surface using Eq. (7.87).
  However, since the laminate plate is not submitted to any moment, $\boldsymbol{M} = \mathbf{0}$. From Eq. (7.85) we can write

$$\begin{bmatrix} \boldsymbol{\varepsilon}^0 \\ \boldsymbol{\kappa} \end{bmatrix} = \begin{bmatrix} \boldsymbol{\alpha} & \boldsymbol{\beta} \\ \boldsymbol{\beta}^T & \boldsymbol{\delta} \end{bmatrix} \cdot \begin{bmatrix} \boldsymbol{N} \\ \boldsymbol{M} = \mathbf{0} \end{bmatrix},$$

which in this case leads to

$$\boldsymbol{\varepsilon}^0 = \boldsymbol{\alpha} \cdot \boldsymbol{N},$$
$$\boldsymbol{\kappa} = \boldsymbol{\beta}^T \cdot \boldsymbol{N}.$$

Thus

$$\begin{bmatrix} \varepsilon_x^0 \\ \varepsilon_y^0 \\ \varepsilon_s^0 \end{bmatrix} = \begin{bmatrix} 9.13 & -2.67 & -3.42 \\ -2.67 & 21.79 & -9.24 \\ -3.42 & -9.24 & 36.27 \end{bmatrix} \cdot 10^{-6} \cdot \begin{bmatrix} N_x = 400 \\ N_y = 250 \\ 0 \end{bmatrix},$$

$$\begin{bmatrix} \kappa_x \\ \kappa_y \\ \kappa_{xy} \end{bmatrix} = \begin{bmatrix} -1.93 & 0.46 & 1.47 \\ 0.46 & 1.01 & 1.47 \\ 1.47 & 1.47 & 1.84 \end{bmatrix} \cdot 10^{-6} \cdot \begin{bmatrix} N_x = 400 \\ N_y = 250 \\ 0 \end{bmatrix},$$

or

$$\begin{bmatrix} \varepsilon_x^0 \\ \varepsilon_y^0 \\ \varepsilon_s^0 \end{bmatrix} = \begin{bmatrix} 2.98 \\ 4.38 \\ -3.68 \end{bmatrix} \cdot 10^{-3},$$

$$\begin{bmatrix} \kappa_x \\ \kappa_y \\ \kappa_{xy} \end{bmatrix} = \begin{bmatrix} -6.58 \\ 4.37 \\ 9.58 \end{bmatrix} \cdot 10^{-4} \text{ mm}^{-1}.$$

- We directly deduce the global strain using Eq. (7.64):

$$\boldsymbol{\varepsilon}^{(x-y)} = \boldsymbol{\varepsilon}^0 + z\,\boldsymbol{\kappa}.$$

Given the linear variation with respect to the coordinate z, we can determine only two values for each ply, namely at the top and at the bottom of each ply. For instance, for the ply #1 at its top, $z = -2$,

$$\begin{bmatrix} \varepsilon_x \\ \varepsilon_y \\ \varepsilon_s \end{bmatrix}_{\text{top, ply \#1}} = \begin{bmatrix} 2.98 \\ 4.38 \\ -3.68 \end{bmatrix} \cdot 10^{-3} - 2 \begin{bmatrix} -6.58 \\ 4.37 \\ 9.58 \end{bmatrix} \cdot 10^{-4}$$

$$= \begin{bmatrix} 4.30 \\ 3.51 \\ -5.60 \end{bmatrix} \cdot 10^{-3}.$$

At the bottom of the ply #1 (or the top of ply #2), $z = -1$,

$$\begin{bmatrix} \varepsilon_x \\ \varepsilon_y \\ \varepsilon_s \end{bmatrix}_{\text{bottom, ply \#1}} = \begin{bmatrix} 2.98 \\ 4.38 \\ -3.68 \end{bmatrix} \cdot 10^{-3} - \begin{bmatrix} -6.58 \\ 4.37 \\ 9.58 \end{bmatrix} \cdot 10^{-4},$$

$$\begin{bmatrix} \varepsilon_x \\ \varepsilon_y \\ \varepsilon_s \end{bmatrix}_{\text{bottom, ply \#1}} = \begin{bmatrix} \varepsilon_x \\ \varepsilon_y \\ \varepsilon_s \end{bmatrix}_{\text{top, ply \#2}} = \begin{bmatrix} 3.64 \\ 3.94 \\ -4.64 \end{bmatrix} \cdot 10^{-3}.$$

Similarly, for the other plies, we obtain:
for $z = 0$,

$$\begin{bmatrix} \varepsilon_x \\ \varepsilon_y \\ \varepsilon_s \end{bmatrix}_{\text{bottom, ply \#2}} = \begin{bmatrix} \varepsilon_x \\ \varepsilon_y \\ \varepsilon_s \end{bmatrix}_{\text{top, ply \#3}} = \begin{bmatrix} 2.98 \\ 4.38 \\ -3.68 \end{bmatrix} \cdot 10^{-3};$$

for $z = +1$,

$$\begin{bmatrix} \varepsilon_x \\ \varepsilon_y \\ \varepsilon_s \end{bmatrix}_{\text{bottom, ply \#3}} = \begin{bmatrix} \varepsilon_x \\ \varepsilon_y \\ \varepsilon_s \end{bmatrix}_{\text{top, ply \#4}} = \begin{bmatrix} 2.33 \\ 4.82 \\ -2.72 \end{bmatrix} \cdot 10^{-3};$$

for $z = +2$,

$$\begin{bmatrix} \varepsilon_x \\ \varepsilon_y \\ \varepsilon_s \end{bmatrix}_{\text{bottom, ply \#4}} = \begin{bmatrix} 1.67 \\ 5.25 \\ -1.76 \end{bmatrix} \cdot 10^{-3}.$$

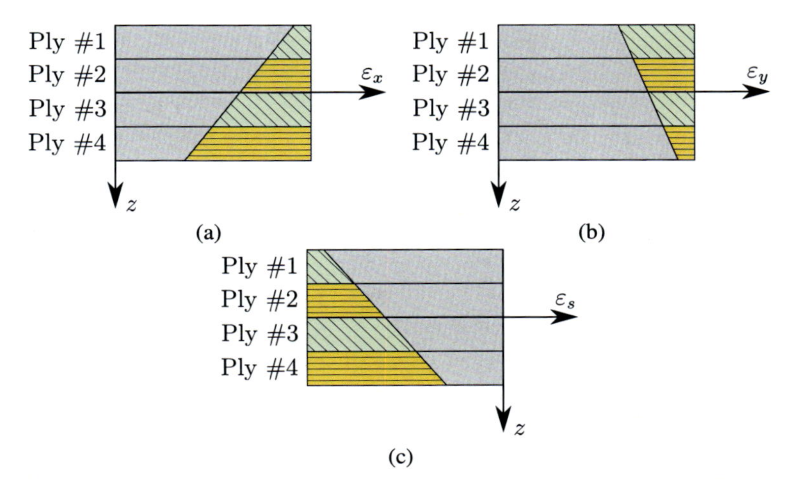

FIGURE 7.27 Distribution through the thickness of (a) ε_x, (b) ε_y, and (c) ε_s.

Fig. 7.27 shows the distributions through the thickness of the global strain components.

- The local strains $\boldsymbol{\varepsilon}^{(1-2)}$ are determined for each ply by transforming the global strains calculated above using the rotation matrix $\boldsymbol{T}_\sigma^T(\vartheta)$ with respect to each lamina angle:

$$\begin{bmatrix} \varepsilon_1 \\ \varepsilon_2 \\ \varepsilon_4 \end{bmatrix}_{\text{top, ply \#1}} = \begin{bmatrix} 1.11 \\ 6.70 \\ -0.80 \end{bmatrix} \cdot 10^{-3},$$

$$\begin{bmatrix} \varepsilon_1 \\ \varepsilon_2 \\ \varepsilon_4 \end{bmatrix}_{\text{bottom, ply \#1}} = \begin{bmatrix} 1.47 \\ 6.11 \\ 0.30 \end{bmatrix} \cdot 10^{-3},$$

$$\begin{bmatrix} \varepsilon_1 \\ \varepsilon_2 \\ \varepsilon_4 \end{bmatrix}_{\text{top, ply \#2}} = \begin{bmatrix} 3.64 \\ 3.94 \\ -4.64 \end{bmatrix} \cdot 10^{-3},$$

$$\begin{bmatrix} \varepsilon_1 \\ \varepsilon_2 \\ \varepsilon_4 \end{bmatrix}_{\text{bottom, ply \#2}} = \begin{bmatrix} 2.98 \\ 4.38 \\ -3.68 \end{bmatrix} \cdot 10^{-3},$$

$$\begin{bmatrix} \varepsilon_1 \\ \varepsilon_2 \\ \varepsilon_4 \end{bmatrix}_{\text{top, ply \#3}} = \begin{bmatrix} 1.84 \\ 5.52 \\ 1.40 \end{bmatrix} \cdot 10^{-3},$$

$$\begin{bmatrix} \varepsilon_1 \\ \varepsilon_2 \\ \varepsilon_4 \end{bmatrix}_{\text{bottom, ply \#3}} = \begin{bmatrix} 2.21 \\ 4.93 \\ 2.49 \end{bmatrix} \cdot 10^{-3},$$

$$\begin{bmatrix} \varepsilon_1 \\ \varepsilon_2 \\ \varepsilon_4 \end{bmatrix}_{\text{top, ply \#4}} = \begin{bmatrix} 2.33 \\ 4.82 \\ -2.72 \end{bmatrix} \cdot 10^{-3},$$

$$\begin{bmatrix} \varepsilon_1 \\ \varepsilon_2 \\ \varepsilon_4 \end{bmatrix}_{\text{bottom, ply \#4}} = \begin{bmatrix} 1.67 \\ 5.25 \\ -1.76 \end{bmatrix} \cdot 10^{-3}.$$

Fig. 7.28 shows the distributions through the thickness of the local strain components.

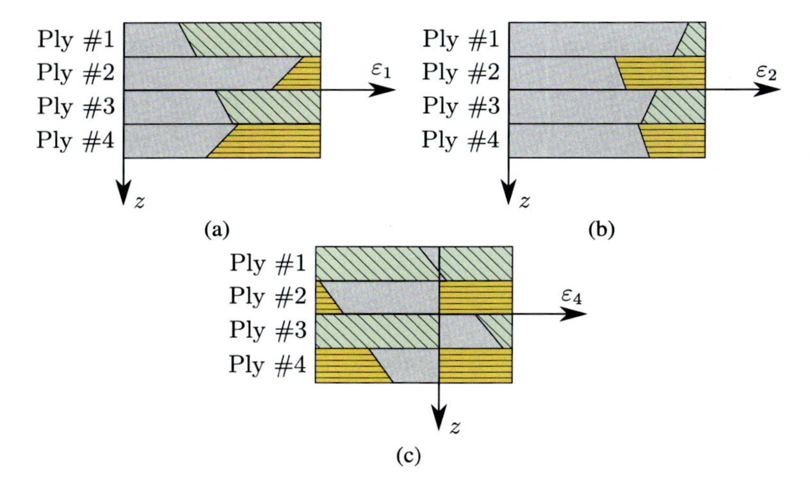

FIGURE 7.28 Distribution through the thickness of (a) ε_1, (b) ε_2 and (c) ε_4.

- The local stresses $\boldsymbol{\sigma}^{(1-2)}$ in each ply (each position) can now be determined using the stress–strain relation (7.28):

$$\begin{bmatrix} \sigma_1 \\ \sigma_2 \\ \sigma_4 \end{bmatrix}_{\text{top, ply \#1}} = \begin{bmatrix} 65.64 \\ 63.59 \\ -3.98 \end{bmatrix} \text{MPa},$$

$$\begin{bmatrix} \sigma_1 \\ \sigma_2 \\ \sigma_4 \end{bmatrix}_{\text{bottom, ply \#1}} = \begin{bmatrix} 81.08 \\ 59.06 \\ 1.50 \end{bmatrix} \text{MPa},$$

$$\begin{bmatrix} \sigma_1 \\ \sigma_2 \\ \sigma_4 \end{bmatrix}_{\text{top, ply \#2}} = \begin{bmatrix} 174.95 \\ 44.23 \\ -23.19 \end{bmatrix} \text{MPa},$$

$$\begin{bmatrix} \sigma_1 \\ \sigma_2 \\ \sigma_4 \end{bmatrix}_{\text{bottom, ply \#2}} = \begin{bmatrix} 145.95 \\ 46.72 \\ -18.40 \end{bmatrix} \text{MPa},$$

$$\begin{bmatrix} \sigma_1 \\ \sigma_2 \\ \sigma_4 \end{bmatrix}_{\text{top, ply \#3}} = \begin{bmatrix} 96.52 \\ 54.52 \\ 6.98 \end{bmatrix} \text{MPa},$$

$$\begin{bmatrix} \sigma_1 \\ \sigma_2 \\ \sigma_4 \end{bmatrix}_{\text{bottom, ply \#3}} = \begin{bmatrix} 111.96 \\ 49.99 \\ 12.46 \end{bmatrix} \text{MPa},$$

$$\begin{bmatrix} \sigma_1 \\ \sigma_2 \\ \sigma_4 \end{bmatrix}_{\text{top, ply \#4}} = \begin{bmatrix} 116.95 \\ 49.21 \\ -13.61 \end{bmatrix} \text{MPa},$$

$$\begin{bmatrix} \sigma_1 \\ \sigma_2 \\ \sigma_4 \end{bmatrix}_{\text{bottom, ply \#4}} = \begin{bmatrix} 87.94 \\ 51.69 \\ -8.82 \end{bmatrix} \text{MPa}.$$

- The local strains and stresses determined above can now be plugged into a chosen lamina failure criterion [4,7] to determine which lamina may fail first (first–ply failure) under the applied biaxial loading. Subsequently, the failed lamina should be withdrawn from the laminate, and the procedure can be repeated until the last–ply failure.

References

[1] R.M. Jones, Mechanics of Composite Materials, second ed., Taylor and Francis, New York, 1999.
[2] S.W. Tsai, N.J. Pagano, Invariant properties of composite materials, Tech. Rep. AD668761, Air Force Materials Laboratory, Ohio, 1968.
[3] S.G. Lekhnitskii, Theory of Elasticity of an Anisotropic Elastic Body, Holden–Day Inc., San Francisco, 1963.
[4] S.W. Tsai, H.T. Hahn, Introduction to Composite Materials, Technomic, Lancaster, PA, 1980.
[5] A.K. Kaw, Mechanics of Composite Materials, CRC Press, Boca Raton, FL, 2006.
[6] N.J. Pagano, R.B. Pipes, The influence of stacking sequence on laminate strength, Journal of Composite Materials 5 (1971) 50–57.
[7] S.W. Tsai, Strength theories of filamentary structures, in: R.T. Schwartz, H.S. Schwartz (Eds.), Fundamental Aspects of Fiber Reinforced Plastic Composites, Wiley–Interscience, New York, 1968.
[8] A. Hauffe, eLamX^2, Source: https://tu-dresden.de/ing/maschinenwesen/ilr/lft/elamx2/elamx, 2020.

PART III

Special topics in homogenization

CHAPTER

8

Composite sphere/cylinder assemblage

OUTLINE

8.1 Composite sphere assemblage

The composite sphere assemblage method (CSA) [1,2] is a homogenization technique that provides effective properties of composites with matrix phase and spherical particles (Fig. 8.1). In this method the RVE consists of two concentric spheres: the inner sphere with radius r_1 represents the particle, whereas the outer sphere with radius r_0 represents the matrix phase (Fig. 8.2_a).

It is well understood that a particulate composite with spherical particles behaves as an isotropic material. Having this in mind, the aim of CSA is to provide two independent effective parameters of the composite material, the bulk modulus $\bar{K}$ and the shear modulus $\bar{\mu}$.

To achieve this goal, Hashin [1] has identified the analytical solutions (in spherical coordinates) for two independent boundary value problems. In the first problem, the composite sphere is subjected to hydrostatic strain

Multiscale Modeling Approaches for Composites
https://doi.org/10.1016/B978-0-12-823143-2.00019-9

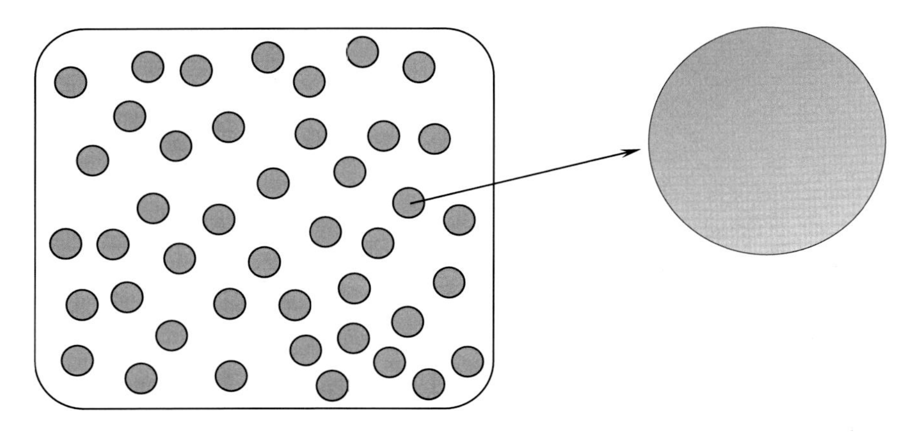

FIGURE 8.1 Cross–section of a particulate composite with spherical particles.

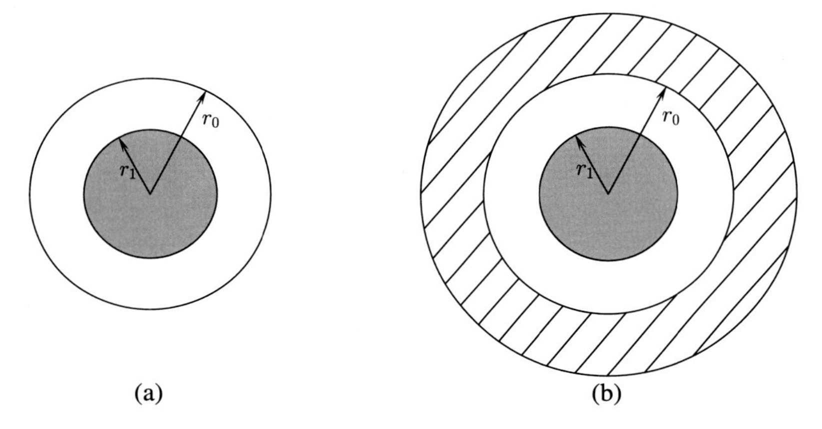

FIGURE 8.2 Composite sphere assemblage method: the considered RVE for (a) the macroscopic bulk modulus and (b) the macroscopic shear modulus.

state, whereas in the second problem the composite sphere is subjected to homogeneous shear displacement. These problems, when solved for the heterogeneous material, provide the strain energy W_c of the composite RVE. Then we can also solve these problems by substituting the phases with hypothetical homogeneous medium and compute the strain energy W_h as a function of the unknown effective properties. The Hill–Mandel principle $W_c = W_h$ allows us to calculate the unknown properties.

The defect of the original CSA method is that whereas the first problem calculates accurately the effective bulk modulus, the second problem provides only bounds for the effective shear modulus. To resolve this issue, Christensen and Lo [2] propose to solve the second boundary value problem using the generalized self–consistent method. In that case the RVE is modified to a problem of three concentric spheres, where the external layer is the unknown effective medium (Fig. 8.2_b). Then, using appropriate energy equivalence principles, we obtain an accurate estimation of the effective shear modulus.

Expressing the problem in spherical coordinates

The spherical form of the inhomogeneities allows us to transform the RVE problem in a spherical coordinate system. In spherical coordinates the axes (x, y, z) are transformed to (r, ϑ, φ). Moreover, at each phase, the strain tensor components are given by the expressions

$$\begin{aligned}
\varepsilon_{rr}^{(q)} &= \frac{\partial u_r^{(q)}}{\partial r}, \\
\varepsilon_{\vartheta\vartheta}^{(q)} &= \frac{1}{r}\frac{\partial u_\vartheta^{(q)}}{\partial \vartheta} + \frac{u_r^{(q)}}{r}, \\
\varepsilon_{\varphi\varphi}^{(q)} &= \frac{1}{r\sin\vartheta}\frac{\partial u_\varphi^{(q)}}{\partial \varphi} + \frac{u_r^{(q)}}{r} + \frac{u_\vartheta^{(q)}\cos\vartheta}{r\sin\vartheta}, \\
2\varepsilon_{r\vartheta}^{(q)} &= \frac{\partial u_\vartheta^{(q)}}{\partial r} + \frac{1}{r}\frac{\partial u_r^{(q)}}{\partial \vartheta} - \frac{u_\vartheta^{(q)}}{r}, \\
2\varepsilon_{r\varphi}^{(q)} &= \frac{\partial u_\varphi^{(q)}}{\partial r} + \frac{1}{r\sin\vartheta}\frac{\partial u_r^{(q)}}{\partial \varphi} - \frac{u_\varphi^{(q)}}{r}, \\
2\varepsilon_{\vartheta\varphi}^{(q)} &= \frac{1}{r}\frac{\partial u_\varphi^{(q)}}{\partial \vartheta} + \frac{1}{r\sin\vartheta}\frac{\partial u_\vartheta^{(q)}}{\partial \varphi} - \frac{u_\varphi^{(q)}\cos\vartheta}{r\sin\vartheta},
\end{aligned} \tag{8.1}$$

whereas the equilibrium equations are written as

$$\begin{aligned}
&\frac{\partial \sigma_{rr}^{(q)}}{\partial r} + \frac{1}{r}\frac{\partial \sigma_{r\vartheta}^{(q)}}{\partial \vartheta} + \frac{\sigma_{r\vartheta}^{(q)}\cos\vartheta}{r\sin\vartheta} + \frac{2\sigma_{rr}^{(q)} - \sigma_{\vartheta\vartheta}^{(q)} - \sigma_{\varphi\varphi}^{(q)}}{r} + \frac{1}{r\sin\vartheta}\frac{\partial \sigma_{r\varphi}^{(q)}}{\partial \varphi} = 0, \\
&\frac{\partial \sigma_{r\vartheta}^{(q)}}{\partial r} + \frac{1}{r}\frac{\partial \sigma_{\vartheta\vartheta}^{(q)}}{\partial \vartheta} + \frac{3\sigma_{r\vartheta}^{(q)}}{r} + \frac{\left[\sigma_{\vartheta\vartheta}^{(q)} - \sigma_{\varphi\varphi}^{(q)}\right]\cos\vartheta}{r\sin\vartheta} + \frac{1}{r\sin\vartheta}\frac{\partial \sigma_{\vartheta\varphi}^{(q)}}{\partial \varphi} = 0, \\
&\frac{\partial \sigma_{r\varphi}^{(q)}}{\partial r} + \frac{1}{r}\frac{\partial \sigma_{\vartheta\varphi}^{(q)}}{\partial \vartheta} + \frac{3\sigma_{r\varphi}^{(q)}}{r} + \frac{2\sigma_{\vartheta\varphi}^{(q)}\cos\vartheta}{r\sin\vartheta} + \frac{1}{r\sin\vartheta}\frac{\partial \sigma_{\varphi\varphi}^{(q)}}{\partial \varphi} = 0.
\end{aligned} \tag{8.2}$$

The superscript q takes the value 1 for the inhomogeneity and 0 for the matrix, whereas for the equivalent homogeneous medium (shear boundary conditions case), it becomes "eq". In isotropic phases the stress and strain tensors are connected through the relation

$$\begin{bmatrix} \sigma_{rr}^{(q)} \\ \sigma_{\vartheta\vartheta}^{(q)} \\ \sigma_{\varphi\varphi}^{(q)} \\ \sigma_{r\vartheta}^{(q)} \\ \sigma_{r\varphi}^{(q)} \\ \sigma_{\vartheta\varphi}^{(q)} \end{bmatrix} = \begin{bmatrix} K_q + \frac{4}{3}\mu_q & K_q - \frac{2}{3}\mu_q & K_q - \frac{2}{3}\mu_q & 0 & 0 & 0 \\ K_q - \frac{2}{3}\mu_q & K_q + \frac{4}{3}\mu_q & K_q - \frac{2}{3}\mu_q & 0 & 0 & 0 \\ K_q - \frac{2}{3}\mu_q & K_q - \frac{2}{3}\mu_q & K_q + \frac{4}{3}\mu_q & 0 & 0 & 0 \\ 0 & 0 & 0 & \mu_q & 0 & 0 \\ 0 & 0 & 0 & 0 & \mu_q & 0 \\ 0 & 0 & 0 & 0 & 0 & \mu_q \end{bmatrix} \cdot \begin{bmatrix} \varepsilon_{rr}^{(q)} \\ \varepsilon_{\vartheta\vartheta}^{(q)} \\ \varepsilon_{\varphi\varphi}^{(q)} \\ 2\varepsilon_{r\vartheta}^{(q)} \\ 2\varepsilon_{r\varphi}^{(q)} \\ 2\varepsilon_{\vartheta\varphi}^{(q)} \end{bmatrix}, \tag{8.3}$$

where K_q and μ_q are the bulk and shear moduli of the q_{th} phase, respectively. In the RVE the inhomogeneity is considered to have the radius $r = r_1$, and the matrix has the external radius r_0 (Fig. 8.2). The ratio

$$c = \frac{r_1^3}{r_0^3} \tag{8.4}$$

corresponds to the inhomogeneity volume fraction. The interface conditions between the inhomogeneity and the matrix are expressed as

$$\begin{aligned} u_r^{(1)}(r_1, \vartheta, \varphi) &= u_r^{(0)}(r_1, \vartheta, \varphi), & \sigma_{rr}^{(1)}(r_1, \vartheta, \varphi) &= \sigma_{rr}^{(0)}(r_1, \vartheta, \varphi), \\ u_\vartheta^{(1)}(r_1, \vartheta, \varphi) &= u_\vartheta^{(0)}(r_1, \vartheta, \varphi), & \sigma_{r\vartheta}^{(1)}(r_1, \vartheta, \varphi) &= \sigma_{r\vartheta}^{(0)}(r_1, \vartheta, \varphi), \\ u_\varphi^{(1)}(r_1, \vartheta, \varphi) &= u_\varphi^{(0)}(r_1, \vartheta, \varphi), & \sigma_{r\varphi}^{(1)}(r_1, \vartheta, \varphi) &= \sigma_{r\varphi}^{(0)}(r_1, \vartheta, \varphi). \end{aligned} \tag{8.5}$$

For the case of shear boundary conditions, the interface conditions between the matrix and the equivalent homogeneous medium (Fig. 8.2_b) are written as

$$\begin{aligned} u_r^{(0)}(r_0, \vartheta, \varphi) &= u_r^{\text{eq}}(r_0, \vartheta, \varphi), & \sigma_{rr}^{(0)}(r_0, \vartheta, \varphi) &= \sigma_{rr}^{\text{eq}}(r_0, \vartheta, \varphi), \\ u_\vartheta^{(0)}(r_0, \vartheta, \varphi) &= u_\vartheta^{\text{eq}}(r_0, \vartheta, \varphi), & \sigma_{r\vartheta}^{(0)}(r_0, \vartheta, \varphi) &= \sigma_{r\vartheta}^{\text{eq}}(r_0, \vartheta, \varphi), \\ u_\varphi^{(0)}(r_0, \vartheta, \varphi) &= u_\varphi^{\text{eq}}(r_0, \vartheta, \varphi), & \sigma_{r\varphi}^{(0)}(r_0, \vartheta, \varphi) &= \sigma_{r\varphi}^{\text{eq}}(r_0, \vartheta, \varphi). \end{aligned} \tag{8.6}$$

The three basis vectors in spherical coordinates are

$$\boldsymbol{n}_r = \begin{bmatrix} \sin\vartheta\cos\varphi \\ \sin\vartheta\sin\varphi \\ \cos\vartheta \end{bmatrix}, \quad \boldsymbol{n}_\vartheta = \begin{bmatrix} \cos\vartheta\cos\varphi \\ \cos\vartheta\cos\varphi \\ -\sin\vartheta \end{bmatrix}, \quad \boldsymbol{n}_\varphi = \begin{bmatrix} -\sin\varphi \\ \cos\varphi \\ 0 \end{bmatrix}. \tag{8.7}$$

Macroscopic bulk modulus

Let us assume that the RVE of Fig. 8.2_a is subjected to the hydrostatic displacement field

$$\boldsymbol{u}^{(0)}(x_{\text{ext}}, y_{\text{ext}}, z_{\text{ext}}) = \begin{bmatrix} \beta x_{\text{ext}} \\ \beta y_{\text{ext}} \\ \beta z_{\text{ext}} \end{bmatrix},$$

which, in spherical coordinates, is expressed as

$$u_r^{(0)}(r_0, \vartheta, \varphi) = \beta r_0, \quad u_\vartheta^{(0)}(r_0, \vartheta, \varphi) = 0, \quad u_\varphi^{(0)}(r_0, \vartheta, \varphi) = 0. \tag{8.8}$$

For this type of boundary conditions, Hashin [1] has demonstrated that, at every phase, the displacement field that satisfies the equilibrium equations (8.2) is given by the analytical expressions

$$u_r^{(q)}(r) = \beta r U_r^{(q)}(r), \quad u_\vartheta^{(q)} = u_\varphi^{(q)} = 0,$$
$$U_r^{(q)}(r) = \Xi_1^{(q)} + \Xi_2^{(q)} \frac{1}{[r/r_1]^3}, \quad q = 0, 1.$$

The unknown constants $\Xi_1^{(0)}$, $\Xi_2^{(0)}$, $\Xi_1^{(1)}$, $\Xi_2^{(1)}$ depend on the volume fraction of the inhomogeneity and the material properties of the phases. Their computation is achieved by using the boundary and interface conditions. Specifically:

- At the center of the RVE the displacement field needs to be finite. Since at $r = 0$ the inhomogeneity phase exists, this leads to the conclusion that $\Xi_2^{(1)} = 0$.
- For the remaining three constants, the interface conditions $(8.5)_{1,2}$ and the boundary condition $(8.8)_1$ provide the linear system

$$\begin{bmatrix} 1 & -1 & -1 \\ 3K_1 & -3K_0 & 4\mu_0 \\ 0 & 1 & c \end{bmatrix} \cdot \begin{bmatrix} \Xi_1^{(1)} \\ \Xi_1^{(0)} \\ \Xi_2^{(0)} \end{bmatrix} = \begin{bmatrix} 0 \\ 0 \\ 1 \end{bmatrix}. \tag{8.9}$$

With the computation of all fields at any position, the next step is computing the average energy in the RVE (inhomogeneity + matrix). In spherical coordinates the surface element in a surface of constant radius r is $\mathrm{d}s_r = r^2 \sin\vartheta \,\mathrm{d}\vartheta \,\mathrm{d}\varphi$. Thus, the required energy is given by

$$U = \frac{1}{2V}\int_{\mathcal{B}} \boldsymbol{\sigma} : \boldsymbol{\varepsilon} \,\mathrm{d}V = \frac{1}{2V}\int_{\mathcal{B}} \operatorname{div}(\boldsymbol{u}\cdot\boldsymbol{\sigma}) \,\mathrm{d}V = \frac{1}{2V}\int_{\partial\mathcal{B}} \boldsymbol{u}\cdot\boldsymbol{\sigma}\cdot\boldsymbol{n} \,\mathrm{d}S$$

$$= \frac{3}{8\pi r_0}\int_0^{2\pi}\int_0^{\pi}\left[\sigma_{rr}^{(0)}u_r^{(0)} + \sigma_{r\vartheta}^{(0)}u_\vartheta^{(0)} + \sigma_{r\varphi}^{(0)}u_\varphi^{(0)}\right]_{r=r_0}\sin\vartheta\,\mathrm{d}\vartheta\,\mathrm{d}\varphi$$

$$= \frac{3\beta^2}{2}\left[3K_0\Xi_1^{(0)} - 4c\mu_0\Xi_2^{(0)}\right]. \tag{8.10}$$

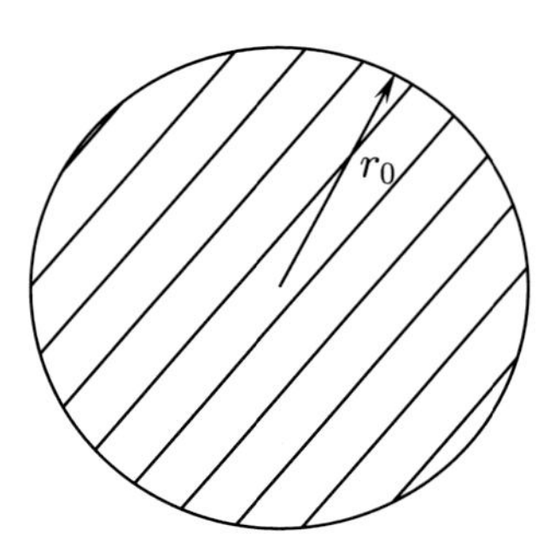

FIGURE 8.3 Hypothetical homogeneous medium for the energy equivalence principle.

According to the Hill–Mandel principle, this energy must be equal to the energy produced by the equivalent homogeneous medium when subjected to the same boundary conditions. If the RVE is substituted by the unknown medium (Fig. 8.3) with bulk modulus $\bar{K}$ and shear modulus $\bar{\mu}$, applying the boundary conditions (8.8) yields that the displacement field satisfies the homogeneous solution

$$\bar{u}_r(r) = \beta r, \quad \bar{u}_\vartheta = \bar{u}_\varphi = 0.$$

The average mechanical energy produced in this hypothetical case is

$$\bar{U} = \frac{3}{8\pi r_0}\int_0^{2\pi}\int_0^{\pi}\left[\bar{\sigma}_{rr}\bar{u}_r + \bar{\sigma}_{r\vartheta}\bar{u}_\vartheta + \bar{\sigma}_{r\varphi}\bar{u}_\varphi\right]_{r=r_2}\sin\vartheta\,\mathrm{d}\vartheta\,\mathrm{d}\varphi = \frac{9\beta^2}{2}\bar{K}. \tag{8.11}$$

Considering $U = \bar{U}$ yields the value of $\bar{K}$. As a conclusion, we have:

BOX 8.1 CSA: Bulk modulus

For an isotropic spherical particulate composite with particle volume fraction c, particle properties K_1, μ_1, and matrix properties K_0, μ_0, the macroscopic bulk modulus, according to the CSA method, is given by

$$\bar{K} = \frac{1}{3}\left[3K_0\Xi_1^{(0)} - 4c\mu_0\Xi_2^{(0)}\right], \tag{8.12}$$

where $\Xi_1^{(0)}$ and $\Xi_2^{(0)}$ are obtained from the solution of the linear system (8.9).

Macroscopic shear modulus

Let us assume the RVE of Fig. 8.2$_b$. At a far distance, this RVE is subjected to the deviatoric displacement field

$$\boldsymbol{u}^{\text{eq}}(x_{\text{ext}}, y_{\text{ext}}, z_{\text{ext}}) = \begin{bmatrix} \beta x_{\text{ext}} \\ -\beta y_{\text{ext}} \\ 0 \end{bmatrix},$$

which, in spherical coordinates, is expressed as

$$\begin{aligned} u_r^{\text{eq}}(r_{\text{ext}}, \vartheta, \varphi) &= \beta r_{\text{ext}} \sin^2\vartheta \cos 2\varphi, \\ u_\vartheta^{\text{eq}}(r_{\text{ext}}, \vartheta, \varphi) &= \beta r_{\text{ext}} \sin\vartheta \cos\vartheta \cos 2\varphi, \\ u_\varphi^{\text{eq}}(r_{\text{ext}}, \vartheta, \varphi) &= -\beta r_{\text{ext}} \sin\vartheta \sin 2\varphi. \end{aligned} \tag{8.13}$$

In the above expressions, $r_{\text{ext}} \to \infty$. For this type of boundary conditions, Christensen and Lo [2] have demonstrated that, at every phase, the displacement field that satisfies the equilibrium equations (8.2) is given by the analytical expressions

$$\begin{aligned} u_r^{(q)}(r, \vartheta, \varphi) &= \beta\, r\, U_r^{(q)}(r)\, \sin^2\vartheta \cos 2\varphi, \\ u_\vartheta^{(q)}(r, \vartheta, \varphi) &= \beta\, r\, U_\vartheta^{(q)}(r)\, \sin\vartheta \cos\vartheta \cos 2\varphi, \\ u_\varphi^{(q)}(r, \vartheta, \varphi) &= -\beta\, r\, U_\vartheta^{(q)}(r)\, \sin\vartheta \sin 2\varphi, \end{aligned}$$

where

$$\begin{aligned} U_r^{(q)}(r) &= \Xi_1^{(q)} + \left[2 - 3\frac{K_q}{\mu_q}\right][r/r_1]^2\,\Xi_2^{(q)} + \frac{3\Xi_3^{(q)}}{[r/r_1]^5} + \left[1 + \frac{K_q}{\mu_q}\right]\frac{3\Xi_4^{(q)}}{[r/r_1]^3}, \\ U_\vartheta^{(q)}(r) &= \Xi_1^{(q)} - \left[\frac{11}{3} + 5\frac{K_q}{\mu_q}\right][r/r_1]^2\,\Xi_2^{(q)} - \frac{2\Xi_3^{(q)}}{[r/r_1]^5} + \frac{2\Xi_4^{(q)}}{[r/r_1]^3}. \end{aligned}$$

For the equivalent medium at the external layer, the displacement field is given by similar expressions. Taking into account the boundary conditions (8.13), we find that

$$\begin{aligned} u_r^{\text{eq}}(r, \vartheta, \varphi) &= \beta\, r\, U_r^{\text{eq}}(r)\, \sin^2\vartheta \cos 2\varphi, \\ u_\vartheta^{\text{eq}}(r, \vartheta, \varphi) &= \beta\, r\, U_\vartheta^{\text{eq}}(r)\, \sin\vartheta \cos\vartheta \cos 2\varphi, \\ u_\varphi^{\text{eq}}(r, \vartheta, \varphi) &= -\beta\, r\, U_\vartheta^{\text{eq}}(r)\, \sin\vartheta \sin 2\varphi, \end{aligned}$$

$$U_r^{\mathrm{eq}}(r) = 1 + \frac{3\Xi_3^{\mathrm{eq}}}{[r/r_0]^5} + \left[1 + \frac{\bar{K}}{\bar{\mu}}\right]\frac{3\Xi_4^{\mathrm{eq}}}{[r/r_0]^3},$$
$$U_\vartheta^{\mathrm{eq}}(r) = 1 - \frac{2\Xi_3^{\mathrm{eq}}}{[r/r_0]^5} + \frac{2\Xi_4^{\mathrm{eq}}}{[r/r_0]^3}.$$

The ten unknown constants ($\Xi_1^{(1)}$, $\Xi_2^{(1)}$, $\Xi_3^{(1)}$, $\Xi_4^{(1)}$, $\Xi_1^{(0)}$, $\Xi_2^{(0)}$, $\Xi_3^{(0)}$, $\Xi_4^{(0)}$, Ξ_3^{eq}, Ξ_4^{eq}) and the macroscopic shear modulus $\bar{\mu}$ are identified as follows:

- At the center of the RVE the displacement field needs to be finite. Since at $r = 0$ the inhomogeneity phase exists, this leads to the conclusion that $\Xi_3^{(1)} = \Xi_4^{(1)} = 0$.
- Considering the effective medium of Fig. 8.3 under the same boundary conditions, the displacement field at every position is given by the homogeneous solution

$$\bar{u}_r(r, \vartheta, \varphi) = \beta r \sin^2\vartheta \cos 2\varphi,$$
$$\bar{u}_\vartheta(r, \vartheta, \varphi) = \beta r \sin\vartheta \cos\vartheta \cos 2\varphi,$$
$$\bar{u}_\varphi(r, \vartheta, \varphi) = -\beta r \sin\vartheta \sin 2\varphi.$$

  From the Eshelby energy principle we find the relation [3]

$$\int_0^{2\pi}\int_0^{\pi} [\sigma_{rr}^{\mathrm{eq}}\bar{u}_r + \sigma_{r\vartheta}^{\mathrm{eq}}\bar{u}_\vartheta + \sigma_{r\varphi}^{\mathrm{eq}}\bar{u}_\varphi - \bar{\sigma}_{rr}u_r^{\mathrm{eq}} - \bar{\sigma}_{r\vartheta}u_\vartheta^{\mathrm{eq}} - \bar{\sigma}_{r\varphi}u_\varphi^{\mathrm{eq}}]_{r=r_0} \sin\vartheta \,\mathrm{d}\vartheta \,\mathrm{d}\varphi = 0. \tag{8.14}$$

  Substituting the stress and displacement fields of the two problems at the position $r = r_0$ into the above integral yields the simple result $\Xi_4^{\mathrm{eq}} = 0$.
- The remaining eight unknowns are identified by the interface conditions (8.5) and (8.6). Since the obtained system of equations is nonlinear, a special treatment is required. Conditions $(8.5)_{1,3,2,4}$ and $(8.6)_{1,3}$ lead to the following system:

$$\boldsymbol{Z} \cdot \begin{bmatrix} \Xi_1^{(1)} \\ \Xi_2^{(1)} \\ \Xi_1^{(0)} \\ \Xi_2^{(0)} \\ \Xi_3^{(0)} \\ \Xi_4^{(0)} \end{bmatrix} = \begin{bmatrix} 0 \\ 0 \\ 0 \\ 0 \\ 1 \\ 1 \end{bmatrix} + \begin{bmatrix} 0 \\ 0 \\ 0 \\ 0 \\ 3 \\ -2 \end{bmatrix} \Xi_3^{\mathrm{eq}}, \tag{8.15}$$

where

$$
\boldsymbol{Z} = \begin{bmatrix}
1 & 2-3\dfrac{K_1}{\mu_1} & -1 & -2+3\dfrac{K_0}{\mu_0} & -3 & -3-3\dfrac{K_0}{\mu_0} \\
1 & -\dfrac{11}{3}-5\dfrac{K_1}{\mu_1} & -1 & \dfrac{11}{3}+5\dfrac{K_0}{\mu_0} & 2 & -2 \\
2\mu_1 & 3K_1-2\mu_1 & -2\mu_0 & -3K_0+2\mu_0 & 24\mu_0 & 18K_0+8\mu_0 \\
2\mu_1 & -16K_1-\dfrac{10}{3}\mu_1 & -2\mu_0 & 16K_0+\dfrac{10}{3}\mu_0 & -16\mu_0 & -6K_0 \\
0 & 0 & 1 & \left[2-3\dfrac{K_0}{\mu_0}\right]c^{-2/3} & 3c^{5/3} & \left[3+3\dfrac{K_0}{\mu_0}\right]c \\
0 & 0 & 1 & -\left[\dfrac{11}{3}+5\dfrac{K_0}{\mu_0}\right]c^{-2/3} & -2c^{5/3} & 2c
\end{bmatrix}.
$$

The solution of thus system is expressed in the form

$$
\begin{bmatrix} \Xi_1^{(1)} \\ \Xi_2^{(1)} \\ \Xi_1^{(0)} \\ \Xi_2^{(0)} \\ \Xi_3^{(0)} \\ \Xi_4^{(0)} \end{bmatrix} = \begin{bmatrix} g_1 \\ g_2 \\ a_1 \\ a_2 \\ a_3 \\ a_4 \end{bmatrix} + \begin{bmatrix} h_1 \\ h_2 \\ b_1 \\ b_2 \\ b_3 \\ b_4 \end{bmatrix} \Xi_3^{\mathrm{eq}}. \tag{8.16}
$$

Using (8.16), we writ the last two conditions $(8.6)_{2,4}$ as

$$
\begin{aligned}
a_5 + b_5\,\Xi_3^{\mathrm{eq}} &= 2\bar{\mu} - 24\bar{\mu}\,\Xi_3^{\mathrm{eq}}, \\
a_6 + b_6\,\Xi_3^{\mathrm{eq}} &= 2\bar{\mu} + 16\bar{\mu}\,\Xi_3^{\mathrm{eq}},
\end{aligned} \tag{8.17}
$$

with

$$
\begin{aligned}
a_5 &= 2\mu_0 a_1 + [3K_0 - 2\mu_0]c^{-2/3}a_2 - 24\mu_0 c^{5/3}a_3 - [18K_0 + 8\mu_0]c\, a_4, \\
a_6 &= 2\mu_0 a_1 - \left[16K_0 + \frac{10}{3}\mu_0\right]c^{-2/3}a_2 + 16\mu_0 c^{5/3}a_3 + 6K_0 c\, a_4, \\
b_5 &= 2\mu_0 b_1 + [3K_0 - 2\mu_0]c^{-2/3}b_2 - 24\mu_0 c^{5/3}b_3 - [18K_0 + 8\mu_0]c\, b_4, \\
b_6 &= 2\mu_0 b_1 - \left[16K_0 + \frac{10}{3}\mu_0\right]c^{-2/3}b_2 + 16\mu_0 c^{5/3}b_3 + 6K_0 c\, b_4.
\end{aligned} \tag{8.18}
$$

Subtracting $(8.17)_1$ from $(8.17)_2$ gives

$$
\Xi_3^{\mathrm{eq}} = \frac{a_6 - a_5}{40\bar{\mu} + b_5 - b_6}.
$$

Substituting the final result into $(8.17)_1$, after some algebra, we obtain the quadratic equation

$$80\bar{\mu}^2 - 2[b_6 - b_5 + 12a_6 + 8a_5]\bar{\mu} + a_5 b_6 - b_5 a_6 = 0.$$

One of two possible solutions is positive, and the other is negative. The positive value is the macroscopic shear modulus.

In conclusion:

BOX 8.2 CSA: Shear modulus

For an isotropic spherical particulate composite with particle volume fraction c, particle properties K_1, μ_1, and matrix properties K_0, μ_0, the macroscopic shear modulus, according to the generalized self–consistent CSA method, is equal to

$$\bar{\mu} = \frac{1}{80}\left[b_6 - b_5 + 12a_6 + 8a_5 + \sqrt{\Delta}\right],$$
$$\Delta = [b_6 - b_5 + 12a_6 + 8a_5]^2 - 80[a_5 b_6 - b_5 a_6]. \tag{8.19}$$

The necessary constants a_i, b_i, $i = 1, ..., 6$, are obtained from the solution of system (8.15) using (8.16) and (8.18).

8.2 Composite cylinder assemblage

The composite cylinder assemblage method (CCA) [4,2] is a homogenization technique that provides the effective properties of unidirectional, cylindrical, long fiber composites (Fig. 8.4). In this method the RVE consists of two concentric cylinders: the inner cylinder with radius r_1 represents the fiber, whereas the outer cylinder with radius r_0 represents the matrix phase.

It is well understood that a unidirectional long fiber composite behaves as a transversely isotropic material. Having this in mind, the aim of CCA is providing five independent parameters of the composite material. The obtained effective properties are the transverse (or in–plane) bulk modulus $\bar{K}^{\text{tr}}$, the axial shear modulus $\bar{\mu}^{\text{ax}}$, the axial Young modulus $\bar{E}^{\text{ax}}$, the axial Poisson ratio $\bar{\nu}^{\text{ax}}$, and the transverse shear modulus $\bar{\mu}^{\text{tr}}$. With these five parameters, we can identify the complete elasticity tensor $\bar{\boldsymbol{L}}$.

To achieve this goal, Hashin and Rosen [4] identified the analytical solutions (in cylindrical coordinates) of four independent boundary value problems. The latter (Fig. 8.5), when solved for the heterogeneous material, provide the strain energy W_c of the composite RVE. Then we can also solve the same problems by substituting the composite cylinder with hypothetical homogeneous medium and compute the strain energy W_h. The

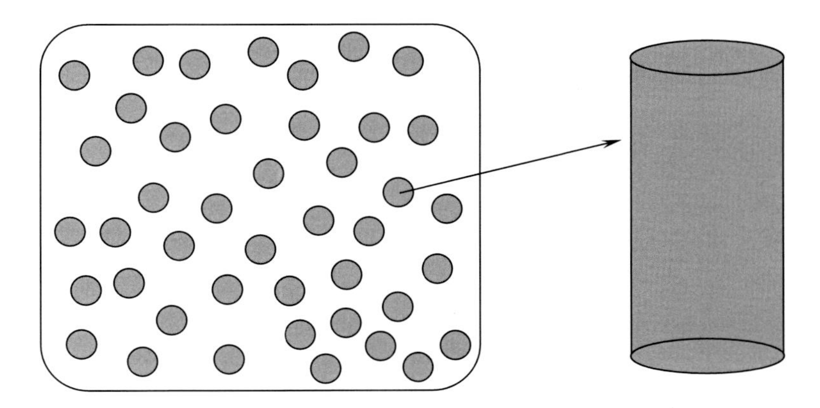

FIGURE 8.4 A unidirectional, long fiber composite.

Hill–Mandel principle $W_c = W_h$ allows us to calculate the unknown effective properties.

Although the obtained four parameters $\bar{K}^{\text{tr}}$, $\bar{\mu}^{\text{ax}}$, $\bar{E}^{\text{ax}}$, and $\bar{\nu}^{\text{ax}}$ represent accurately the effective medium, the fifth boundary value problem of the original CCA method provides only a bound on the $\bar{\mu}^{\text{tr}}$. To resolve this issue, Christensen and Lo [2] propose to solve the fourth boundary value problem using the generalized self–consistent method. In that case the RVE is modified to a problem of three concentric cylinders, where the external layer is the unknown effective medium (Fig. 8.5_d). Then, using appropriate energy equivalence principles, we obtain an accurate estimation of the transverse shear modulus.

Expressing the problem in cylindrical coordinates

The cylindrical form of the inhomogeneities allows us to transform the RVE problem in a cylindrical coordinate system. In cylindrical coordinates the axes (x, y, z) are transformed to (r, ϑ, z). Moreover, at each phase, the strain tensor components are given by the expressions

$$
\begin{aligned}
\varepsilon_{rr}^{(q)} &= \frac{\partial u_r^{(q)}}{\partial r}, \\
\varepsilon_{\vartheta\vartheta}^{(q)} &= \frac{1}{r}\frac{\partial u_\vartheta^{(q)}}{\partial \vartheta} + \frac{u_r^{(q)}}{r}, \\
\varepsilon_{zz}^{(q)} &= \frac{\partial u_z^{(q)}}{\partial z}, \\
2\varepsilon_{r\vartheta}^{(q)} &= \frac{\partial u_\vartheta^{(q)}}{\partial r} + \frac{1}{r}\frac{\partial u_r^{(q)}}{\partial \vartheta} - \frac{u_\vartheta^{(q)}}{r},
\end{aligned}
$$

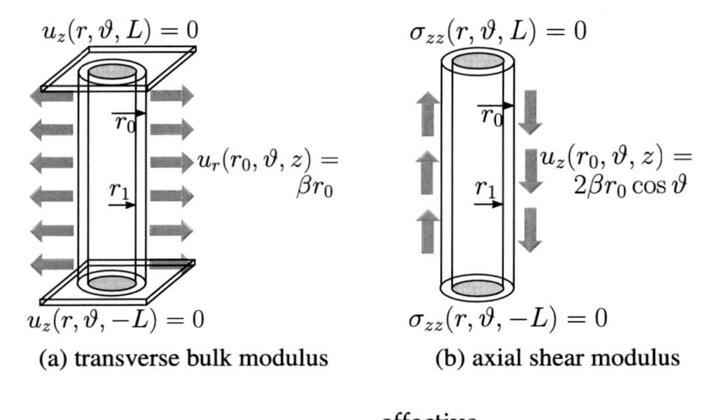

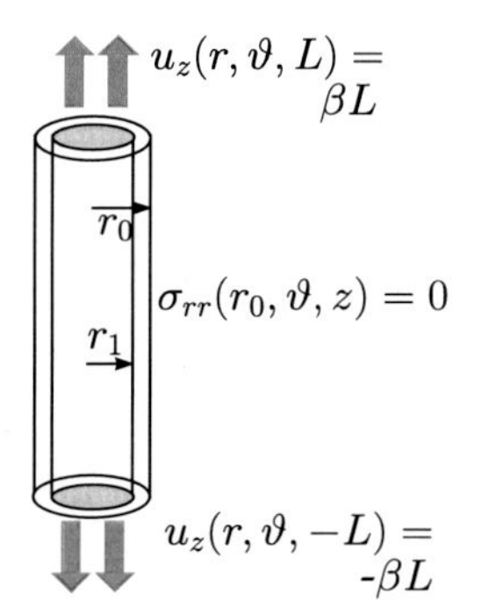

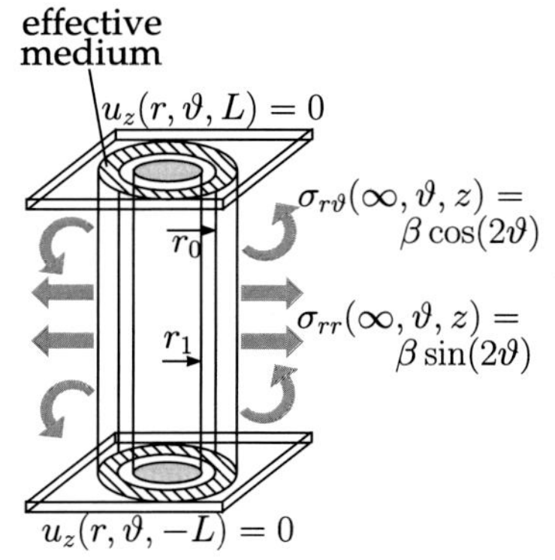

FIGURE 8.5 The four boundary value problems of composite cylinder assemblage method.

$$
\begin{aligned}
2\varepsilon_{rz}^{(q)} &= \frac{\partial u_z^{(q)}}{\partial r} + \frac{\partial u_r^{(q)}}{\partial z}, \\
2\varepsilon_{\vartheta z}^{(q)} &= \frac{1}{r}\frac{\partial u_z^{(q)}}{\partial \vartheta} + \frac{\partial u_\vartheta^{(q)}}{\partial z},
\end{aligned} \tag{8.20}
$$

whereas the equilibrium equations are written as

$$
\begin{aligned}
\frac{\partial \sigma_{rr}^{(q)}}{\partial r} + \frac{1}{r}\frac{\partial \sigma_{r\vartheta}^{(q)}}{\partial \vartheta} + \frac{\sigma_{rr}^{(q)} - \sigma_{\vartheta\vartheta}^{(q)}}{r} + \frac{\partial \sigma_{rz}^{(q)}}{\partial z} &= 0, \\
\frac{\partial \sigma_{r\vartheta}^{(q)}}{\partial r} + \frac{1}{r}\frac{\partial \sigma_{\vartheta\vartheta}^{(q)}}{\partial \vartheta} + \frac{2\sigma_{r\vartheta}^{(q)}}{r} + \frac{\partial \sigma_{\vartheta z}^{(q)}}{\partial z} &= 0,
\end{aligned}
$$

$$\frac{\partial \sigma_{rz}^{(q)}}{\partial r} + \frac{1}{r}\frac{\partial \sigma_{\vartheta z}^{(q)}}{\partial \vartheta} + \frac{\sigma_{rz}^{(q)}}{r} + \frac{\partial \sigma_{zz}^{(q)}}{\partial z} = 0. \tag{8.21}$$

The superscript q takes the value 1 for the inhomogeneity and 0 for the matrix, whereas for the equivalent homogeneous medium (case of Fig. 8.5_d), it becomes "eq". In transversely isotropic phases (where the axis of symmetry is the z–direction) the stress and strain tensors are connected through the relation

$$\begin{bmatrix} \sigma_{rr}^{(q)} \\ \sigma_{\vartheta\vartheta}^{(q)} \\ \sigma_{zz}^{(q)} \\ \sigma_{r\vartheta}^{(q)} \\ \sigma_{rz}^{(q)} \\ \sigma_{\vartheta z}^{(q)} \end{bmatrix} = \begin{bmatrix} K_q^{\text{tr}} + \mu_q^{\text{tr}} & K_q^{\text{tr}} - \mu_q^{\text{tr}} & l_q & 0 & 0 & 0 \\ K_q^{\text{tr}} - \mu_q^{\text{tr}} & K_q^{\text{tr}} + \mu_q^{\text{tr}} & l_q & 0 & 0 & 0 \\ l_q & l_q & n_q & 0 & 0 & 0 \\ 0 & 0 & 0 & \mu_q^{\text{tr}} & 0 & 0 \\ 0 & 0 & 0 & 0 & \mu_q^{\text{ax}} & 0 \\ 0 & 0 & 0 & 0 & 0 & \mu_q^{\text{ax}} \end{bmatrix} \cdot \begin{bmatrix} \varepsilon_{rr}^{(q)} \\ \varepsilon_{\vartheta\vartheta}^{(q)} \\ \varepsilon_{zz}^{(q)} \\ 2\varepsilon_{r\vartheta}^{(q)} \\ 2\varepsilon_{rz}^{(q)} \\ 2\varepsilon_{\vartheta z}^{(q)} \end{bmatrix}, \tag{8.22}$$

where K_q^{tr}, μ_q^{tr}, μ_q^{ax}, l_q, and n_q are the five elastic constants of the q_{th} phase. Two of these constants (l_q and n_q) can be interchanged with the axial Young modulus E_q^{ax} and the axial Poisson ratio ν_q^{ax} with the help of the relations

$$l_q = 2\nu_q^{\text{ax}} K_q^{\text{tr}}, \qquad n_q = E_q^{\text{ax}} + 2\nu_q^{\text{ax}} l_q. \tag{8.23}$$

In the RVE the inhomogeneity is considered to have radius $r = r_1$, and the matrix has external radius r_0 (Fig. 8.5). The ratio

$$c = \frac{r_1^2}{r_0^2} \tag{8.24}$$

corresponds to the inhomogeneity volume fraction. The interface conditions between the inhomogeneity and the matrix are expressed as

$$\begin{aligned} u_r^{(1)}(r_1, \vartheta, z) &= u_r^{(0)}(r_1, \vartheta, z), & \sigma_{rr}^{(1)}(r_1, \vartheta, z) &= \sigma_{rr}^{(0)}(r_1, \vartheta, z), \\ u_\vartheta^{(1)}(r_1, \vartheta, z) &= u_\vartheta^{(0)}(r_1, \vartheta, z), & \sigma_{r\vartheta}^{(1)}(r_1, \vartheta, z) &= \sigma_{r\vartheta}^{(0)}(r_1, \vartheta, z), \\ u_z^{(1)}(r_1, \vartheta, z) &= u_z^{(0)}(r_1, \vartheta, z), & \sigma_{rz}^{(1)}(r_1, \vartheta, z) &= \sigma_{rz}^{(0)}(r_1, \vartheta, z). \end{aligned} \tag{8.25}$$

For the case of shear boundary conditions, the interface conditions between the matrix and the equivalent homogeneous medium (Fig. 8.5_d) are written as

$$u_r^{(0)}(r_0,\vartheta,z)=u_r^{\mathrm{eq}}(r_0,\vartheta,z),\qquad \sigma_{rr}^{(0)}(r_0,\vartheta,z)=\sigma_{rr}^{\mathrm{eq}}(r_0,\vartheta,z),$$
$$u_\vartheta^{(0)}(r_0,\vartheta,z)=u_\vartheta^{\mathrm{eq}}(r_0,\vartheta,z),\qquad \sigma_{r\vartheta}^{(0)}(r_0,\vartheta,z)=\sigma_{r\vartheta}^{\mathrm{eq}}(r_0,\vartheta,z),$$
$$u_z^{(0)}(r_0,\vartheta,z)=u_z^{\mathrm{eq}}(r_0,\vartheta,z),\qquad \sigma_{rz}^{(0)}(r_0,\vartheta,z)=\sigma_{rz}^{\mathrm{eq}}(r_0,\vartheta,z). \tag{8.26}$$

The three basis vectors in cylindrical coordinates are

$$\boldsymbol{n}_r=\begin{bmatrix}\cos\vartheta\\ \sin\vartheta\\ 0\end{bmatrix},\qquad \boldsymbol{n}_\vartheta=\begin{bmatrix}-\sin\vartheta\\ \cos\vartheta\\ 0\end{bmatrix},\qquad \boldsymbol{n}_z=\begin{bmatrix}0\\ 0\\ 1\end{bmatrix}. \tag{8.27}$$

With the help of these vectors, the displacements of the phases are transformed to the Cartesian coordinate system through the formula

$$\boldsymbol{u}^{(q)}=u_r^{(q)}\,\boldsymbol{n}_r+u_\vartheta^{(q)}\,\boldsymbol{n}_\vartheta+u_z^{(q)}\,\boldsymbol{n}_z. \tag{8.28}$$

With regard to integrals in cylindrical coordinates, the surface element in a surface of constant radius r (a vertical cylinder) is $\mathrm{d}s_r=r\mathrm{d}\vartheta\mathrm{d}z$, and the surface element in a surface of constant z is $\mathrm{d}s_z=r\mathrm{d}r\mathrm{d}\vartheta$. For a cylinder of radius r_0 and length $2L$, the average energy is expressed as

$$\begin{aligned}U&=\frac{1}{2V}\int_{\mathcal{B}}\boldsymbol{\sigma}:\boldsymbol{\varepsilon}\,\mathrm{d}V=\frac{1}{2V}\int_{\mathcal{B}}\mathrm{div}(\boldsymbol{u}\cdot\boldsymbol{\sigma})\,\mathrm{d}V=\frac{1}{2V}\int_{\partial\mathcal{B}}\boldsymbol{u}\cdot\boldsymbol{\sigma}\cdot\boldsymbol{n}\,\mathrm{d}S\\
&=\frac{1}{4L\pi r_0}\int_{-L}^{L}\int_0^{2\pi}\left[\sigma_{rr}^{(0)}u_r^{(0)}+\sigma_{r\vartheta}^{(0)}u_\vartheta^{(0)}+\sigma_{rz}^{(0)}u_z^{(0)}\right]_{r=r_0}\mathrm{d}\vartheta\mathrm{d}z\\
&+\frac{1}{4L\pi r_0^2}\int_0^{2\pi}\int_0^{r_0}\left[\sigma_{rz}^{(0)}u_r^{(0)}+\sigma_{\vartheta z}^{(0)}u_\vartheta^{(0)}+\sigma_{zz}^{(0)}u_z^{(0)}\right]_{z=L}r\mathrm{d}r\mathrm{d}\vartheta\\
&-\frac{1}{4L\pi r_0^2}\int_0^{2\pi}\int_0^{r_0}\left[\sigma_{rz}^{(0)}u_r^{(0)}+\sigma_{\vartheta z}^{(0)}u_\vartheta^{(0)}+\sigma_{zz}^{(0)}u_z^{(0)}\right]_{z=-L}r\mathrm{d}r\mathrm{d}\vartheta.\end{aligned} \tag{8.29}$$

The average strain in the same cylinder is given by the expression

$$\begin{aligned}\langle\boldsymbol{\varepsilon}\rangle&=\frac{1}{V}\int_{\mathcal{B}}\boldsymbol{\varepsilon}\,\mathrm{d}V=\frac{1}{2V}\int_{\partial\mathcal{B}}[\boldsymbol{u}\otimes\boldsymbol{n}+\boldsymbol{n}\otimes\boldsymbol{u}]\,\mathrm{d}S\\
&=\frac{1}{4L\pi r_0}\int_{-L}^{L}\int_0^{2\pi}\left[\boldsymbol{u}^{(0)}\otimes\boldsymbol{n}_r+\boldsymbol{n}_r\otimes\boldsymbol{u}^{(0)}\right]_{r=r_0}\mathrm{d}\vartheta\mathrm{d}z\\
&+\frac{1}{4L\pi r_0^2}\int_0^{2\pi}\int_0^{r_0}[\boldsymbol{u}\otimes\boldsymbol{n}_z+\boldsymbol{n}_z\otimes\boldsymbol{u}]_{z=L}\,r\mathrm{d}r\mathrm{d}\vartheta\\
&-\frac{1}{4L\pi r_0^2}\int_0^{2\pi}\int_0^{r_0}[\boldsymbol{u}\otimes\boldsymbol{n}_z+\boldsymbol{n}_z\otimes\boldsymbol{u}]_{z=-L}\,r\mathrm{d}r\mathrm{d}\vartheta.\end{aligned} \tag{8.30}$$

Macroscopic transverse bulk modulus

Let us consider the boundary value problem of Fig. 8.5_a. For such conditions, the displacement field with general form per phase [4]

$$u_r^{(q)}(r) = \beta r U_r^{(q)}(r), \quad u_\vartheta^{(q)} = u_z^{(q)} = 0,$$
$$U_r^{(q)}(r) = \Xi_1^{(q)} + \Xi_2^{(q)} \frac{1}{[r/r_1]^2}, \quad q = 0, 1,$$

produces stresses that satisfy the equilibrium equations (8.21). The four unknown constants $\Xi_1^{(1)}$, $\Xi_2^{(1)}$, $\Xi_1^{(0)}$, and $\Xi_2^{(0)}$ are identified as follows:

- At the center of the RVE the displacement field needs to be finite. Since at $r = 0$ the inhomogeneity phase exists, this leads to the conclusion that $\Xi_2^{(1)} = 0$.
- For the remaining three constants, the interface conditions $(8.25)_{1,2}$ and the boundary condition

$$u_r^{(0)} = \beta r_0 \tag{8.31}$$

provide the linear system

$$\begin{bmatrix} 1 & -1 & -1 \\ 2K_1^{\text{tr}} & -2K_0^{\text{tr}} & 2\mu_0^{\text{tr}} \\ 0 & 1 & c \end{bmatrix} \cdot \begin{bmatrix} \Xi_1^{(1)} \\ \Xi_1^{(0)} \\ \Xi_2^{(0)} \end{bmatrix} = \begin{bmatrix} 0 \\ 0 \\ 1 \end{bmatrix}. \tag{8.32}$$

With the computation of all fields at any position, the next step is computing the average energy in the RVE (inhomogeneity + matrix) using expression (8.29). For the specific boundary value problem, the average energy is equal to

$$U = \beta^2 \left[2K_0^{\text{tr}} \Xi_1^{(0)} - 2c\mu_0^{\text{tr}} \Xi_2^{(0)}\right]. \tag{8.33}$$

According to the Hill–Mandel principle, this energy must be equal to the energy produced by the equivalent homogeneous medium when subjected to the same boundary conditions. If the RVE is substituted by the unknown medium (Fig. 8.6) with transverse bulk modulus $\bar{K}^{\text{tr}}$, applying the boundary conditions (8.31) yields that the displacement field satisfies the homogeneous solution

$$\bar{u}_r(r) = \beta r, \quad \bar{u}_\vartheta = \bar{u}_z = 0.$$

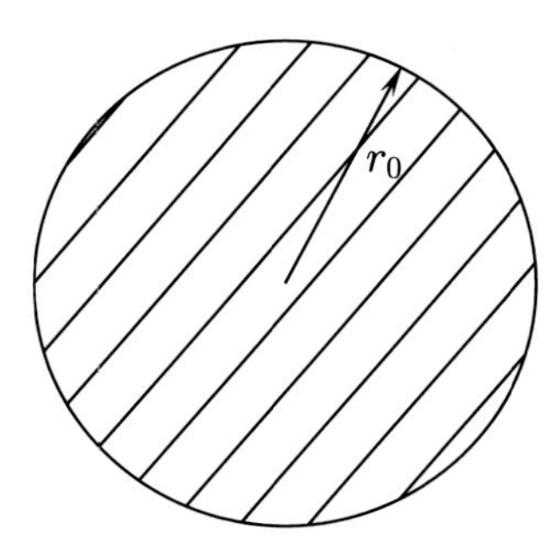

FIGURE 8.6 Hypothetical homogeneous medium for the energy equivalence principle.

The average mechanical energy produced in this hypothetical case is

$$\bar{U} = \frac{1}{2V}\int_{\partial\mathcal{B}} \bar{\boldsymbol{u}}\cdot\bar{\boldsymbol{\sigma}}\cdot\boldsymbol{n}\,\mathrm{d}S = 2\beta^2\bar{K}^{\mathrm{tr}}. \tag{8.34}$$

Considering $U = \bar{U}$ yields the value of $\bar{K}^{\mathrm{tr}}$. As a conclusion, we have:

BOX 8.3 CCA: Transverse bulk modulus

Consider a transversely isotropic long fiber composite with fiber volume fraction c, fiber properties K_1^{tr}, μ_1^{tr}, μ_1^{ax}, l_1, n_1, and matrix properties K_0^{tr}, μ_0^{tr}, μ_0^{ax}, l_0, n_0. The macroscopic transverse bulk modulus, according to the CCA method, is given by

$$\bar{K}^{\mathrm{tr}} = K_0^{\mathrm{tr}}\Xi_1^{(0)} - c\mu_0^{\mathrm{tr}}\Xi_2^{(0)}, \tag{8.35}$$

where $\Xi_1^{(0)}$ and $\Xi_2^{(0)}$ are obtained from the solution of the linear system (8.32).

Macroscopic axial shear modulus

Let us consider the boundary value problem of Fig. 8.5$_b$. For such conditions, the displacement field with general form per phase [4]

$$u_z^{(q)}(r,\vartheta) = \beta r U_z^{(q)}(r)\cos\vartheta, \qquad u_r^{(q)} = u_\vartheta^{(q)} = 0,$$
$$U_z^{(q)}(r) = \Xi_1^{(q)} + \Xi_2^{(q)}\frac{1}{[r/r_1]^2}, \quad q = 0, 1,$$

produces stresses that satisfy the equilibrium equations (8.21). The four unknown constants $\Xi_1^{(1)}$, $\Xi_2^{(1)}$, $\Xi_1^{(0)}$, and $\Xi_2^{(0)}$ are identified as follows:

- At the center of the RVE the displacement field needs to be finite. Since at $r = 0$ the inhomogeneity phase exists, this leads to the conclusion that $\Xi_2^{(1)} = 0$.
- For the remaining three constants, the interface conditions $(8.25)_{5,6}$ and the boundary condition

$$u_z^{(0)} = 2\beta r_0 \cos\vartheta \tag{8.36}$$

provide the linear system

$$\begin{bmatrix} 1 & -1 & -1 \\ \mu_1^{\text{ax}} & -\mu_0^{\text{ax}} & \mu_0^{\text{ax}} \\ 0 & 1 & c \end{bmatrix} \cdot \begin{bmatrix} \Xi_1^{(1)} \\ \Xi_1^{(0)} \\ \Xi_2^{(0)} \end{bmatrix} = \begin{bmatrix} 0 \\ 0 \\ 2 \end{bmatrix}. \tag{8.37}$$

With the computation of all fields at any position, the next step is computing the average energy in the RVE (inhomogeneity + matrix), using expression (8.29). For the specific boundary value problem, the average energy is equal to

$$U = \mu_0^{\text{ax}} \beta^2 \left[\Xi_1^{(0)} - c\,\Xi_2^{(0)} \right]. \tag{8.38}$$

According to the Hill–Mandel principle, this energy must be equal to the energy produced by the equivalent homogeneous medium when subjected to the same boundary conditions. If the RVE is substituted by the unknown medium (Fig. 8.6) with axial shear modulus $\bar{\mu}^{\text{ax}}$, applying the boundary conditions (8.36) yields that the displacement field satisfies the homogeneous solution

$$\bar{u}_z(r,\vartheta) = 2\beta r \cos\vartheta, \quad \bar{u}_r = \bar{u}_\vartheta = 0.$$

The average mechanical energy produced in this hypothetical case is

$$\bar{U} = \frac{1}{2V} \int_{\partial\mathcal{B}} \bar{\boldsymbol{u}} \cdot \bar{\boldsymbol{\sigma}} \cdot \boldsymbol{n} \, \mathrm{d}S = 2\beta^2 \bar{\mu}^{\text{ax}}. \tag{8.39}$$

Considering $U = \bar{U}$ yields the value of $\bar{\mu}^{\text{ax}}$. As a conclusion, we have:

BOX 8.4 CCA: Axial shear modulus

Consider a transversely isotropic long fiber composite with fiber volume fraction c, fiber properties K_1^{tr}, μ_1^{tr}, μ_1^{ax}, l_1, n_1, and matrix properties K_0^{tr}, μ_0^{tr}, μ_0^{ax}, l_0, n_0. The macroscopic axial shear modulus, according to the CCA method, is given by

$$\bar{\mu}^{\mathrm{ax}} = \frac{\mu_0^{\mathrm{ax}}}{2}\left[\Xi_1^{(0)} - c\,\Xi_2^{(0)}\right], \tag{8.40}$$

where $\Xi_1^{(0)}$ and $\Xi_2^{(0)}$ are obtained from the solution of the linear system (8.37).

Macroscopic axial Young's modulus/Poisson's ratio

Let us consider the boundary value problem of Fig. 8.5$_c$. For such conditions, the displacement field with general form per phase [4]

$$u_r^{(q)}(r) = \beta r U_r^{(q)}(r), \quad u_\vartheta^{(q)} = 0, \quad u_z^{(q)} = \beta z,$$
$$U_r^{(q)}(r) = \Xi_1^{(q)} + \Xi_2^{(q)}\frac{1}{[r/r_1]^2}, \quad q = 0, 1,$$

produces stresses that satisfy the equilibrium equations (8.21). The four unknown constants $\Xi_1^{(1)}$, $\Xi_2^{(1)}$, $\Xi_1^{(0)}$, and $\Xi_2^{(0)}$ are identified as follows:

- At the center of the RVE the displacement field needs to be finite. Since at $r = 0$ the inhomogeneity phase exists, this leads to the conclusion that $\Xi_2^{(1)} = 0$.
- For the remaining three constants, the interface conditions $(8.25)_{1,2}$ and the boundary condition

$$\sigma_{rr}^{(0)} = 0 \tag{8.41}$$

provide the linear system

$$\begin{bmatrix} 1 & -1 & -1 \\ 2K_1^{\mathrm{tr}} & -2K_0^{\mathrm{tr}} & 2\mu_0^{\mathrm{tr}} \\ 0 & 2K_0^{\mathrm{tr}} & -2c\mu_0^{\mathrm{tr}} \end{bmatrix} \cdot \begin{bmatrix} \Xi_1^{(1)} \\ \Xi_1^{(0)} \\ \Xi_2^{(0)} \end{bmatrix} = \begin{bmatrix} 0 \\ l_0 - l_1 \\ -l_0 \end{bmatrix}. \tag{8.42}$$

With the computation of all fields at any position, the next step is computing the average energy in the RVE (inhomogeneity + matrix) using expression (8.29). For the specific boundary value problem, the average energy is equal to

$$U = \frac{\beta^2}{2}\left[c\left[2l_1\Xi_1^{(1)} + n_1\right] + [1-c]\left[2l_0\Xi_1^{(0)} + n_0\right]\right]. \tag{8.43}$$

In addition, the average strain tensor in the RVE is given by (8.30):

$$\langle \boldsymbol{\varepsilon} \rangle = \beta \begin{bmatrix} \Xi_1^{(0)} + c\Xi_2^{(0)} \\ \Xi_1^{(0)} + c\Xi_2^{(0)} \\ 1 \\ 0 \\ 0 \\ 0 \end{bmatrix}. \tag{8.44}$$

According to the Hill–Mandel principle, this energy must be equal to the energy produced by the equivalent homogeneous medium when subjected to the same boundary conditions. If the RVE is substituted by the unknown medium (Fig. 8.6), applying the boundary conditions (8.41) yields that the displacement field satisfies the homogeneous solution

$$u_r^{(q)}(r) = -\beta r \frac{\bar{l}}{2\bar{K}^{\text{tr}}} = -\beta r \bar{\nu}^{\text{ax}}, \quad u_\vartheta^{(q)} = 0, \quad u_z^{(q)} = \beta z.$$

The average mechanical energy produced in this hypothetical case is

$$\bar{U} = \frac{1}{2V} \int_{\partial \mathcal{B}} \bar{\boldsymbol{u}} \cdot \bar{\boldsymbol{\sigma}} \cdot \boldsymbol{n} \, \mathrm{d}S = \frac{\beta^2}{2} \left[-2\bar{l}\bar{\nu}^{\text{ax}} + \bar{n} \right] = \frac{\beta^2}{2} \bar{E}^{\text{ax}}. \tag{8.45}$$

In addition, the average strain tensor in the hypothetical medium is equal to

$$\langle \bar{\boldsymbol{\varepsilon}} \rangle = \beta \begin{bmatrix} -\bar{\nu}^{\text{ax}} \\ -\bar{\nu}^{\text{ax}} \\ 1 \\ 0 \\ 0 \\ 0 \end{bmatrix}. \tag{8.46}$$

Considering $U = \bar{U}$ yields the value of $\bar{E}^{\text{ax}}$, whereas comparing the average strains of the two systems yields the value of $\bar{\nu}^{\text{ax}}$. As a conclusion, we have:

BOX 8.5 CCA: Axial Young's modulus and Poisson's ratio

Consider a transversely isotropic, long fiber composite with fiber volume fraction c, fiber properties K_1^{tr}, μ_1^{tr}, μ_1^{ax}, l_1, n_1, and matrix properties K_0^{tr}, μ_0^{tr}, μ_0^{ax}, l_0, n_0. According to the CCA method, the macroscopic axial Young modulus and Poisson ratio are given by

$$\bar{E}^{\mathrm{ax}} = c\left[2l_1\Xi_1^{(1)} + n_1\right] + [1-c]\left[2l_0\Xi_1^{(0)} + n_0\right],$$
$$\bar{\nu}^{\mathrm{ax}} = -\left[\Xi_1^{(0)} + c\,\Xi_2^{(0)}\right]. \tag{8.47}$$

The constants $\Xi_1^{(1)}$, $\Xi_1^{(0)}$, and $\Xi_2^{(0)}$ are obtained from the solution of the linear system (8.42).

Macroscopic transverse shear modulus

Let us consider the boundary value problem of Fig. 8.5_d. For such conditions, the displacement field with general form [4]

$$u_r^{(q)}(r,\vartheta) = \beta r U_r^{(q)}(r)\sin 2\vartheta, \quad u_\vartheta^{(q)}(r,\vartheta) = \beta r U_\vartheta^{(q)}(r)\cos 2\vartheta, \quad u_z^{(q)} = 0,$$

where

$$U_r^{(q)}(r) = \frac{K_q^{\mathrm{tr}} - \mu_q^{\mathrm{tr}}}{2K_q^{\mathrm{tr}} + \mu_q^{\mathrm{tr}}}[r/r_1]^2\Xi_1^{(q)} + \Xi_2^{(q)} - \frac{\Xi_3^{(q)}}{[r/r_1]^4} + \frac{K_q^{\mathrm{tr}} + \mu_q^{\mathrm{tr}}}{\mu_q^{\mathrm{tr}}}\frac{\Xi_4^{(q)}}{[r/r_1]^2},$$
$$U_\vartheta^{(q)}(r) = [r/r_1]^2\Xi_1^{(q)} + \Xi_2^{(q)} + \frac{\Xi_3^{(q)}}{[r/r_1]^4} + \frac{\Xi_4^{(q)}}{[r/r_1]^2},$$

satisfies the equilibrium equations. Christensen and Lo [2] have considered the addition of an extra layer, made by the unknown equivalent medium, whose displacement field is expressed as

$$u_r^{\mathrm{eq}}(r,\vartheta) = \frac{\beta r}{2\bar{\mu}^{\mathrm{tr}}}U_r^{\mathrm{eq}}(r)\sin 2\vartheta, \quad u_\vartheta^{\mathrm{eq}}(r,\vartheta) = \frac{\beta r}{2\bar{\mu}^{\mathrm{tr}}}U_\vartheta^{\mathrm{eq}}(r)\cos 2\vartheta, \quad u_z^{\mathrm{eq}} = 0,$$

with

$$U_r^{\mathrm{eq}}(r) = 1 - \frac{\Xi_3^{\mathrm{eq}}}{[r/r_0]^4} + \frac{\bar{K}^{\mathrm{tr}} + \bar{\mu}^{\mathrm{tr}}}{\bar{\mu}^{\mathrm{tr}}}\frac{\Xi_4^{\mathrm{eq}}}{[r/r_0]^2},$$
$$U_\vartheta^{\mathrm{eq}}(r) = 1 + \frac{\Xi_3^{\mathrm{eq}}}{[r/r_0]^4} + \frac{\Xi_4^{\mathrm{eq}}}{[r/r_0]^2}.$$

In the last expressions the boundary conditions shown in Fig. 8.5_d are taken into account. The ten unknown constants ($\Xi_1^{(1)}$, $\Xi_2^{(1)}$, $\Xi_3^{(1)}$, $\Xi_4^{(1)}$, $\Xi_1^{(0)}$, $\Xi_2^{(0)}$, $\Xi_3^{(0)}$, $\Xi_4^{(0)}$, Ξ_3^{eq}, Ξ_4^{eq}) and the macroscopic transverse shear modulus $\bar{\mu}^{\mathrm{tr}}$ are identified as follows:

- At the center of the RVE the displacement field needs to be finite. Since at $r = 0$ the inhomogeneity phase exists, this leads to the conclusion that $\Xi_3^{(1)} = \Xi_4^{(1)} = 0$.
- Considering the effective medium of Fig. 8.6 under the same boundary conditions, the displacement field at every position is given by the homogeneous solution

$$\bar{u}_r(r,\vartheta) = \frac{\beta r}{2\bar{\mu}^{\text{tr}}}\sin 2\vartheta, \quad \bar{u}_\vartheta(r,\vartheta) = \frac{\beta r}{2\bar{\mu}^{\text{tr}}}\cos 2\vartheta, \quad \bar{u}_z = 0.$$

From the Eshelby energy principle we find the relation [3]

$$\int_{-L}^{L}\int_0^{2\pi}[\sigma_{rr}^{\text{eq}}\bar{u}_r + \sigma_{r\vartheta}^{\text{eq}}\bar{u}_\vartheta + \sigma_{rz}^{\text{eq}}\bar{u}_z - \bar{\sigma}_{rr}u_r^{\text{eq}} \\ -\bar{\sigma}_{r\vartheta}u_\vartheta^{\text{eq}} - \bar{\sigma}_{rz}u_z^{\text{eq}}]_{r=r_0}\mathrm{d}\vartheta\,\mathrm{d}z = 0\,. \tag{8.48}$$

Substituting the stress and displacement fields of the two problems at the position $r = r_0$ into the above integral yields the simple result $\Xi_4^{\text{eq}} = 0$.
- The remaining eight unknowns are identified by the interface conditions (8.25) and (8.26). Since the obtained system of equations is nonlinear, a special treatment is required. Conditions $(8.25)_{1,3,2,4}$ and $(8.26)_{2,4}$ lead to the following system:

$$\boldsymbol{Z}\cdot\begin{bmatrix}\Xi_1^{(1)}\\ \Xi_2^{(1)}\\ \Xi_1^{(0)}\\ \Xi_2^{(0)}\\ \Xi_3^{(0)}\\ \Xi_4^{(0)}\end{bmatrix} = \begin{bmatrix}0\\0\\0\\0\\1\\1\end{bmatrix} + \begin{bmatrix}0\\0\\0\\0\\3\\-3\end{bmatrix}\Xi_3^{\text{eq}}, \tag{8.49}$$

where

$$\mathbf{Z} = \begin{bmatrix} \frac{K_1^{\mathrm{tr}} - \mu_1^{\mathrm{tr}}}{2K_1^{\mathrm{tr}} + \mu_1^{\mathrm{tr}}} & 1 & -\frac{K_0^{\mathrm{tr}} - \mu_0^{\mathrm{tr}}}{2K_0^{\mathrm{tr}} + \mu_0^{\mathrm{tr}}} & -1 & 1 & -\frac{K_0^{\mathrm{tr}} + \mu_0^{\mathrm{tr}}}{\mu_0^{\mathrm{tr}}} \\ 1 & 1 & -1 & -1 & -1 & -1 \\ 0 & 2\mu_1^{\mathrm{tr}} & 0 & -2\mu_0^{\mathrm{tr}} & -6\mu_0^{\mathrm{tr}} & 4K_0^{\mathrm{tr}} \\ \frac{6K_1^{\mathrm{tr}}\mu_1^{\mathrm{tr}}}{2K_1^{\mathrm{tr}} + \mu_1^{\mathrm{tr}}} & 2\mu_1^{\mathrm{tr}} & -\frac{6K_0^{\mathrm{tr}}\mu_0^{\mathrm{tr}}}{2K_0^{\mathrm{tr}} + \mu_0^{\mathrm{tr}}} & -2\mu_0^{\mathrm{tr}} & 6\mu_0^{\mathrm{tr}} & -2K_0^{\mathrm{tr}} \\ 0 & 0 & 0 & 2\mu_0^{\mathrm{tr}} & 6c^2\mu_0^{\mathrm{tr}} & -4cK_0^{\mathrm{tr}} \\ 0 & 0 & \frac{1}{c}\frac{6K_0^{\mathrm{tr}}\mu_0^{\mathrm{tr}}}{2K_0^{\mathrm{tr}} + \mu_0^{\mathrm{tr}}} & 2\mu_0^{\mathrm{tr}} & -6c^2\mu_0^{\mathrm{tr}} & 2cK_0^{\mathrm{tr}} \end{bmatrix}.$$

The solution of this system is expressed in the form

$$\begin{bmatrix} \Xi_1^{(1)} \\ \Xi_2^{(1)} \\ \Xi_1^{(0)} \\ \Xi_2^{(0)} \\ \Xi_3^{(0)} \\ \Xi_4^{(0)} \end{bmatrix} = \begin{bmatrix} g_1 \\ g_2 \\ a_1 \\ a_2 \\ a_3 \\ a_4 \end{bmatrix} + \begin{bmatrix} h_1 \\ h_2 \\ b_1 \\ b_2 \\ b_3 \\ b_4 \end{bmatrix} \Xi_3^{\mathrm{eq}}. \tag{8.50}$$

Using (8.50), we write the last two conditions $(8.26)_{1,3}$ as

$$\begin{aligned} a_5 + b_5 \Xi_3^{\mathrm{eq}} &= \frac{1}{2\overline{\mu}^{\mathrm{tr}}}\left[1 - \Xi_3^{\mathrm{eq}}\right], \\ a_6 + b_6 \Xi_3^{\mathrm{eq}} &= \frac{1}{2\overline{\mu}^{\mathrm{tr}}}\left[1 + \Xi_3^{\mathrm{eq}}\right], \end{aligned} \tag{8.51}$$

with

$$\begin{aligned} a_5 &= \frac{1}{c}\frac{K_0^{\mathrm{tr}} - \mu_0^{\mathrm{tr}}}{2K_0^{\mathrm{tr}} + \mu_0^{\mathrm{tr}}} a_1 + a_2 - c^2 a_3 + c\frac{K_0^{\mathrm{tr}} + \mu_0^{\mathrm{tr}}}{\mu_0^{\mathrm{tr}}} a_4, \\ a_6 &= \frac{1}{c} a_1 + a_2 + c^2 a_3 + c a_4, \\ b_5 &= \frac{1}{c}\frac{K_0^{\mathrm{tr}} - \mu_0^{\mathrm{tr}}}{2K_0^{\mathrm{tr}} + \mu_0^{\mathrm{tr}}} b_1 + b_2 - c^2 b_3 + c\frac{K_0^{\mathrm{tr}} + \mu_0^{\mathrm{tr}}}{\mu_0^{\mathrm{tr}}} b_4, \end{aligned}$$

$$b_6 = \frac{1}{c} b_1 + b_2 + c^2 b_3 + c b_4. \tag{8.52}$$

Subtracting $(8.51)_1$ from $(8.51)_2$ gives

$$\Xi_3^{\text{eq}} = \frac{\bar{\mu}^{\text{tr}}[a_6 - a_5]}{\bar{\mu}^{\text{tr}}[b_5 - b_6] + 1}.$$

Substituting the final result into $(8.51)_1$, after some algebra, we obtain the quadratic equation

$$2[a_6 b_5 - a_5 b_6]\left[\bar{\mu}^{\text{tr}}\right]^2 + [a_5 + a_6 + b_6 - b_5]\bar{\mu}^{\text{tr}} - 1 = 0.$$

One of two possible solutions is positive, and the other is negative. The positive value is the macroscopic shear modulus.

In conclusion:

BOX 8.6 CCA: Transverse shear modulus

Consider a transversely isotropic long fiber composite with fiber volume fraction c, fiber properties K_1^{tr}, μ_1^{tr}, μ_1^{ax}, l_1, n_1, and matrix properties K_0^{tr}, μ_0^{tr}, μ_0^{ax}, l_0, n_0. The macroscopic transverse shear modulus, according to the generalized self–consistent CCA method, is given by

$$\bar{\mu}^{\text{tr}} = \max\left\{\frac{-w + \sqrt{\Delta}}{4[a_6 b_5 - a_5 b_6]}, \frac{-w - \sqrt{\Delta}}{4[a_6 b_5 - a_5 b_6]}\right\},$$

$$w = a_5 + a_6 + b_6 - b_5, \qquad \Delta = w^2 + 8[a_6 b_5 - a_5 b_6]. \tag{8.53}$$

The necessary constants a_i, b_i, $i = 1, ..., 6$, are obtained from the solution of system (8.49) and (8.50) and (8.52).

8.3 Eshelby's energy principle

Before closing the theoretical discussion on the CSA and CCA methods, it is important to mention the Eshelby energy principle, a powerful tool that has been utilized in the development of the generalized self–consistent approach. The proof presented here follows the methodology described by Christensen [3].

Let us consider a composite consisting of a matrix phase and reinforcement. According to the self–consistent approach, we can visualize this composite as a three–phase medium, where the matrix and reinforcement occupy the space $\mathcal{B}_0$, and the homogenized (effective) medium is added as an external layer extended until the far distance boundary surface $\partial\mathcal{B}$

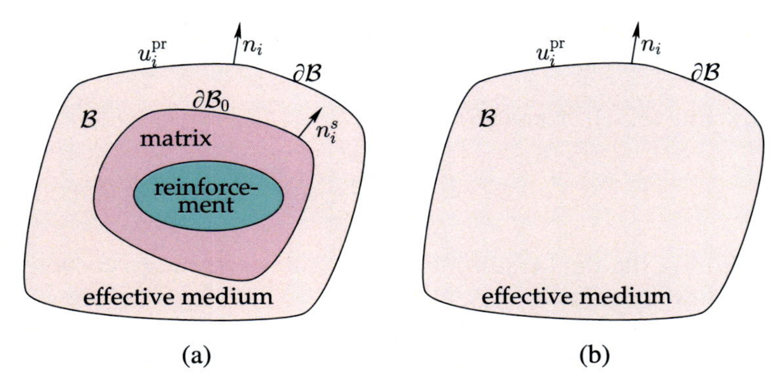

FIGURE 8.7 (a) Three–phase model, including the reinforcement, the matrix, and the effective medium. (b) Hypothetical homogenized medium. The two cases are subjected to the same displacement boundary conditions.

(Fig. 8.7$_a$). At $\partial\mathcal{B}$ a prescribed displacement field u_i^{pr} is applied. The elasticity tensor of the effective medium is denoted as $\bar{L}_{ijkl}$, whereas the elasticity tensor in the matrix–reinforcement system is taken as a position–dependent function $L_{ijkl}(\boldsymbol{x})$. The system of equations that describe the mechanical response of this medium are the following:

$$
\begin{aligned}
\varepsilon_{ij} &= \frac{1}{2}\left[\frac{\partial u_i}{\partial x_j} + \frac{\partial u_j}{\partial x_i}\right] \quad \text{in } \mathcal{B}, \\
\frac{\partial \sigma_{ij}}{\partial x_j} &= 0 \quad \text{in } \mathcal{B}, \\
\sigma_{ij} &= \begin{cases} L_{ijkl}(\boldsymbol{x})\varepsilon_{kl}, & \boldsymbol{x} \in \mathcal{B}_0, \\ \bar{L}_{ijkl}\varepsilon_{kl}, & \boldsymbol{x} \in \mathcal{B} - \mathcal{B}_0, \end{cases} \\
u_i^{\text{pr}} & \text{ known on } \partial\mathcal{B}.
\end{aligned} \tag{8.54}
$$

Using the divergence theorem and the equilibrium equation $(8.54)_2$, we write the energy of this composite medium as

$$
\begin{aligned}
U &= \frac{1}{2}\int_{\mathcal{B}} \sigma_{ij}\varepsilon_{ij}\,\mathrm{d}V = \frac{1}{2}\int_{\mathcal{B}} \frac{\partial}{\partial x_j}\left(u_i\sigma_{ij}\right)\mathrm{d}V - \frac{1}{2}\int_{\mathcal{B}} \frac{\partial \sigma_{ij}}{\partial x_j} u_i\,\mathrm{d}V \\
&= \frac{1}{2}\int_{\partial\mathcal{B}} t_i u_i^{\text{pr}}\,\mathrm{d}S,
\end{aligned} \tag{8.55}
$$

where $t_i = \sigma_{ij}n_j$ is the traction vector on the boundary surface $\partial\mathcal{B}$, and n_i is the unit normal vector to the boundary.

The aim of the homogenization is to substitute all the phases into the $\mathcal{B}$ space with one single homogeneous material with elastic properties $\bar{L}_{ijkl}$

(Fig. 8.7_b). The homogenized body is described by the system of equations

$$\begin{aligned}\bar{\varepsilon}_{ij} &= \frac{1}{2}\left[\frac{\partial \bar{u}_i}{\partial x_j} + \frac{\partial \bar{u}_j}{\partial x_i}\right] \quad \text{in } \mathcal{B},\\ \frac{\partial \bar{\sigma}_{ij}}{\partial x_j} &= 0 \quad \text{in } \mathcal{B},\\ \bar{\sigma}_{ij} &= \bar{L}_{ijkl}\bar{\varepsilon}_{kl} \quad \text{in } \mathcal{B},\\ u_i^{\text{pr}} & \quad \text{known on } \partial\mathcal{B}.\end{aligned} \tag{8.56}$$

Following the same calculation steps, the energy of the homogenized medium is given by

$$\bar{U} = \frac{1}{2}\int_{\mathcal{B}} \bar{\sigma}_{ij}\bar{\varepsilon}_{ij}\,\mathrm{d}V = \frac{1}{2}\int_{\partial\mathcal{B}} \bar{t}_i u_i^{\text{pr}}\,\mathrm{d}S\,. \tag{8.57}$$

Since the two bodies of Figs. 8.7_a and 8.7_b are assumed equivalent, the total mechanical energies produced by their mechanical fields must be equal, that is, $U = \bar{U}$ or

$$\int_{\partial\mathcal{B}} \left[t_i u_i^{\text{pr}} - \bar{t}_i u_i^{\text{pr}}\right] \mathrm{d}S = 0. \tag{8.58}$$

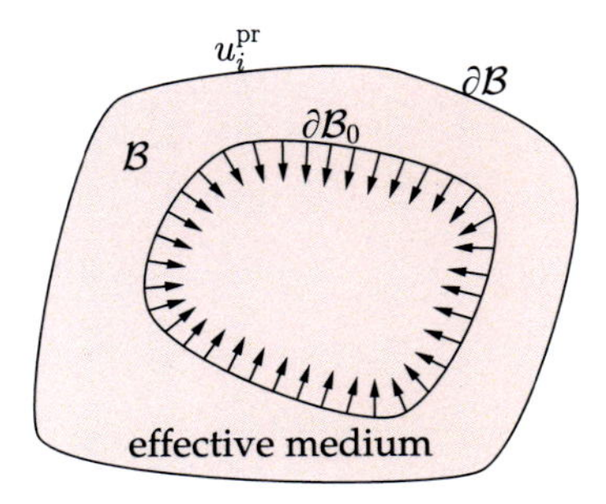

FIGURE 8.8 Hypothetical medium, in which the matrix-reinforcement system of Fig. 8.7_a is substituted by the effective medium and certain spatially dependent body forces. The mechanical fields produced in the space $\mathcal{B} - \mathcal{B}_0$ in this hypothetical medium and the actual medium of Fig. 8.7_a are assumed identical.

At this point, we introduce a new hypothetical medium (Fig. 8.8), in which the matrix–reinforcement system occupying the space $\mathcal{B}_0$ of Fig. 8.7_a is substituted by the effective medium and certain spatially dependent body forces. The body forces are chosen in such way that the displacements, strains, and stresses produced in the space $\mathcal{B} - \mathcal{B}_0$ are identical with those produced in the actual medium of 8.7_a. Using the symbol $\breve{}$

above a variable to denote fields of the new hypothetical medium, the latter condition is written as

$$\breve{u}_i = u_i, \quad \breve{\varepsilon}_{ij} = \varepsilon_{ij}, \quad \breve{\sigma}_{ij} = \sigma_{ij} \quad \text{in } \mathcal{B} - \mathcal{B}_0. \tag{8.59}$$

The system of equations describing the mechanical response of the hypothetical medium of Fig. 8.8 are

$$\begin{aligned} \breve{\varepsilon}_{ij} &= \frac{1}{2}\left[\frac{\partial \breve{u}_i}{\partial x_j} + \frac{\partial \breve{u}_j}{\partial x_i}\right] \quad \text{in } \mathcal{B}, \\ \frac{\partial \breve{\sigma}_{ij}}{\partial x_j} &+ \breve{b}_i(\boldsymbol{x}) = 0 \quad \text{in } \mathcal{B}_0, \\ \frac{\partial \breve{\sigma}_{ij}}{\partial x_j} &= 0 \quad \text{in } \mathcal{B} - \mathcal{B}_0, \\ \breve{\sigma}_{ij} &= \bar{L}_{ijkl}\breve{\varepsilon}_{kl} \quad \text{in } \mathcal{B}, \\ u_i^{\text{pr}} & \text{ known on } \partial\mathcal{B}. \end{aligned} \tag{8.60}$$

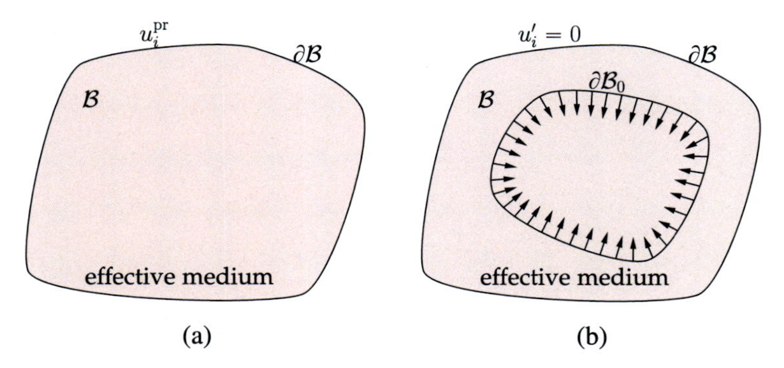

FIGURE 8.9 Decomposition of the boundary value problem of Fig. 8.8 into two simpler ones.

Due to the principle of superposition that holds in elasticity, the latter mechanical response can be decomposed into two simpler boundary value problems:

1. In the first one the body forces are removed, whereas the boundary displacements remain the same (Fig. 8.9$_a$). This boundary value problem is identical with that of Fig. 8.7$_b$, leading to the same Eqs. (8.56).
2. In the second one the body forces remain, whereas the body is displacement free at the boundary (Fig. 8.9$_b$). Denoting by $'$ the fields related to this problem, the set of equations that characterize the behavior of the

second medium is

$$
\begin{aligned}
\varepsilon'_{ij} &= \frac{1}{2}\left[\frac{\partial u'_i}{\partial x_j} + \frac{\partial u'_j}{\partial x_i}\right] \quad \text{in } \mathcal{B}, \\
\frac{\partial \sigma'_{ij}}{\partial x_j} &+ b'_i(\boldsymbol{x}) = 0 \quad \text{in } \mathcal{B}_0, \\
\frac{\partial \sigma'_{ij}}{\partial x_j} &= 0 \quad \text{in } \mathcal{B} - \mathcal{B}_0, \\
\sigma'_{ij} &= \bar{L}_{ijkl}\varepsilon'_{kl} \quad \text{in } \mathcal{B}, \\
u'_i &= 0 \quad \text{on } \partial\mathcal{B}.
\end{aligned} \tag{8.61}
$$

According to the superposition principle, we have the relations

$$
\breve{\sigma}_{ij} = \bar{\sigma} + \sigma'_{ij}, \quad \breve{\varepsilon}_{ij} = \bar{\varepsilon}_{ij} + \varepsilon'_{ij}, \quad \breve{u}_i = \bar{u} + u'_i . \tag{8.62}
$$

Taking into account conditions (8.59), we can implement the field decompositions (8.62) in (8.58) and obtain

$$
\int_{\partial\mathcal{B}} t'_i u_i^{\mathrm{pr}} \, \mathrm{d}S = 0, \quad t'_i = \sigma'_{ij} n_j . \tag{8.63}
$$

The elastic energy of the medium of Fig. 8.8 is given by

$$
\begin{aligned}
\breve{U} &= \frac{1}{2}\int_{\mathcal{B}} \breve{\sigma}_{ij}\breve{\varepsilon}_{ij} \, \mathrm{d}V = \frac{1}{2}\int_{\mathcal{B}} [\bar{\sigma}_{ij} + \sigma'_{ij}][\bar{\varepsilon}_{ij} + \varepsilon'_{ij}] \, \mathrm{d}V \\
&= \frac{1}{2}\int_{\mathcal{B}} \bar{\sigma}_{ij}\bar{\varepsilon}_{ij} \, \mathrm{d}V + \frac{1}{2}\int_{\mathcal{B}} \sigma'_{ij}\varepsilon'_{ij} \, \mathrm{d}V + \frac{1}{2}\int_{\mathcal{B}} [\sigma'_{ij}\bar{\varepsilon}_{ij} + \bar{\sigma}_{ij}\varepsilon'_{ij}] \, \mathrm{d}V .
\end{aligned} \tag{8.64}
$$

In (8.64) the last term is the interaction energy U_{INT}. The symmetries of the stress, the strain, and the macroscopic elasticity tensor $\bar{L}_{ijkl}$ allow us to write

$$
\sigma'_{ij}\bar{\varepsilon}_{ij} = [\bar{L}_{ijkl}\varepsilon'_{kl}]\bar{\varepsilon}_{ij} = [\bar{L}_{ijkl}\bar{\varepsilon}_{kl}]\varepsilon'_{ij} = \bar{\sigma}_{ij}\varepsilon'_{ij} . \tag{8.65}
$$

So the interaction energy can be written in the forms

$$
\begin{aligned}
U_{\mathrm{INT}} &= \frac{1}{2}\int_{\mathcal{B}} [\sigma'_{ij}\bar{\varepsilon}_{ij} + \bar{\sigma}_{ij}\varepsilon'_{ij}] \, \mathrm{d}V \\
&= \int_{\mathcal{B}} \bar{\sigma}_{ij}\varepsilon'_{ij} \, \mathrm{d}V = \int_{\mathcal{B}_0} \bar{\sigma}_{ij}\varepsilon'_{ij} \, \mathrm{d}V + \int_{\mathcal{B}-\mathcal{B}_0} \sigma'_{ij}\bar{\varepsilon}_{ij} \, \mathrm{d}V .
\end{aligned} \tag{8.66}
$$

By the divergence theorem and Eq. $(8.56)_2$ the second form of (8.66) yields

$$
U_{\mathrm{INT}} = \int_{\mathcal{B}} \bar{\sigma}_{ij}\varepsilon'_{ij} \, \mathrm{d}V = \int_{\partial\mathcal{B}} \bar{t}_i u'_i \, \mathrm{d}S = 0, \tag{8.67}
$$

since $u_i' = 0$ on $\partial\mathcal{B}$. Implementing the divergence theorem in the third form of (8.66) and taking into account (8.56)$_2$, (8.61)$_3$, and the continuity of displacements on the surface $\partial\mathcal{B}_0$ yield

$$U_{\text{INT}} = \int_{\partial\mathcal{B}_0} \bar{t}_i u_i' \,\mathrm{d}S - \int_{\partial\mathcal{B}_0} t_i' \bar{u}_i \,\mathrm{d}S + \int_{\partial\mathcal{B}} t_i' u_i^{\text{pr}} \,\mathrm{d}S\,, \tag{8.68}$$

where $t_i' = \sigma_{ij} n_j^s$, and n_i^s is the unit normal vector of the surface $\partial\mathcal{B}_0$ (Fig. 8.7$_a$). Using (8.63) and (8.67) in the last result gives

$$\int_{\partial\mathcal{B}_0} \left[t_i' \bar{u}_i - \bar{t}_i u_i'\right] \mathrm{d}S = 0 \tag{8.69}$$

or, by (8.62),

$$\int_{\partial\mathcal{B}_0} \left[\breve{t}_i \bar{u}_i - \bar{t}_i \breve{u}_i\right] \mathrm{d}S = 0\,, \tag{8.70}$$

where $\breve{t}_i = \breve{\sigma}_{ij} n_j^s$. Using conditions (8.59) and the continuity of tractions on the surface $\partial\mathcal{B}_0$, we obtain the expression

$$\int_{\partial\mathcal{B}_0} \left[t_i \bar{u}_i - \bar{t}_i u_i\right] \mathrm{d}S = 0\,. \tag{8.71}$$

The final formula is called Eshelby's energy principle. It is important to mention that we achieve the same result even if applied tractions instead of displacements are considered on the boundary $\partial\mathcal{B}$ [3]. Eq. (8.71) has been utilized in the computation of the shear moduli in the generalized self–consistent composite sphere and composite cylinder assemblage approaches (Eqs. (8.14) and (8.48), respectively).

8.4 Universal relations for fiber composites

As already mentioned in Chapter 2, the elasticity tensor of a transversely isotropic medium can be described by five independent material parameters. A composite reinforced with unidirectional long cylindrical fibers presents transversely isotropic behavior, but as Hill [5] and Dvorak [6] have illustrated, three parameters are sufficient to express the elasticity tensor of this composite.

To demonstrate this finding, let us assume that the matrix phase and the fibers of the reference composite are at most transversely isotropic media whose elasticity tensors are given by the relations

$$\boldsymbol{L}_q = \begin{bmatrix} K_q^{\text{tr}} + \mu_q^{\text{tr}} & K_q^{\text{tr}} - \mu_q^{\text{tr}} & l_q & 0 & 0 & 0 \\ K_q^{\text{tr}} - \mu_q^{\text{tr}} & K_q^{\text{tr}} + \mu_q^{\text{tr}} & l_q & 0 & 0 & 0 \\ l_q & l_q & n_q & 0 & 0 & 0 \\ 0 & 0 & 0 & \mu_q^{\text{tr}} & 0 & 0 \\ 0 & 0 & 0 & 0 & \mu_q^{\text{ax}} & 0 \\ 0 & 0 & 0 & 0 & 0 & \mu_q^{\text{ax}} \end{bmatrix}, \tag{8.72}$$

where $q = 0$ denotes the matrix, and $q = 1$ refers to the fibers. The composite macroscopic elasticity tensor is expressed in a similar form as

$$\bar{\boldsymbol{L}} = \begin{bmatrix} \bar{K}^{\text{tr}} + \bar{\mu}^{\text{tr}} & \bar{K}^{\text{tr}} - \bar{\mu}^{\text{tr}} & \bar{l} & 0 & 0 & 0 \\ \bar{K}^{\text{tr}} - \bar{\mu}^{\text{tr}} & \bar{K}^{\text{tr}} + \bar{\mu}^{\text{tr}} & \bar{l} & 0 & 0 & 0 \\ \bar{l} & \bar{l} & \bar{n} & 0 & 0 & 0 \\ 0 & 0 & 0 & \bar{\mu}^{\text{tr}} & 0 & 0 \\ 0 & 0 & 0 & 0 & \bar{\mu}^{\text{ax}} & 0 \\ 0 & 0 & 0 & 0 & 0 & \bar{\mu}^{\text{ax}} \end{bmatrix}. \tag{8.73}$$

Upon a triaxial state of macroscopic deformation without shearing (i.e., $\bar{\varepsilon}_{12} = \bar{\varepsilon}_{13} = \bar{\varepsilon}_{23} = 0$), the stress–strain relation for the composite can be described by the following two equations:

$$\frac{1}{2}\left[\bar{\sigma}_{11} + \bar{\sigma}_{22}\right] = \bar{K}^{\text{tr}}\bar{e} + \bar{l}\bar{\varepsilon}_{33},$$
$$\bar{\sigma}_{33} = \bar{l}\bar{e} + \bar{n}\bar{\varepsilon}_{33}, \tag{8.74}$$

with

$$\bar{e} = \bar{\varepsilon}_{11} + \bar{\varepsilon}_{22}. \tag{8.75}$$

As Hill [5] pointed out, under such a loading, we expect that *"all transverse sections remain plane and the axial strain component is independent of position"*. In other words, both the matrix and the fibers present average stress and average strain states that obey the relations

$$\frac{1}{2}\left[\sigma_{11}^r + \sigma_{22}^r\right] = K_r^{\text{tr}} e^r + l_r\bar{\varepsilon}_{33},$$

$$\sigma_{33}^r = l_r e^r + n_r \bar{\varepsilon}_{33}, \tag{8.76}$$

with

$$e^r = \varepsilon_{11}^r + \varepsilon_{22}^r. \tag{8.77}$$

The quantities σ_{11}^r, σ_{22}^r, ε_{11}^r, ε_{22}^r $(r = 0, 1)$ denote the average values of the corresponding stress and strain components at the phase r in the sense of expressions (3.23) of Chapter 3. If c is the volume fraction of the fibers, then the macroscopic stresses and strains and their microscopic counterparts are connected through the classical averaging expressions

$$\begin{aligned}\frac{1}{2}\left[\bar{\sigma}_{11} + \bar{\sigma}_{22}\right] &= \frac{1-c}{2}\left[\sigma_{11}^0 + \sigma_{22}^0\right] + \frac{c}{2}\left[\sigma_{11}^1 + \sigma_{22}^1\right],\\ \bar{\sigma}_{33} &= [1-c]\sigma_{33}^0 + c\sigma_{33}^1,\\ \bar{e} &= [1-c]e^0 + ce^1.\end{aligned} \tag{8.78}$$

Substituting (8.78) and (8.76) into (8.74) after small algebra yields

$$\begin{aligned}[1-c]e^0[\bar{K}^{\text{tr}} - K_0^{\text{tr}}] + ce^1[\bar{K}^{\text{tr}} - K_1^{\text{tr}}] + \bar{\varepsilon}_{33}\left[\bar{l} - [1-c]l_0 - cl_1\right] &= 0,\\ [1-c]e^0[\bar{l} - l_0] + ce^1[\bar{l} - l_1] + \bar{\varepsilon}_{33}\left[\bar{n} - [1-c]n_0 - cn_1\right] &= 0.\end{aligned} \tag{8.79}$$

The last expressions hold for arbitrary choices of macroscopic strains $\bar{\varepsilon}_{11}$, $\bar{\varepsilon}_{22}$, and $\bar{\varepsilon}_{33}$. Choosing $\bar{\varepsilon}_{33} = 0$, a combination of the two relations of (8.79) gives

$$\frac{\bar{K}^{\text{tr}} - K_0^{\text{tr}}}{\bar{l} - l_0} = \frac{\bar{K}^{\text{tr}} - K_1^{\text{tr}}}{\bar{l} - l_1}. \tag{8.80}$$

On the other hand, it is possible to identify such values for the three macroscopic components that lead to $e^1 = 0$. Under such conditions, a combination of the two relations of (8.79) gives

$$\frac{\bar{K}^{\text{tr}} - K_0^{\text{tr}}}{\bar{l} - l_0} = \frac{\bar{l} - [1-c]l_0 - cl_1}{\bar{n} - [1-c]n_0 - cn_1}. \tag{8.81}$$

The so–called universal relations (8.80) and (8.81) imply that the knowledge of only one of the parameters $\bar{K}^{\text{tr}}$, $\bar{l}$, and $\bar{n}$ is sufficient for identifying the other two.

With the help of some cumbersome algebraic calculations we can show that the composite cylinder assemblage renders macroscopic properties that respect the universal relations.

8.5 Examples

Example 1: particulate composite effective properties

Consider a bi–phase composite consisting of an epoxy matrix and glass reinforcement in the form of spherical particles. The material properties (bulk modulus K and shear modulus μ) of the two phases are summarized below:

matrix	glass particles
$K_0 = 2.5$ GPa	$K_1 = 44$ GPa
$\mu_0 = 1.15$ GPa	$\mu_1 = 30$ GPa

Identify the macroscopic elastic properties of this composite for reinforcement volume fraction between 20% and 80% using a) the CSA and b) the Mori–Tanaka method.

Solution:

For the Mori–Tanaka method, the analytical solution of the problem is provided in Example 1 of Chapter 5. For the CSA and the generalized self–consistent CSA (GSCCSA) approaches, the formulas of Section 8.1 are utilized. The obtained results for the macroscopic bulk and shear moduli are illustrated in Fig. 8.10. As we can observe, the two methods provide exactly the same macroscopic bulk modulus, and they have small difference in the macroscopic shear modulus. For higher volume fractions, CSA predicts higher values for $\bar{\mu}$.

The corresponding python script for applying the composite sphere assemblage method to the studied composite is the following:

```
import numpy as np
# Data
K0=2.5
mu0=1.15
K1=44.0
mu1=30.0

tableA=np.zeros((61,3))

for n in range(61):
   c=0.01*n+0.2

# Bulk modulus
   B1=np.matrix([[1., -1., -1],\
                 [3*K1, -3*K0, 4*mu0],\
                 [0., 1., c]])
   F1=np.matrix([[0.], [0.], [1.]])
```

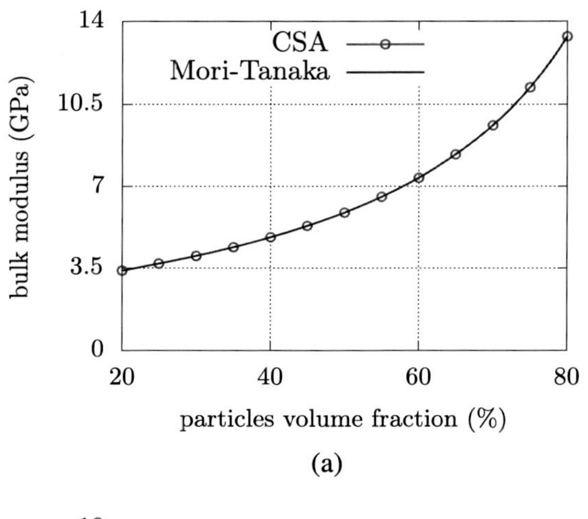

(a)

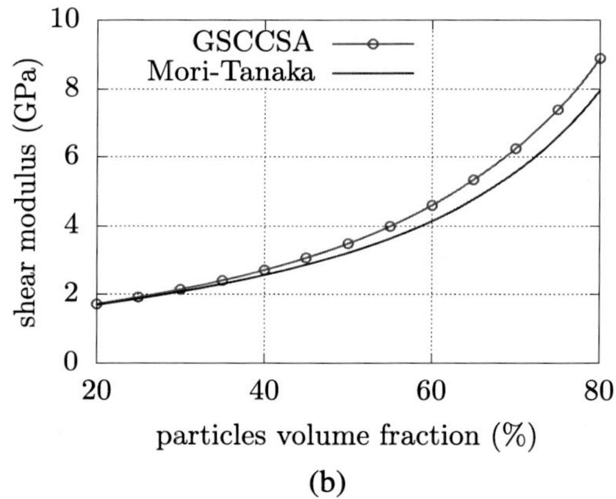

(b)

FIGURE 8.10 Macroscopic properties versus particles volume fraction using CSA (or GSCCSA) and Mori–Tanaka methods: (a) bulk modulus and (b) shear modulus.

```
    U1=(B1.I)*F1
    Keff=(3*K0*U1[1]-4*c*mu0*U1[2])/3

# Shear modulus
    B2=np.matrix([[1., 2.-3*K1/mu1, -1.,\
        -2.+3*K0/mu0, -3., -3.-3*K0/mu0],\
          [1., -11./3.-5*K1/mu1, -1.,\
         11./3.+5*K0/mu0, 2., -2.],\
          [2*mu1, 3*K1-2*mu1, -2*mu0,\
        -3*K0+2*mu0, 24*mu0, 18*K0+8*mu0],\
          [2*mu1, -16*K1-10*mu1/3, -2*mu0,\
         16*K0+10*mu0/3, -16*mu0, -6*K0],\
```

```
        [0., 0., 1.,\
         (2.-3*K0/mu0)*(c**(-2./3.)),\
      3*(c**(5./3.)), (3+3*K0/mu0)*c],\
        [0., 0., 1.,\
         -(11./3.+5*K0/mu0)*(c**(-2./3.)),\
      -2*(c**(5./3.)), 2*c]])
  F2=np.matrix([[0., 0.], [0., 0.], [0., 0.],\
               [0., 0.], [1., 3.], [1., -2.]])
  U2=(B2.I)*F2
  a5=2*mu0*U2[2,0]\
     +(3*K0-2*mu0)*c**(-2./3.)*U2[3,0]\
     -24*mu0*c**(5./3.)*U2[4,0]\
     -(18*K0+8*mu0)*c*U2[5,0]
  a6=2*mu0*U2[2,0]\
     -(16*K0+10*mu0/3)*c**(-2./3.)*U2[3,0]\
     +16*mu0*c**(5./3.)*U2[4,0]\
     +6*K0*c*U2[5,0]
  b5=2*mu0*U2[2,1]\
     +(3*K0-2*mu0)*c**(-2./3.)*U2[3,1]\
     -24*mu0*c**(5./3.)*U2[4,1]\
     -(18*K0+8*mu0)*c*U2[5,1]
  b6=2*mu0*U2[2,1]\
     -(16*K0+10*mu0/3)*c**(-2./3.)*U2[3,1]\
     +16*mu0*c**(5./3.)*U2[4,1]\
     +6*K0*c*U2[5,1]
  Delta=(b6-b5+12*a6+8*a5)**2-80*(a5*b6-b5*a6)
  mueff=(b6-b5+12*a6+8*a5+np.sqrt(Delta))/80

  tableA[n,:]=[c, Keff, mueff]

np.savetxt('K_mu_CSA.txt',tableA,fmt='%1.5f')
```

Example 2: long fiber composite effective properties

In the previous example, substitute the particles with long fibers and compute the macroscopic transverse shear modulus using the generalized self–consistent composite cylinder assemblage (GSCCCA), Mori–Tanaka, and Voigt–Reuss methods.

Solution:

The computation of the macroscopic properties according to the Mori–Tanaka method has been discussed in Example 3 of Chapter 5. The overall behavior according to the Voigt–Reuss approach can be obtained using formulas (4.16) of Chapter 4. For the generalized self–consistent CCA (GSCCCA) approach, the formulas of Section 8.2 are utilized. The obtained results from both methods for macroscopic transverse shear modulus are illustrated in Fig. 8.11. As we can observe, for higher volume fractions, CCA predicts higher values for $\bar{\mu}^{tr}$. The Voigt–Reuss approach clearly underestimates the macroscopic shear modulus.

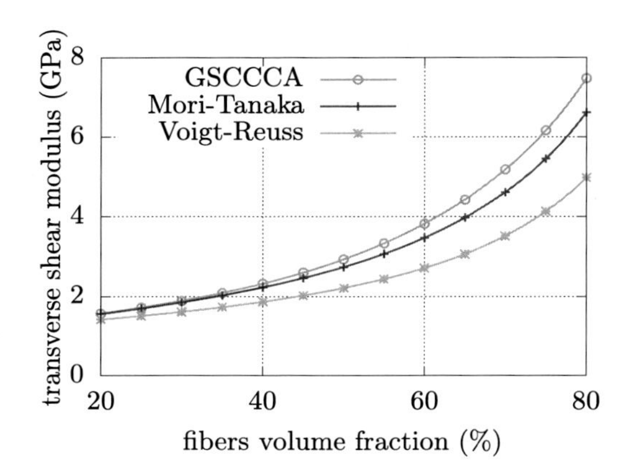

FIGURE 8.11 Macroscopic transverse shear modulus versus fiber volume fraction using GSCCCA, Mori–Tanaka and Voigt–Reuss methods.

The Python script for the Mori–Tanaka approach is similar to that described in Example 3 of Chapter 5. For completeness, we give the Python script for obtaining all the macroscopic properties using the composite cylinders assemblage method:

```
import numpy as np
# Data
K0=2.5
mu0=1.15
K1=44.0
mu1=30.0

Ktr0=K0+mu0/3
mutr0=mu0
muax0=mu0
n0=Ktr0+mutr0
l0=Ktr0-mutr0
Ktr1=K1+mu1/3
mutr1=mu1
muax1=mu1
n1=Ktr1+mutr1
l1=Ktr1-mutr1
```

```
tableA=np.zeros((61,6))

for n in range(61):
   c=0.01*n+0.2

# Transverse bulk modulus
   B1=np.matrix([[1., -1., -1.],\
                 [2*Ktr1, -2*Ktr0, 2*mutr0],\
                 [0., 1., c]])
   F1=np.matrix([[0.], [0.], [1.]])
   U1=(B1.I)*F1
   Ktreff=Ktr0*U1[1]-c*mutr0*U1[2]

# Axial shear modulus
   B2=np.matrix([[1., -1., -1.],\
                 [muax1, -muax0, muax0],\
                 [0., 1., c]])
   F2=np.matrix([[0.], [0.], [2.]])
   U2=(B2.I)*F2
   muaxeff=(U2[1]-c*U2[2])*muax0/2

# Axial Young's modulus and Poisson's ratio
   B3=np.matrix([[1., -1., -1.],\
                 [2*Ktr1, -2*Ktr0, 2*mutr0],\
                 [0., 2*Ktr0, -2*c*mutr0]])
   F3=np.matrix([[0.], [l0-l1], [-l0]])
   U3=(B3.I)*F3
   Eaxeff=c*(2*l1*U3[0]+n1)\
         +(1-c)*(2*l0*U3[1]+n0)
   vaxeff=-(U3[1]+c*U3[2])

# Transverse shear modulus
   B5=np.matrix([[(Ktr1-mutr1)/(2*Ktr1+mutr1), 1.,\
         -(Ktr0-mutr0)/(2*Ktr0+mutr0), -1.,\
         1., -(Ktr0+mutr0)/mutr0],\
      [1., 1., -1., -1., -1., -1.],\
      [0., 2*mutr1, 0., -2*mutr0, -6*mutr0, 4*Ktr0],\
      [6*Ktr1*mutr1/(2*Ktr1+mutr1), 2*mutr1,\
       -6*Ktr0*mutr0/(2*Ktr0+mutr0),\
       -2*mutr0, 6*mutr0, -2*Ktr0],\
      [0., 0., 0., 2*mutr0, 6*c*c*mutr0, -4*c*Ktr0],\
      [0., 0., (6/c)*Ktr0*mutr0/(2*Ktr0+mutr0),\
       2*mutr0, -6*c*c*mutr0, 2*c*Ktr0]])
   F5=np.matrix([[0., 0.], [0., 0.], [0., 0.],\
                 [0., 0.], [1., 3.], [1., -3.]])
```

```
    U5=(B5.I)*F5
    a5=(U5[2,0]/c)*(Ktr0-mutr0)/(2*Ktr0+mutr0)\
       +U5[3,0]-c*c*U5[4,0]\
       +c*U5[5,0]*(Ktr0+mutr0)/mutr0
    a6=(U5[2,0]/c)+U5[3,0]+c*c*U5[4,0]+c*U5[5,0]
    b5=(U5[2,1]/c)*(Ktr0-mutr0)/(2*Ktr0+mutr0)\
       +U5[3,1]-c*c*U5[4,1]\
       +c*U5[5,1]*(Ktr0+mutr0)/mutr0
    b6=(U5[2,1]/c)+U5[3,1]+c*c*U5[4,1]+c*U5[5,1]
    w=a5+a6+b6-b5
    Delta=w*w+8*(a6*b5-a5*b6)
    mutreff=np.amax(\
              [(-w+np.sqrt(Delta))/(4*(a6*b5-a5*b6)),\
               (-w-np.sqrt(Delta))/(4*(a6*b5-a5*b6))])

    tableA[n,:]=[c, Ktreff, muaxeff, Eaxeff,\
                 vaxeff, mutreff]

np.savetxt('prop_CCA.txt',tableA,fmt='%1.5f')
```

Example 3: assessment of universal relations

Let us consider a unidirectional fiber composite whose RVE consists of a matrix material (phase 0) and a long fiber (phase 1). Both phases exhibit transversely isotropic behavior, and their material parameters are detailed below:

matrix	fibers
$K_0^{\text{tr}} = 4$ GPa	$K_1^{\text{tr}} = 30$ GPa
$\mu_0^{\text{tr}} = 1.5$ GPa	$\mu_1^{\text{tr}} = 10$ GPa
$\mu_0^{\text{ax}} = 1.7$ GPa	$\mu_1^{\text{ax}} = 12$ GPa
$l_0 = 2.3$ GPa	$l_1 = 19$ GPa
$n_0 = 7.3$ GPa	$n_1 = 85$ GPa

The volume fraction of fibers in the composite is 30%. Identify the overall properties $\bar{K}^{\text{tr}}, \bar{l}$, and $\bar{n}$ using the CCA approach and demonstrate that they satisfy the universal relations discussed in Section 8.4.

Solution:

Using a Python script similar to that described in the previous example, we can compute the macroscopic properties. With the help of expressions (8.23), the obtained values up to two significant digits are $\bar{K}^{\text{tr}} = 5.81$ GPa, $\bar{l} = 3.46$ GPa, and $\bar{n} = 28.14$ GPa. The universal relations in this case give

$$\frac{\bar{K}^{\text{tr}} - K_0^{\text{tr}}}{\bar{l} - l_0} = \frac{5.81 - 4}{3.46 - 2.3} = 1.56,$$

$$\frac{\bar{K}^{\text{tr}} - K_1^{\text{tr}}}{\bar{l} - l_1} = \frac{5.81 - 30}{3.46 - 19} = 1.56,$$

$$\frac{\bar{l} - [1 - c]l_0 - cl_1}{\bar{n} - [1 - c]n_0 - cn_1} = \frac{3.46 - 0.7 \cdot 2.3 - 0.3 \cdot 19}{28.14 - 0.7 \cdot 7.3 - 0.3 \cdot 85} = 1.56.$$

References

[1] Z. Hashin, The elastic moduli of heterogeneous materials, Journal of Applied Mechanics 29 (1) (1962) 143–150.

[2] R.M. Christensen, K.H. Lo, Solutions for effective shear properties in three phase sphere and cylinder models, Journal of the Mechanics and Physics of Solids 27 (1979) 315–330.

[3] R.M. Christensen, Mechanics of Composite Materials, Dover, New York, 1979.

[4] Z. Hashin, B.W. Rosen, The elastic moduli of fiber–reinforced materials, Journal of Applied Mechanics 31 (1964) 223–232.

[5] R. Hill, Theory of mechanical properties of fibre–strengthened materials: I. Elastic behaviour, Journal of the Mechanics and Physics of Solids 12 (1964) 199–212.

[6] G.J. Dvorak, On uniform fields in heterogeneous media, Proceedings of the Royal Society A. Mathematical, Physical and Engineering Sciences 431 (1881) (1990) 89–110.

CHAPTER

9

Green's tensor

OUTLINE

9.1 Preliminaries

9.1.1 Fourier transform

The Fourier transform is a very useful tool of mathematics frequently used to solve differential equations.

For simplicity in the expressions that follow, we represent the triple integral in an infinite space (from $-\infty$ to ∞ in all directions) as

$$\int_{-\infty}^{\infty}\int_{-\infty}^{\infty}\int_{-\infty}^{\infty} \mathrm{d}x_1 \mathrm{d}x_2 \mathrm{d}x_3 = \int_{-\infty}^{\infty} \mathrm{d}V(\boldsymbol{x}). \tag{9.1}$$

The notations $\mathrm{d}V(\boldsymbol{x})$ and $\mathrm{d}S(\boldsymbol{x})$ in this chapter are used to identify that respectively the volume and surface integrals are computed in terms of the position vector $\boldsymbol{x}$. The Fourier transform of a function $g(\boldsymbol{x})$, denoted as

Multiscale Modeling Approaches for Composites
https://doi.org/10.1016/B978-0-12-823143-2.00020-5

$\mathcal{G}(g(\boldsymbol{x}))$, is another function $\widehat{g}(\boldsymbol{\xi})$ defined as

$$\mathcal{G}(g(\boldsymbol{x})) = \widehat{g}(\boldsymbol{\xi}) = \frac{1}{8\pi^3}\int_{-\infty}^{\infty} g(\boldsymbol{x})e^{-\mathbb{i}\boldsymbol{\xi}\cdot\boldsymbol{x}}\,\mathrm{d}V(\boldsymbol{x}) \tag{9.2}$$

with $\mathbb{i} = \sqrt{-1}$. The inverse of the Fourier transform gives the function g as

$$g(\boldsymbol{x}) = \int_{-\infty}^{\infty} \widehat{g}(\boldsymbol{\xi})e^{\mathbb{i}\boldsymbol{\xi}\cdot\boldsymbol{x}}\,\mathrm{d}V(\boldsymbol{\xi}). \tag{9.3}$$

A very nice property that follows from this definition is that

$$\mathcal{G}\left(\frac{\partial g(\boldsymbol{x})}{\partial x_i}\right) = \mathbb{i}\xi_i\widehat{g}(\boldsymbol{\xi}). \tag{9.4}$$

A function linked to the Fourier transform is the Dirac delta $\delta(\boldsymbol{x})$. Mathematically, it is defined as

$$\delta(\boldsymbol{x}) = \frac{1}{8\pi^3}\int_{-\infty}^{\infty} e^{-\mathbb{i}\boldsymbol{\xi}\cdot\boldsymbol{x}}\,\mathrm{d}V(\boldsymbol{\xi}) = \begin{cases} +\infty, & \boldsymbol{x} = \mathbf{0}, \\ 0, & \boldsymbol{x} \neq \mathbf{0}, \end{cases} \tag{9.5}$$

and it has the following property for an arbitrary function g:

$$\int_{-\infty}^{\infty} g(\boldsymbol{y})\delta(\boldsymbol{x}-\boldsymbol{y})\,\mathrm{d}V(\boldsymbol{y}) = \int_{-\infty}^{\infty} g(\boldsymbol{y})\delta(\boldsymbol{y}-\boldsymbol{x})\,\mathrm{d}V(\boldsymbol{y}) = g(\boldsymbol{x}). \tag{9.6}$$

9.1.2 Betti's reciprocal theorem

The Betti's reciprocal theorem has a wide range of applications in structural engineering (e.g., in the development of the Castigliano method).

Let us consider a body made by a homogeneous elastic material $\boldsymbol{L}'$, occupying a space $\mathcal{B}$ with boundary surface $\partial\mathcal{B}$, whose unit normal vector is denoted by $\boldsymbol{n}$. Two cases of loading conditions are considered, named as 1 and 2. In each case the body is subjected to body force $\boldsymbol{b}$ at the volume and prescribed displacements $\boldsymbol{u}$ and/or tractions $\boldsymbol{t} = \boldsymbol{\sigma}\cdot\boldsymbol{n}$ at the boundary. These boundary value problems are expressed in indicial notation by the following set of equations:

$$\begin{aligned} \frac{\partial\sigma_{ij}^r}{\partial x_j} + b_i^r &= 0 \quad \text{in } \mathcal{B}, \\ \sigma_{ij}^r &= L'_{ijkl}\varepsilon_{kl}^r \quad \text{in } \mathcal{B}, \\ \varepsilon_{ij}^r &= \frac{1}{2}\left[\frac{\partial u_i^r}{\partial x_j} + \frac{\partial u_j^r}{\partial x_i}\right] \quad \text{in } \mathcal{B}, \end{aligned}$$

$$t_i^r = \sigma_{ij}^r n_j \text{ known on } \partial\mathcal{B}^t,$$
$$u_i^r \text{ known on } \partial\mathcal{B}^u, \tag{9.7}$$

for

$$\partial\mathcal{B}^t \cup \partial\mathcal{B}^u = \partial\mathcal{B}, \quad \partial\mathcal{B}^t \cap \partial\mathcal{B}^u = \emptyset, \quad r = 1, 2. \tag{9.8}$$

The Betti reciprocal theorem expresses equivalences between energies produced by mixing the fields of the two problems. Specifically, it states that

$$\int_{\partial\mathcal{B}} t_i^1 u_i^2 \,\mathrm{d}S + \int_{\mathcal{B}} b_i^1 u_i^2 \,\mathrm{d}V = \int_{\partial\mathcal{B}} t_i^2 u_i^1 \,\mathrm{d}S + \int_{\mathcal{B}} b_i^2 u_i^1 \,\mathrm{d}V. \tag{9.9}$$

Proof. Taking into account the major symmetries of $\boldsymbol{L}'$, we have that

$$\sigma_{ij}^1 \varepsilon_{ij}^2 = [L'_{ijkl}\varepsilon_{kl}^1]\varepsilon_{ij}^2 = [L'_{ijkl}\varepsilon_{kl}^2]\varepsilon_{ij}^1 = \sigma_{ij}^2 \varepsilon_{ij}^1. \tag{9.10}$$

Taking the volume integral in the last equalities yields

$$\int_{\mathcal{B}} \sigma_{ij}^1 \varepsilon_{ij}^2 \,\mathrm{d}V = \int_{\mathcal{B}} \sigma_{ij}^2 \varepsilon_{ij}^1 \,\mathrm{d}V. \tag{9.11}$$

The divergence theorem and Eqs. (9.7) allow us to write the left–hand side of (9.11) as

$$\begin{aligned}
\int_{\mathcal{B}} \sigma_{ij}^1 \varepsilon_{ij}^2 \,\mathrm{d}V &= \int_{\mathcal{B}} \sigma_{ij}^1 \frac{\partial u_i^2}{\partial x_j} \,\mathrm{d}V \\
&= \int_{\partial\mathcal{B}} \sigma_{ij}^1 u_i^2 n_j \,\mathrm{d}V - \int_{\mathcal{B}} \frac{\partial \sigma_{ij}^1}{\partial x_j} u_i^2 \,\mathrm{d}V \\
&= \int_{\partial\mathcal{B}} t_i^1 u_i^2 \,\mathrm{d}V + \int_{\mathcal{B}} b_i^1 u_i^2 \,\mathrm{d}V.
\end{aligned} \tag{9.12}$$

The right–hand side of (9.11) is similarly written as

$$\int_{\mathcal{B}} \sigma_{ij}^2 \varepsilon_{ij}^1 \,\mathrm{d}V = \int_{\partial\mathcal{B}} t_i^2 u_i^1 \,\mathrm{d}V + \int_{\mathcal{B}} b_i^2 u_i^1 \,\mathrm{d}V. \tag{9.13}$$

Substituting (9.12) and (9.13) into (9.11) yields the Betti identity (9.9). □

9.2 Definition and properties

The presentation of the Green's tensor on this section follows the discussion in [1].

Definition

Assume a body made by homogeneous elastic material $\boldsymbol{L}'$ that lies on an infinite space $\mathcal{B}$. At the far distance boundary $\partial\mathcal{B}$ the body is displacement free. At a specific position $\boldsymbol{y}$ of space $\mathcal{B}$ a concentrated force $\boldsymbol{F}$ is applied (Fig. 9.1). Under these conditions, what is the displacement field generated at every position $\boldsymbol{x}$ of the body?

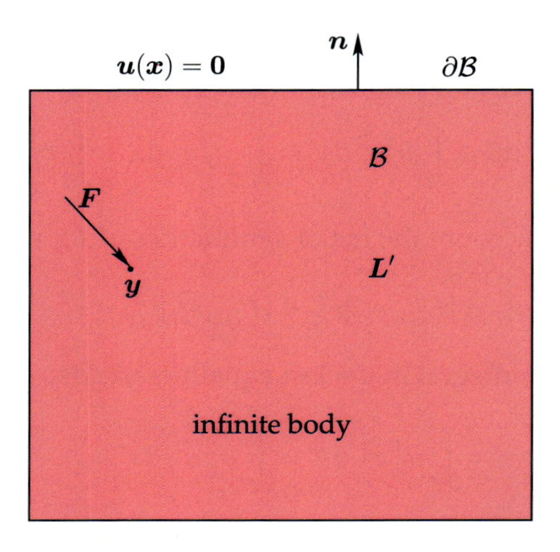

FIGURE 9.1 Homogeneous medium with constant properties $\boldsymbol{L}'$, occupying an infinite space $\mathcal{B}$ and being displacement free at $\partial\mathcal{B}$. At the position $\boldsymbol{y}$ a concentrated force $\boldsymbol{F}$ is applied.

The equilibrium equation and the boundary conditions for this mechanical problem are expressed in indicial notation as

$$L'_{ijkl}\frac{\partial^2 u_k(\boldsymbol{x})}{\partial x_j \partial x_l} + F_i\delta(\boldsymbol{x}-\boldsymbol{y}) = 0 \quad \text{in } \mathcal{B}, \tag{9.14}$$

$$u_i(\boldsymbol{x}) \to 0 \quad \text{as } x_1^2 + x_2^2 + x_3^2 \to \infty, \tag{9.15}$$

where $\delta(\boldsymbol{x}-\boldsymbol{y})$ is the three–dimensional Dirac delta function, whose definition is given in (9.5). This boundary value problem can be solved with the help of the Fourier transform method (see Subsection 9.1.1). Applying the Fourier transform to Eq. (9.14) and using properties (9.4) and (9.6) yield

$$L'_{ijkl}[\mathbb{i}\xi_j][\mathbb{i}\xi_l]\widehat{u}_k(\boldsymbol{\xi}) + \frac{1}{8\pi^3}F_i e^{-\mathbb{i}\boldsymbol{\xi}\cdot\boldsymbol{y}} = 0, \tag{9.16}$$

where $\widehat{u}_i(\boldsymbol{\xi})$ is the Fourier transform of $u_i(\boldsymbol{x})$. Thus we obtain the linear equation

$$L'_{ijkl}\xi_j\xi_l\widehat{u}_k(\boldsymbol{\xi}) = \frac{1}{8\pi^3}F_i e^{-\mathrm{i}\boldsymbol{\xi}\cdot\boldsymbol{y}}. \tag{9.17}$$

The second–order tensor $L'_{ijkl}\xi_j\xi_l$ is symmetric. Its inverse[1]

$$N_{ik} = \left[L'_{ijkl}\xi_j\xi_l\right]^{-1} \tag{9.18}$$

is also a symmetric tensor. Using it in (9.17) yields

$$\widehat{u}_i(\boldsymbol{\xi}) = \frac{1}{8\pi^3}N_{ij}F_j e^{-\mathrm{i}\boldsymbol{\xi}\cdot\boldsymbol{y}}. \tag{9.19}$$

Applying the inverse Fourier transform, we express the displacement in the real space as

$$u_i(\boldsymbol{x}) = G^{\infty}_{ij}(\boldsymbol{x}-\boldsymbol{y})F_j, \tag{9.20}$$

where

$$G^{\infty}_{ij}(\boldsymbol{x}-\boldsymbol{y}) = \frac{1}{8\pi^3}\int_{-\infty}^{\infty} N_{ij}e^{\mathrm{i}\boldsymbol{\xi}\cdot[\boldsymbol{x}-\boldsymbol{y}]}\,\mathrm{d}V(\boldsymbol{\xi}), \tag{9.21}$$

is called the Green's tensor.

BOX 9.1 Physical meaning of Green's tensor

Consider an infinite body, displacement free at the boundary, made by a homogeneous material. Assume that at the position $\boldsymbol{y}$ of the body a concentrated force $\boldsymbol{F}$ is applied. The displacement field $\boldsymbol{u}$ in the position $\boldsymbol{x}$ of the body, produced by the presence of the concentrated force, is equal to the single contraction product of the Green's tensor $\boldsymbol{G}^{\infty}$ with the $\boldsymbol{F}$. Alternatively, we can say that the Green's tensor is the displacement at the position $\boldsymbol{x}$ generated by a unit concentrated force at the position $\boldsymbol{y}$.

Substituting (9.20) into (9.14) yields

$$\left[L'_{ijkl}\frac{\partial^2}{\partial x_j \partial x_l}G^{\infty}_{km}(\boldsymbol{x}-\boldsymbol{y}) + \delta_{im}\delta(\boldsymbol{x}-\boldsymbol{y})\right]F_m = 0. \tag{9.22}$$

The latter equation needs to hold for arbitrary concentrated force F_i. In addition, due to the boundary condition (9.15), the Green's tensor must be zero at infinity. Summarizing, we have the following definition:

[1]The inverse A^{-1}_{ij} of a second–order tensor A_{ij} denotes the usual 3×3 matrix inverse that satisfies the relation $A^{-1}_{ik}A_{kj} = A_{ik}A^{-1}_{kj} = \delta_{ij}$.

BOX 9.2 Green's tensor definition

Mathematically, the Green's tensor of an infinite body with elasticity tensor $\boldsymbol{L}'$ is given by

$$G_{ij}^{\infty}(\boldsymbol{x}-\boldsymbol{y}) = \frac{1}{8\pi^3}\int_{-\infty}^{\infty} N_{ij} e^{\mathrm{i}\boldsymbol{\xi}\cdot[\boldsymbol{x}-\boldsymbol{y}]}\,\mathrm{d}V(\boldsymbol{\xi}), \tag{9.23}$$

where

$$N_{ij} = N_{ji} = \left[L'_{ikjl}\xi_k\xi_l\right]^{-1}. \tag{9.24}$$

The Green's tensor satisfies the differential equation

$$L'_{ijkl}\frac{\partial^2}{\partial x_j \partial x_l}G_{km}^{\infty}(\boldsymbol{x}-\boldsymbol{y}) + \delta_{im}\delta(\boldsymbol{x}-\boldsymbol{y}) = 0 \tag{9.25}$$

with boundary condition

$$G_{km}^{\infty}(\boldsymbol{x}-\boldsymbol{y}) \to 0 \ \text{ as } x_1^2 + x_2^2 + x_3^2 \to \infty. \tag{9.26}$$

Properties

For the Green's tensor, we have the following symmetries:

$$G_{ij}^{\infty}(\boldsymbol{x}-\boldsymbol{y}) = G_{ji}^{\infty}(\boldsymbol{x}-\boldsymbol{y}) = G_{ij}^{\infty}(\boldsymbol{y}-\boldsymbol{x}) = G_{ji}^{\infty}(\boldsymbol{y}-\boldsymbol{x}). \tag{9.27}$$

Proof. Considering the symmetry of the tensor N_{ij}, from (9.23) it immediately follows that

$$G_{ij}^{\infty}(\boldsymbol{x}-\boldsymbol{y}) = G_{ji}^{\infty}(\boldsymbol{x}-\boldsymbol{y}) \ \text{ and } \ G_{ij}^{\infty}(\boldsymbol{y}-\boldsymbol{x}) = G_{ji}^{\infty}(\boldsymbol{y}-\boldsymbol{x}).$$

To complete the equalities, we utilize the Betti's reciprocal theorem. Let us consider a body that occupies an infinite space $\mathcal{B}$. For this body, two different loading cases are assumed:

1. A concentrated load F_i^1 in the position $\boldsymbol{x}$, which at every position $\boldsymbol{z}$ leads to the body force $F_i^1\delta(\boldsymbol{z}-\boldsymbol{x})$ and causes the displacement $G_{ij}^{\infty}(\boldsymbol{z}-\boldsymbol{x})F_j^1$.
2. A concentrated load F_i^2 in the position $\boldsymbol{y}$, which at every position $\boldsymbol{z}$ leads to the body force $F_i^2\delta(\boldsymbol{z}-\boldsymbol{y})$ and causes the displacement $G_{ij}^{\infty}(\boldsymbol{z}-\boldsymbol{y})F_j^2$.

In both cases the displacement at the far distance boundary is zero. According to Betti's reciprocal theorem (9.9), we have the following relation:

$$\int_{\mathcal{B}} F_i^1\delta(\boldsymbol{z}-\boldsymbol{x})G_{ij}^{\infty}(\boldsymbol{z}-\boldsymbol{y})F_j^2\,\mathrm{d}V(\boldsymbol{z})$$

$$= \int_{\mathcal{B}} F_j^2 \delta(z - y) G_{ji}^{\infty}(z - x) F_i^1 \, dV(z).$$

The surface integrals of the theorem vanish due to the displacement–free boundary condition. Using the Dirac delta property (9.6), we write the last equality as

$$G_{ij}^{\infty}(x - y) F_i^1 F_j^2 = G_{ji}^{\infty}(y - x) F_i^1 F_j^2.$$

This relation must hold for arbitrary choice of concentrated forces F_i^1 and F_j^2. Thus we obtain the expression

$$G_{ij}^{\infty}(x - y) = G_{ji}^{\infty}(y - x).$$

This completes the proof. □

From the structure of the Green's tensor (9.23) we can easily verify that

$$\frac{\partial}{\partial x_l} G_{km}^{\infty}(x - y) = -\frac{\partial}{\partial y_l} G_{km}^{\infty}(x - y). \tag{9.28}$$

For future reference, the modified Green's tensor $\Gamma_{ijkl}^{\infty}(x - y)$ is defined according to the formulas

$$\begin{aligned} \Gamma_{ijkl}^{\infty}(x - y) = & \frac{1}{4} \frac{\partial^2}{\partial x_j \partial y_l} G_{ki}^{\infty}(x - y) + \frac{1}{4} \frac{\partial^2}{\partial x_i \partial y_l} G_{kj}^{\infty}(x - y) \\ & + \frac{1}{4} \frac{\partial^2}{\partial x_j \partial y_k} G_{li}^{\infty}(x - y) + \frac{1}{4} \frac{\partial^2}{\partial x_i \partial y_k} G_{lj}^{\infty}(x - y). \end{aligned} \tag{9.29}$$

The two alternative forms

$$\begin{aligned} \Gamma_{ijkl}^{\infty}(x - y) = & -\frac{1}{4} \frac{\partial^2}{\partial x_j \partial x_l} G_{ki}^{\infty}(x - y) - \frac{1}{4} \frac{\partial^2}{\partial x_i \partial x_l} G_{kj}^{\infty}(x - y) \\ & - \frac{1}{4} \frac{\partial^2}{\partial x_j \partial x_k} G_{li}^{\infty}(x - y) - \frac{1}{4} \frac{\partial^2}{\partial x_i \partial x_k} G_{lj}^{\infty}(x - y) \\ = & -\frac{1}{4} \frac{\partial^2}{\partial y_j \partial y_l} G_{ki}^{\infty}(x - y) - \frac{1}{4} \frac{\partial^2}{\partial y_i \partial y_l} G_{kj}^{\infty}(x - y) \\ & - \frac{1}{4} \frac{\partial^2}{\partial y_j \partial y_k} G_{li}^{\infty}(x - y) - \frac{1}{4} \frac{\partial^2}{\partial y_i \partial y_k} G_{lj}^{\infty}(x - y) \end{aligned} \tag{9.30}$$

are due to identity (9.28). From the symmetries of G_{ij}^{∞} and the structure of Γ_{ijkl}^{∞} it becomes clear that the following symmetries hold:

$$\begin{aligned} & \Gamma_{ijkl}^{\infty}(x - y) = \Gamma_{jikl}^{\infty}(x - y) = \Gamma_{ijlk}^{\infty}(x - y) = \Gamma_{klij}^{\infty}(x - y) \\ & = \Gamma_{ijkl}^{\infty}(y - x) = \Gamma_{jikl}^{\infty}(y - x) = \Gamma_{ijlk}^{\infty}(y - x) = \Gamma_{klij}^{\infty}(y - x). \end{aligned} \tag{9.31}$$

9.3 Applications of Green's tensor

We illustrate the usefulness of the Green's tensor in micromechanics in the following special applications motivated by mean–field homogenization theories.

9.3.1 Infinite homogeneous body with varying eigenstresses

The following special boundary value problem has been utilized for the derivation of the generalized form of the Hashin–Shtrikman bounds.

A homogeneous body with elasticity modulus $\boldsymbol{L}'$ occupies the space $\mathcal{B}$ and is subjected to: i) zero displacement field $\boldsymbol{u}(\boldsymbol{x}) = \mathbf{0}$ at the boundary surface $\partial\mathcal{B}$ with unit vector $\boldsymbol{n}$ and ii) a spatially varying eigenstress $\boldsymbol{\sigma}^p(\boldsymbol{x})$, as shown in Fig. 9.2. The space $\mathcal{B}$ is assumed to be infinite.

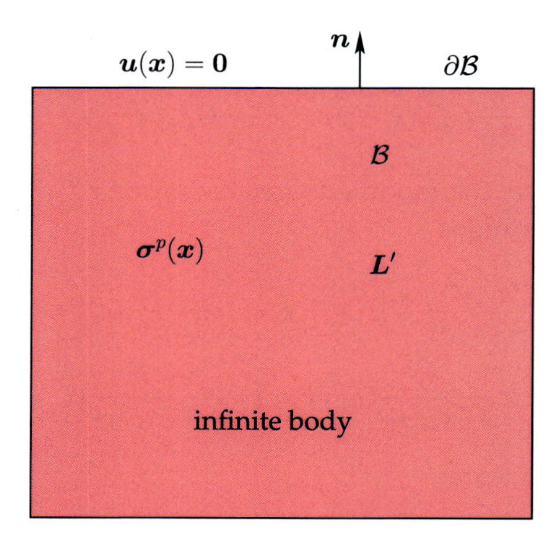

FIGURE 9.2 Hypothetical homogeneous medium with constant properties $\boldsymbol{L}'$ and spatially varying eigenstress $\boldsymbol{\sigma}^p(\boldsymbol{x})$. The material occupies an infinite space $\mathcal{B}$ and is subjected to zero displacement at $\partial\mathcal{B}$.

This boundary value problem can be described in indicial notation by the differential equation

$$\frac{\partial \sigma_{ij}^p}{\partial x_j} + L'_{ijkl}\frac{\partial \varepsilon_{kl}}{\partial x_j} = 0 \quad \text{in } \mathcal{B} \tag{9.32}$$

and the boundary condition

$$u_i = 0 \quad \text{on } \partial\mathcal{B}, \tag{9.33}$$

where ε_{ij} denotes the strain tensor. The Green's tensor corresponding to the infinite medium satisfies relations (9.25) and (9.26). Multiplying (9.25) by $u_i(\boldsymbol{x})$ and integrating over $\mathcal{B}$ yield

$$\int_{\mathcal{B}} u_i(\boldsymbol{x}) L'_{ijkl} \frac{\partial^2}{\partial x_j \partial x_l} G^{\infty}_{km}(\boldsymbol{x}-\boldsymbol{y})\, \mathrm{d}V(\boldsymbol{x}) + u_m(\boldsymbol{y}) = 0, \tag{9.34}$$

which by the divergence theorem gives

$$\begin{aligned} u_m(\boldsymbol{y}) &= \int_{\mathcal{B}} \frac{\partial u_i(\boldsymbol{x})}{\partial x_j} L'_{ijkl} \frac{\partial}{\partial x_l} G^{\infty}_{km}(\boldsymbol{x}-\boldsymbol{y})\, \mathrm{d}V(\boldsymbol{x}) \\ &\quad - \int_{\partial\mathcal{B}} u_i(\boldsymbol{x}) L'_{ijkl} \frac{\partial}{\partial x_l} G^{\infty}_{km}(\boldsymbol{x}-\boldsymbol{y}) n_j\, \mathrm{d}S(\boldsymbol{x}) \\ &= \int_{\mathcal{B}} \varepsilon_{ij}(\boldsymbol{x}) L'_{ijkl} \frac{\partial}{\partial x_l} G^{\infty}_{km}(\boldsymbol{x}-\boldsymbol{y})\, \mathrm{d}V(\boldsymbol{x}) \end{aligned} \tag{9.35}$$

due to (9.33). Similarly, multiplying (9.32) by $G^{\infty}_{im}(\boldsymbol{x}-\boldsymbol{y})$ and integrating over $\mathcal{B}$ give

$$\int_{\mathcal{B}} G^{\infty}_{im}(\boldsymbol{x}-\boldsymbol{y}) \frac{\partial \sigma^{p}_{ij}(\boldsymbol{x})}{\partial x_j}\, \mathrm{d}V(\boldsymbol{x}) + \int_{\mathcal{B}} G^{\infty}_{im}(\boldsymbol{x}-\boldsymbol{y}) L'_{ijkl} \frac{\partial \varepsilon_{kl}(\boldsymbol{x})}{\partial x_j}\, \mathrm{d}V(\boldsymbol{x}) = 0. \tag{9.36}$$

Using the divergence theorem in both terms yields

$$\begin{aligned} &\int_{\partial\mathcal{B}} G^{\infty}_{im}(\boldsymbol{x}-\boldsymbol{y}) \sigma^{p}_{ij}(\boldsymbol{x}) n_j\, \mathrm{d}S(\boldsymbol{x}) \\ &\quad - \int_{\mathcal{B}} \frac{\partial}{\partial x_j} G^{\infty}_{im}(\boldsymbol{x}-\boldsymbol{y}) \sigma^{p}_{ij}(\boldsymbol{x})\, \mathrm{d}V(\boldsymbol{x}) \\ &\quad + \int_{\partial\mathcal{B}} G^{\infty}_{im}(\boldsymbol{x}-\boldsymbol{y}) L'_{ijkl} \varepsilon_{kl}(\boldsymbol{x}) n_j\, \mathrm{d}S(\boldsymbol{x}) \\ &\quad - \int_{\mathcal{B}} \frac{\partial}{\partial x_j} G^{\infty}_{im}(\boldsymbol{x}-\boldsymbol{y}) L'_{ijkl} \varepsilon_{kl}(\boldsymbol{x})\, \mathrm{d}V(\boldsymbol{x}) = 0. \end{aligned} \tag{9.37}$$

The surface integrals (first and third terms) vanish since $\partial\mathcal{B}$ is very far away, as (9.26) implies. Due to the symmetry of $\boldsymbol{L}'$, the last term is equal to the right–hand side of (9.35). Consequently,

$$u_m(\boldsymbol{y}) = -\int_{\mathcal{B}} \frac{\partial}{\partial x_j} G^{\infty}_{im}(\boldsymbol{x}-\boldsymbol{y}) \sigma^{p}_{ij}(\boldsymbol{x})\, \mathrm{d}V(\boldsymbol{x}). \tag{9.38}$$

The latter integral does not converge for arbitrary choice of eigenstress. To fix this issue, the tensor σ^{p}_{ij} is substituted by $\sigma^{p}_{ij} - \bar{\sigma}^{p}_{ij}$, where $\bar{\sigma}^{p}_{ij}$ is the constant equal to the average eigenstress. With this modification, (9.32) is still satisfied, whereas the average of $\sigma_{ij}^{P} - \bar{\sigma}_{ij}^{P}$ vanishes, erasing the convergence issues [2–4]. Thus the displacement field takes the final form

$$u_m(\boldsymbol{y}) = -\int_{\mathcal{B}} \frac{\partial}{\partial x_j} G_{im}^{\infty}(\boldsymbol{x}-\boldsymbol{y})[\sigma_{ij}^{P}(\boldsymbol{x}) - \bar{\sigma}_{ij}^{P}]\,\mathrm{d}V(\boldsymbol{x}). \tag{9.39}$$

Due to the symmetry of the eigenstresses, we can write

$$\frac{\partial}{\partial x_j} G_{im}^{\infty}(\boldsymbol{x}-\boldsymbol{y})[\sigma_{ij}^{P}(\boldsymbol{x}) - \bar{\sigma}_{ij}^{P}] = \frac{1}{2}\left[\frac{\partial}{\partial x_j} G_{im}^{\infty}(\boldsymbol{x}-\boldsymbol{y}) + \frac{\partial}{\partial x_i} G_{jm}^{\infty}(\boldsymbol{x}-\boldsymbol{y})\right][\sigma_{ij}^{P}(\boldsymbol{x}) - \bar{\sigma}_{ij}^{P}].$$

With this relation in mind, differentiating expression (9.39) and taking the symmetric part yield

$$\varepsilon_{mn}(\boldsymbol{y}) = -\int_{\mathcal{B}} \Gamma_{mnij}^{\infty}(\boldsymbol{x}-\boldsymbol{y})\left[\sigma_{ij}^{P}(\boldsymbol{x}) - \bar{\sigma}_{ij}^{P}\right]\mathrm{d}V(\boldsymbol{x}), \tag{9.40}$$

where $\Gamma_{ijkl}^{\infty}(\boldsymbol{x}-\boldsymbol{y})$ is the modified Green's tensor (9.29). Considering the symmetries (9.31), we can write the last expression in the form

$$\varepsilon_{mn}(\boldsymbol{x}) = -\int_{\mathcal{B}} \Gamma_{mnij}^{\infty}(\boldsymbol{x}-\boldsymbol{y})\left[\sigma_{ij}^{P}(\boldsymbol{y}) - \bar{\sigma}_{ij}^{P}\right]\mathrm{d}V(\boldsymbol{y}). \tag{9.41}$$

9.3.2 Eshelby's inclusion problem

The boundary value problem concerning the Eshelby inclusion case study has been already described in Subsection 5.1.1 of Chapter 5. It is actually similar to the previous boundary value problem if we set the eigenstress as

$$\sigma_{ij}^{P}(\boldsymbol{x}) = \begin{cases} -L'_{ijkl}\varepsilon_{kl}^{*} & \text{if } \boldsymbol{x} \in \Omega, \\ 0 & \text{if } \boldsymbol{x} \in \mathcal{B}-\Omega, \end{cases} \tag{9.42}$$

where we recall that Ω is the space that the inclusion occupies and ε_{ij}^{*} is the eigenstrain. Due to the infinite size of $\mathcal{B}$, $\bar{\sigma}_{ij}^{P} \to 0$. Thus, according to Eq. (9.41), the strain at every position of the medium is given by

$$\varepsilon_{mn}(\boldsymbol{x}) = -\int_{\mathcal{B}} \Gamma_{mnij}^{\infty}(\boldsymbol{x}-\boldsymbol{y})\sigma_{ij}^{P}(\boldsymbol{y})\,\mathrm{d}V(\boldsymbol{y}).$$

Since the eigenstress is nonzero only inside the inclusion, we write the last equation as

$$\varepsilon_{mn}(\boldsymbol{x}) = L'_{ijkl}\varepsilon_{kl}^{*}\int_{\Omega} \Gamma_{mnij}^{\infty}(\boldsymbol{x}-\boldsymbol{y})\,\mathrm{d}V(\boldsymbol{y}).$$

By setting

$$P_{ijkl}(\boldsymbol{x}) = \int_{\Omega} \Gamma^{\infty}_{ijkl}(\boldsymbol{x}-\boldsymbol{y})\,\mathrm{d}V(\boldsymbol{y})$$

the strain at every position is given by

$$\varepsilon_{ij}(\boldsymbol{x}) = P_{ijkl}(\boldsymbol{x}) L'_{mnkl} \varepsilon^*_{kl}.$$

$P_{ijmn}(\boldsymbol{x})$ is often called the polarization tensor and exhibits minor and major symmetries. Substituting it with the product

$$P_{ijmn}(\boldsymbol{x}) = S_{ijpq}(\boldsymbol{x})[L'_{pqmn}]^{-1},$$

where S_{ijpq} denotes the Eshelby tensor, we obtain the final expression

$$\varepsilon_{ij}(\boldsymbol{x}) = S_{ijkl}(\boldsymbol{x})\varepsilon^*_{kl}. \tag{9.43}$$

Eshelby [5] has proven that P_{ijkl} (and, consequently, S_{ijkl}) is constant if the inclusion is of ellipsoidal shape, leading to the final relation (5.2). The special form (5.3) of the Eshelby tensor is obtained with the help of proper coordinate transformations [6,7].

9.4 Examples

Example 1

Assume an infinite homogeneous isotropic body with shear modulus $\mu' = 1150$ MPa and Poisson's ratio $\nu' = 0.3$. At the position $\boldsymbol{y} = [0\ 0\ 0]^T$ the concentrated force $\boldsymbol{F} = [1\ 0\ 0]^T$ MN is applied. Compute the produced displacement component u_1 on the plane $x_1 = 10$ mm.

Solution:

For an isotropic material with shear modulus μ' and Poisson's ratio ν', the Green's tensor of the infinite body is given by the analytical formula [1]

$$G^{\infty}_{ij}(\boldsymbol{x}-\boldsymbol{y}) = \frac{1}{16\pi\mu'[1-\nu']|\boldsymbol{x}-\boldsymbol{y}|}\left[[3-4\nu']\delta_{ij} + \frac{[x_i-y_i][x_j-y_j]}{|\boldsymbol{x}-\boldsymbol{y}|^2}\right], \tag{9.44}$$

where

$$|\boldsymbol{x}-\boldsymbol{y}| = \sqrt{[x_1-y_1]^2+[x_2-y_2]^2+[x_3-y_3]^2}. \tag{9.45}$$

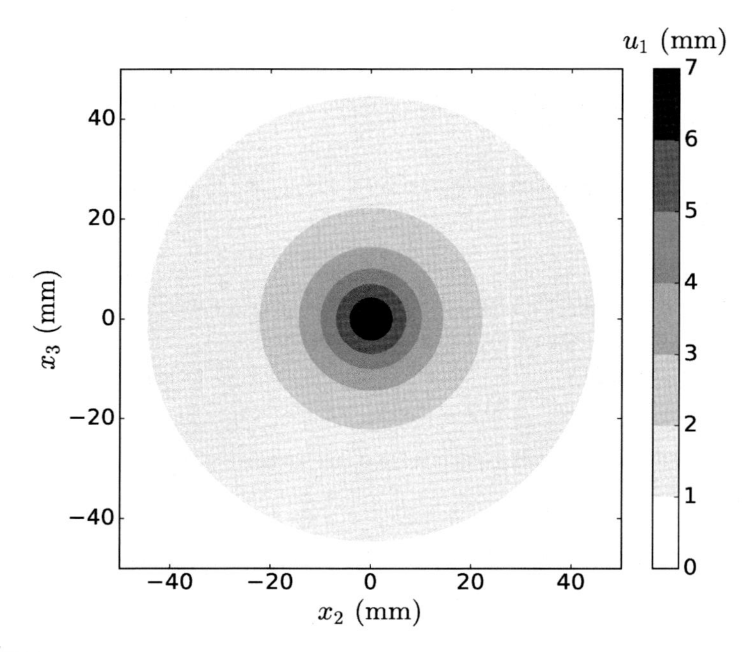

FIGURE 9.3 Displacement component u_1 on the plane $x_1 = 10$ mm due to the concentrated force $\boldsymbol{F}$.

The displacement component u_1 is given by (9.20) as

$$u_1 = G^{\infty}_{1j}(\boldsymbol{x} - \boldsymbol{y})F_j = G^{\infty}_{11}(\boldsymbol{x} - \boldsymbol{y})F_1.$$

Fig. 9.3 illustrates the distribution of u_1 on the plane $x_1 = 10$ mm. As expected, the displacement presents its highest value at the position $(x_2, x_3) = (0, 0)$, which is the closest to the position $\boldsymbol{y}$, and it decays at spatial positions further to the concentrated load.

References

[1] C. Weinberger, W. Cai, D. Barnett, ME340B Lecture Notes, Elasticity of Microscopic Structures, Stanford University, 2005.

[2] Z. Hashin, S. Shtrikman, A variational approach to the theory of the elastic behaviour of multiphase materials, Journal of the Mechanics and Physics of Solids 11 (2) (1963) 127–140.

[3] Z. Hashin, On elastic behaviour of fibre reinforced materials of arbitrary transverse phase geometry, Journal of the Mechanics and Physics of Solids 13 (1965) 119–134.

[4] J.R. Willis, J.R. Acton, The overall elastic moduli of a dilute suspension of spheres, Quarterly Journal of Mechanics and Applied Mathematics 29 (1976) 163–177.

[5] J.D. Eshelby, The determination of the elastic field of an ellipsoidal inclusion, and related problems, Proceedings of the Royal Society of London. Series A, Mathematical and Physical Sciences 241 (1226) (1957) 376–396.

[6] J. Qu, M. Cherkaoui, Fundamentals of Micromechanics of Solids, Wiley, New Jersey, 2006.
[7] T. Mura, Micromechanics of Defects in Solids, second revised ed., Kluwer Academic Publishers, Dordrecht, 1987.

CHAPTER 10

Hashin–Shtrikman bounds

OUTLINE

10.1 Preliminaries

Due to the excessive use of both indicial and tensorial notation in this chapter, the indices indicating the material phase are difficult to handle. For this reason, we consider the following convention: A vector $\boldsymbol{v}$, a second–order tensor $\boldsymbol{a}$, and a fourth–order tensor $\boldsymbol{A}$ related to the material phase r will be denoted with superscript when indicial notation is chosen and with subscript when tensorial notation is chosen, that is,

$$v_i^r, \ a_{ij}^r, \text{ and } A_{ijkl}^r \quad \text{or} \quad \boldsymbol{v}_r, \ \boldsymbol{a}_r, \text{ and } \boldsymbol{A}_r.$$

10.1.1 Positive and negative definite matrices

As already discussed in the previous chapters, a fourth–order tensor with minor symmetries can be represented as a 6×6 matrix. Thus any definition related to symmetric matrices can be extended to the fourth–order tensors. To simplify the computations in this section, we adopt the matrix notation using the symbol $\cdot$ for the classical matrix multiplication.

https://doi.org/10.1016/B978-0-12-823143-2.00021-7

The positive and negative definiteness of a matrix is a very useful notion for mechanics, often utilized in computations related to maximization/minimization of functionals and existence/uniqueness of solutions in various problems.

- A positive definite $n \times n$ matrix $\boldsymbol{A}$ has all its eigenvalues positive and satisfies the relation

$$\boldsymbol{v}^T \cdot \boldsymbol{A} \cdot \boldsymbol{v} > 0 \tag{10.1}$$

  for every vector $\boldsymbol{v}$. If $\boldsymbol{A}$ is symmetric and positive definite, then $\boldsymbol{A}^{-1}$ exists and is also symmetric and positive definite. For simplicity in the notation, the positive definiteness is often represented by the notion

$$\boldsymbol{A} \succ 0. \tag{10.2}$$

- A negative definite $n \times n$ matrix $\boldsymbol{A}$ has all its eigenvalues negative and satisfies the relation

$$\boldsymbol{v}^T \cdot \boldsymbol{A} \cdot \boldsymbol{v} < 0 \tag{10.3}$$

  for every $n \times 1$ vector $\boldsymbol{v}$. If $\boldsymbol{A}$ is negative definite, then $\boldsymbol{A}^{-1}$ exists and is also negative definite. For simplicity of notation, the negative definiteness is often denoted as

$$\boldsymbol{A} \prec 0. \tag{10.4}$$

Two important identities are the following:

Identity 1:

Consider two $n \times n$, symmetric positive definite matrices $\mathbf{A}$ and $\boldsymbol{B}$. If $\mathbf{A} - \boldsymbol{B}$ is negative definite, then $\boldsymbol{B}^{-1} - \boldsymbol{A}^{-1}$ is negative definite and vice versa.

Proof. For the proof, we need certain important properties of symmetric matrices and two lemmas.

- For every symmetric matrix $\boldsymbol{B}$, there are a diagonal matrix $\boldsymbol{D}$ and an orthogonal matrix $\boldsymbol{G}$ (i.e., $\boldsymbol{G}^{-1} = \boldsymbol{G}^T$) such that

$$\boldsymbol{B} = \boldsymbol{G} \cdot \boldsymbol{D} \cdot \boldsymbol{G}^T. \tag{10.5}$$

  The diagonal terms of $\boldsymbol{D}$ are the eigenvalues of $\boldsymbol{B}$. If $\boldsymbol{B}$ is also positive definite, then all terms of $\boldsymbol{D}$ are non–zero, and thus $\boldsymbol{B}$ is invertible with

$$\boldsymbol{B}^{-1} = \boldsymbol{G} \cdot \boldsymbol{D}^{-1} \cdot \boldsymbol{G}^T. \tag{10.6}$$

- For every symmetric positive definite matrix $\boldsymbol{B}$, there are matrices $\boldsymbol{B}^{1/2}$ and $\boldsymbol{B}^{-1/2}$ such that

$$\boldsymbol{B}^{1/2} = \boldsymbol{G} \cdot \boldsymbol{D}^{1/2} \cdot \boldsymbol{G}^T, \quad \boldsymbol{B}^{-1/2} = \boldsymbol{G} \cdot \boldsymbol{D}^{-1/2} \cdot \boldsymbol{G}^T. \tag{10.7}$$

  The $\boldsymbol{D}$ to the power of $\pm 1/2$ denotes the square root (or inverse square root) of the diagonal elements. Clearly, $\boldsymbol{B}^{1/2}$ and $\boldsymbol{B}^{-1/2}$ are symmetric. Moreover,

$$\boldsymbol{B}^{1/2} \cdot \boldsymbol{B}^{1/2} = \boldsymbol{B}, \quad \boldsymbol{B}^{-1/2} \cdot \boldsymbol{B}^{-1/2} = \boldsymbol{B}^{-1}. \tag{10.8}$$

- *Lemma 1:* If $\boldsymbol{A} \prec \boldsymbol{B}$, then $\boldsymbol{C} \cdot \boldsymbol{A} \cdot \boldsymbol{C} \prec \boldsymbol{C} \cdot \boldsymbol{B} \cdot \boldsymbol{C}$ for any conformable symmetric matrix $\boldsymbol{C}$.
  The proof of this lemma is quite obvious, since

$$\begin{aligned} \boldsymbol{v}^T \cdot \boldsymbol{C} \cdot [\boldsymbol{A} - \boldsymbol{B}] \cdot \boldsymbol{C} \cdot \boldsymbol{v} &= \boldsymbol{v}^T \cdot \boldsymbol{C}^T \cdot [\boldsymbol{A} - \boldsymbol{B}] \cdot \boldsymbol{C} \cdot \boldsymbol{v} \\ &= [\boldsymbol{C} \cdot \boldsymbol{v}]^T \cdot [\boldsymbol{A} - \boldsymbol{B}] \cdot [\boldsymbol{C} \cdot \boldsymbol{v}] < 0 \end{aligned}$$

  for any $\boldsymbol{v}$ and $\boldsymbol{C}$.
- *Lemma 2:* If $\boldsymbol{B}$ is symmetric, $\boldsymbol{I}$ is the $n \times n$ identity matrix, and

$$\boldsymbol{I} \prec \boldsymbol{B}, \tag{10.9}$$

  then $\boldsymbol{B}$ is positive definite and invertible with

$$\boldsymbol{B}^{-1} \prec \boldsymbol{I}. \tag{10.10}$$

  The proof of this lemma is as follows:

$$\boldsymbol{B} = \boldsymbol{G} \cdot \boldsymbol{D} \cdot \boldsymbol{G}^T \quad \Rightarrow \quad \boldsymbol{D} = \boldsymbol{G}^T \cdot \boldsymbol{B} \cdot \boldsymbol{G} \succ \boldsymbol{G}^T \cdot \boldsymbol{I} \cdot \boldsymbol{G} = \boldsymbol{I} \tag{10.11}$$

  due to Lemma 1. Thus all eigenvalues of $\boldsymbol{B}$ are greater than 1, which means that $\boldsymbol{B}$ is invertible. Consequently, we can write

$$\boldsymbol{B}^{-1} = \boldsymbol{B}^{-1/2} \cdot \boldsymbol{B}^{-1/2} = \boldsymbol{B}^{-1/2} \cdot \boldsymbol{I} \cdot \boldsymbol{B}^{-1/2} \prec \boldsymbol{B}^{-1/2} \cdot \boldsymbol{B} \cdot \boldsymbol{B}^{-1/2} = \boldsymbol{I}.$$

With the help of the above definitions and lemmas, we can pass to the main proof of the identity. The relation $\boldsymbol{A} - \boldsymbol{B} \prec 0$ allows us to write

$$\begin{aligned} \boldsymbol{B} - \boldsymbol{A} \succ \boldsymbol{0} \;&\Leftrightarrow\; \boldsymbol{A}^{-1/2} \cdot [\boldsymbol{B} - \boldsymbol{A}] \cdot \boldsymbol{A}^{-1/2} \succ \boldsymbol{0} \;\Leftrightarrow \\ \boldsymbol{A}^{-1/2} \cdot \boldsymbol{B} \cdot \boldsymbol{A}^{-1/2} \succ \boldsymbol{I} \;&\Leftrightarrow\; \boldsymbol{A}^{1/2} \cdot \boldsymbol{B}^{-1} \cdot \boldsymbol{A}^{1/2} \prec \boldsymbol{I} \;\Leftrightarrow \\ \boldsymbol{A}^{-1/2} \cdot \boldsymbol{A}^{1/2} \cdot \boldsymbol{B}^{-1} \cdot \boldsymbol{A}^{1/2} \cdot \boldsymbol{A}^{-1/2} &\prec \boldsymbol{A}^{-1/2} \cdot \boldsymbol{A}^{-1/2} \;\Leftrightarrow \\ &\boldsymbol{B}^{-1} \prec \boldsymbol{A}^{-1}. \end{aligned} \tag{10.12}$$

The latter concludes the proof of identity 1. □

Identity 2:

Consider two $n \times n$ symmetric positive definite matrices $\boldsymbol{A}$ and $\boldsymbol{B}$. If $[\boldsymbol{A} - \boldsymbol{B}]^{-1} + \boldsymbol{B}^{-1}$ is negative definite, then $\boldsymbol{A} - \boldsymbol{B}$ is negative definite and vice versa.

Proof. Let $\boldsymbol{I}$ be the $n \times n$ identity matrix. Initially, we have that

$$[\boldsymbol{A} - \boldsymbol{B}]^{-1} \cdot \boldsymbol{B} = \left[\boldsymbol{B}^{-1} \cdot [\boldsymbol{A} - \boldsymbol{B}]\right]^{-1} = \left[\boldsymbol{B}^{-1} \cdot \boldsymbol{A} - \boldsymbol{I}\right]^{-1}$$

and

$$\begin{aligned}
&\boldsymbol{B}^{-1} \cdot \left[\boldsymbol{B}^{-1} - \boldsymbol{A}^{-1}\right]^{-1} - \boldsymbol{I} \\
&\quad = \left[[\boldsymbol{B}^{-1} - \boldsymbol{A}^{-1}] \cdot \boldsymbol{B}\right]^{-1} - \boldsymbol{I} = \left[\boldsymbol{I} - \boldsymbol{A}^{-1} \cdot \boldsymbol{B}\right]^{-1} - \boldsymbol{I} \\
&\quad = \left[\boldsymbol{I} - \boldsymbol{A}^{-1} \cdot \boldsymbol{B}\right]^{-1} - \left[\boldsymbol{I} - \boldsymbol{A}^{-1} \cdot \boldsymbol{B}\right] \cdot \left[\boldsymbol{I} - \boldsymbol{A}^{-1} \cdot \boldsymbol{B}\right]^{-1} \\
&\quad = \boldsymbol{A}^{-1} \cdot \boldsymbol{B} \cdot \left[\boldsymbol{I} - \boldsymbol{A}^{-1} \cdot \boldsymbol{B}\right]^{-1} \\
&\quad = \left[\boldsymbol{B}^{-1} \cdot \boldsymbol{A} - \boldsymbol{A}^{-1} \cdot \boldsymbol{B} \cdot \boldsymbol{B}^{-1} \cdot \boldsymbol{A}\right]^{-1} = \left[\boldsymbol{B}^{-1} \cdot \boldsymbol{A} - \boldsymbol{I}\right]^{-1},
\end{aligned}$$

which leads to the identity

$$[\boldsymbol{A} - \boldsymbol{B}]^{-1} \cdot \boldsymbol{B} = \boldsymbol{B}^{-1} \cdot \left[\boldsymbol{B}^{-1} - \boldsymbol{A}^{-1}\right]^{-1} - \boldsymbol{I}. \tag{10.13}$$

Let $\boldsymbol{\eta}$ and $\boldsymbol{\tau}$ be arbitrary $n \times 1$ vectors connected through the relation

$$\boldsymbol{\tau} = \boldsymbol{B} \cdot \boldsymbol{\eta}. \tag{10.14}$$

Using these two vectors and the symmetry of the matrices yield

$$\boldsymbol{\tau}^T \cdot \boldsymbol{B}^{-1} \cdot \boldsymbol{\tau} = \boldsymbol{\eta}^T \cdot \boldsymbol{B} \cdot \boldsymbol{B}^{-1} \cdot \boldsymbol{B} \cdot \boldsymbol{\eta} = \boldsymbol{\eta}^T \cdot \boldsymbol{B} \cdot \boldsymbol{\eta} \tag{10.15}$$

and, by (10.13),

$$\begin{aligned}
\boldsymbol{\tau}^T \cdot [\boldsymbol{A} - \boldsymbol{B}]^{-1} \cdot \boldsymbol{\tau} &= \boldsymbol{\eta}^T \cdot \boldsymbol{B} \cdot [\boldsymbol{A} - \boldsymbol{B}]^{-1} \cdot \boldsymbol{B} \cdot \boldsymbol{\eta} \\
&= \boldsymbol{\eta}^T \cdot \boldsymbol{B} \cdot \boldsymbol{B}^{-1} \cdot \left[\boldsymbol{B}^{-1} - \boldsymbol{A}^{-1}\right]^{-1} \cdot \boldsymbol{\eta} - \boldsymbol{\eta}^T \cdot \boldsymbol{B} \cdot \boldsymbol{\eta} \\
&= \boldsymbol{\eta}^T \cdot \left[\boldsymbol{B}^{-1} - \boldsymbol{A}^{-1}\right]^{-1} \cdot \boldsymbol{\eta} - \boldsymbol{\eta}^T \cdot \boldsymbol{B} \cdot \boldsymbol{\eta}.
\end{aligned} \tag{10.16}$$

Adding (10.15) and (10.16) yields

$$\boldsymbol{\tau}^T \cdot [[\boldsymbol{A} - \boldsymbol{B}]^{-1} + \boldsymbol{B}^{-1}] \cdot \boldsymbol{\tau} = \boldsymbol{\eta}^T \cdot \left[\boldsymbol{B}^{-1} - \boldsymbol{A}^{-1}\right]^{-1} \cdot \boldsymbol{\eta}. \tag{10.17}$$

The last equation shows that if $[\boldsymbol{A} - \boldsymbol{B}]^{-1} + \boldsymbol{B}^{-1}$ is a negative definite tensor, then the tensor $\left[\boldsymbol{B}^{-1} - \boldsymbol{A}^{-1}\right]^{-1}$ (and, consequently, its inverse) is also negative definite and vice versa. According to identity 1, if $\boldsymbol{B}^{-1} - \boldsymbol{A}^{-1}$ is negative definite, then, equivalently, $\boldsymbol{A} - \boldsymbol{B}$ is negative definite. □

10.1.2 Calculus of variations

The appearance of functionals in mechanics is very frequent. In many cases, we minimize or maximize a functional to identify appropriate conservation laws or variational principles. To achieve that, first and second variations of the functionals are required.

Let us consider a functional I of two spatially depended functions $y(x)$ and $z(x)$, written in the form

$$I(y(x), z(x)) = \int_{\mathcal{B}} F(y, z)\,\mathrm{d}V. \tag{10.18}$$

Assuming a small parameter ϵ, the functions y and z are altered according to the formulas

$$\widetilde{y} = y + \delta y = y + \epsilon\psi, \qquad \widetilde{z} = z + \delta z = z + \epsilon\zeta. \tag{10.19}$$

Using the Taylor expansion, the functional I for the new functions $\widetilde{y}(x)$ and $\widetilde{z}(x)$ is expressed as

$$\begin{aligned} I(\widetilde{y}(x), \widetilde{z}(x)) &= \int_{\mathcal{B}} F(y + \epsilon\psi, z + \epsilon\zeta)\,\mathrm{d}V \\ &= I(y(x), z(x)) + \epsilon \int_{\mathcal{B}} \left[\frac{\partial F}{\partial y}\psi + \frac{\partial F}{\partial z}\zeta\right] \mathrm{d}V \\ &\quad + \frac{\epsilon^2}{2}\int_{\mathcal{B}} \left[\frac{\partial^2 F}{\partial y^2}\psi^2 + \frac{\partial^2 F}{\partial z^2}\zeta^2 + \frac{\partial^2 F}{\partial y \partial z}\psi\zeta\right] \mathrm{d}V + \cdots . \end{aligned} \tag{10.20}$$

From the final expression the first variation δI and the second variation $\delta^2 I$ of the functional I are defined as

$$\delta I = \int_{\mathcal{B}} \left[\frac{\partial F}{\partial y}\delta y + \frac{\partial F}{\partial z}\delta z\right] \mathrm{d}V \tag{10.21}$$

and

$$\delta^2 I = \frac{1}{2}\int_{\mathcal{B}} \left[\frac{\partial^2 F}{\partial y^2}\delta y \delta y + \frac{\partial^2 F}{\partial z^2}\delta z \delta z + \frac{\partial^2 F}{\partial y \partial z}\delta y \delta z\right] \mathrm{d}V. \tag{10.22}$$

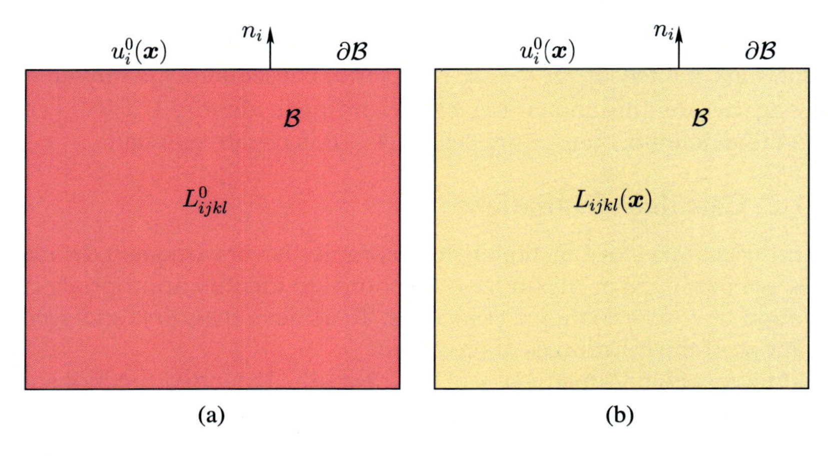

FIGURE 10.1 (a) Reference and (b) heterogeneous medium defined for the Hashin–Shtrikman variational principle.

10.2 Hashin–Shtrikman variational principle

For the derivation of their bounds, Hashin and Shtrikman [1] have identified an appropriate variational principle considering two different materials, homogeneous (reference) and heterogeneous (actual composite).

Reference medium:

Let us consider a reference material (index 0) with elasticity tensor L^0_{ijkl} occupying a space $\mathcal{B}$ with volume V and boundary surface $\partial\mathcal{B}$ (Fig. 10.1_a). The normal vector to $\partial\mathcal{B}$ is denoted as n_i. For this body, the displacement field is denoted as u_i^0. On $\partial\mathcal{B}$, we prescribe the displacement, that is,

$$u_i^0(\boldsymbol{x}), \quad \forall \boldsymbol{x} \in \partial\mathcal{B} \quad \text{known.} \tag{10.23}$$

The kinematics, equilibrium, and constitutive law that hold for this reference material in the complete space $\mathcal{B}$ are written as

$$\varepsilon^0_{ij} = \frac{1}{2}\left[\frac{\partial u_i^0}{\partial x_j} + \frac{\partial u_j^0}{\partial x_i}\right], \tag{10.24}$$

$$\frac{\partial \sigma^0_{ij}}{\partial x_j} = 0, \tag{10.25}$$

$$\sigma^0_{ij} = L^0_{ijkl}\varepsilon^0_{kl} \quad \text{or} \quad \varepsilon^0_{ij} = M^0_{ijkl}\sigma^0_{kl}, \tag{10.26}$$

where M^0_{ijkl} is the inverse of L^0_{ijkl}. Its elastic energy is defined as

$$W^0 = \frac{1}{2V} \int_{\mathcal{B}} \sigma^0_{ij} \varepsilon^0_{ij} \, \mathrm{d}V. \tag{10.27}$$

Heterogeneous medium:

In the same space $\mathcal{B}$ the reference material is substituted by a heterogeneous material with elasticity tensor $L_{ijkl}(\boldsymbol{x})$ (Fig. 10.1$_b$). For this body, the displacement field is denoted as u_i. On $\partial\mathcal{B}$ the displacement field is the same as that for the reference material, that is,

$$u_i(\boldsymbol{x}) = u^0_i(\boldsymbol{x}), \quad \forall \boldsymbol{x} \in \partial\mathcal{B}. \tag{10.28}$$

The kinematics, equilibrium, and constitutive law that hold for this heterogeneous medium in the complete space $\mathcal{B}$ are written as

$$\varepsilon_{ij} = \frac{1}{2}\left[\frac{\partial u_i}{\partial x_j} + \frac{\partial u_j}{\partial x_i}\right], \tag{10.29}$$

$$\frac{\partial \sigma_{ij}}{\partial x_j} = 0, \tag{10.30}$$

$$\sigma_{ij} = L_{ijkl}\varepsilon_{kl} \quad \text{or} \quad \varepsilon_{ij} = M_{ijkl}\sigma_{kl}, \tag{10.31}$$

where M_{ijkl} is the inverse of L_{ijkl}. Its elastic energy is defined as

$$W = \frac{1}{2V} \int_{\mathcal{B}} \sigma_{ij} \varepsilon_{ij} \, \mathrm{d}V. \tag{10.32}$$

Once the two different bodies, the reference and the heterogeneous, are defined, we can proceed to the definition of the fluctuation strain and polarization stress.

Let us identify a displacement field u'_i for the whole space $\mathcal{B}$, which describes the difference between the displacements of the heterogeneous and the reference body

$$u'_i = u_i - u^0_i. \tag{10.33}$$

At the boundary $\partial\mathcal{B}$,

$$u'_i(\boldsymbol{x}) = u_i(\boldsymbol{x}) - u^0_i(\boldsymbol{x}) = 0, \quad \forall \boldsymbol{x} \in \partial\mathcal{B}. \tag{10.34}$$

The corresponding strain, called the fluctuation strain, is given by

$$\varepsilon'_{ij} = \varepsilon_{ij} - \varepsilon^0_{ij} = \frac{1}{2}\left[\frac{\partial u'_i}{\partial x_j} + \frac{\partial u'_j}{\partial x_i}\right]. \tag{10.35}$$

Besides the fluctuation strain, we can also define the polarization stress τ_{ij} as

$$\tau_{ij} = \sigma_{ij} - L^0_{ijkl}\varepsilon_{kl} = [L_{ijkl} - L^0_{ijkl}]\varepsilon_{kl} = \sigma_{ij} - \sigma^0_{ij} - L^0_{ijkl}\varepsilon'_{kl}. \tag{10.36}$$

Combining the last expression with the equilibrium equations (10.25) and (10.30) yields

$$\frac{\partial \tau_{ij}}{\partial x_j} + L^0_{ijkl}\frac{\partial \varepsilon'_{kl}}{\partial x_j} = 0. \tag{10.37}$$

In terms of elastic energies, it is shown that

$$\int_{\mathcal{B}} \sigma_{ij}\varepsilon'_{ij}\,\mathrm{d}V = \int_{\mathcal{B}} \sigma_{ij}\frac{\partial u'_i}{\partial x_j}\,\mathrm{d}V = \int_{\partial\mathcal{B}} \sigma_{ij}n_j u'_i\,\mathrm{d}S - \int_{\mathcal{B}} u'_i\frac{\partial \sigma_{ij}}{\partial x_j}\,\mathrm{d}V = 0 \tag{10.38}$$

due to (10.34) and (10.30). Similarly,

$$\int_{\mathcal{B}} \sigma^0_{ij}\varepsilon'_{ij}\,\mathrm{d}V = 0. \tag{10.39}$$

Using (10.35), (10.36), (10.38), and (10.39) and the symmetry of L^0_{ijkl} yield the following relation between the energies of the heterogeneous medium W and the reference medium W^0:

$$\begin{aligned}
W &= \frac{1}{2V}\int_{\mathcal{B}} \sigma_{ij}\varepsilon_{ij}\,\mathrm{d}V = \frac{1}{2V}\int_{\mathcal{B}} \sigma_{ij}\varepsilon^0_{ij}\,\mathrm{d}V \\
&= \frac{1}{2V}\int_{\mathcal{B}} \left[\sigma^0_{ij} + \tau_{ij} + L^0_{ijkl}\varepsilon'_{kl}\right]\varepsilon^0_{ij}\,\mathrm{d}V \\
&= W^0 + \frac{1}{2V}\int_{\mathcal{B}} \tau_{ij}\varepsilon^0_{ij}\,\mathrm{d}V + \frac{1}{2V}\int_{\mathcal{B}} \sigma^0_{ij}\varepsilon'_{ij}\,\mathrm{d}V \\
&= W^0 + \frac{1}{2V}\int_{\mathcal{B}} \tau_{ij}\varepsilon^0_{ij}\,\mathrm{d}V.
\end{aligned} \tag{10.40}$$

With the help of the above definitions, now we can present the following variational principle:

BOX 10.1 Hashin–Shtrikman variational principle

Consider a homogeneous material with elasticity tensor L^0_{ijkl} and a heterogeneous with elasticity tensor $L_{ijkl}(\boldsymbol{x})$. Under the subsidiary condition

$$\frac{\partial \tau_{ij}}{\partial x_j} + L^0_{ijkl}\frac{\partial \varepsilon'_{kl}}{\partial x_j} = 0 \text{ in } \mathcal{B} \tag{10.41}$$

and the boundary condition

$$u'_i(\boldsymbol{x}) = 0 \ \text{ on } \partial\mathcal{B}, \tag{10.42}$$

the Hashin–Shtrikman functional

$$U^*(\tau_{ij}, \varepsilon'_{ij}) = W^0 - \frac{1}{2V}\int_{\mathcal{B}} \left[\tau_{ij} H_{ijkl}\tau_{kl} - \tau_{ij}\varepsilon'_{ij} - 2\tau_{ij}\varepsilon^0_{ij}\right] \mathrm{d}V \tag{10.43}$$

with $H_{ijkl} = [L_{ijkl} - L^0_{ijkl}]^{-1}$ is stationary when

$$H_{ijkl}\tau_{kl} = \varepsilon^0_{ij} + \varepsilon'_{ij} = \varepsilon_{ij}. \tag{10.44}$$

Proof. By rewriting (10.41) in the form

$$\frac{\partial \lambda_{ij}}{\partial x_j} = 0, \qquad \lambda_{ij} = \tau_{ij} + L^0_{ijkl}\varepsilon'_{kl} \tag{10.45}$$

we express its variation as

$$\frac{\partial \delta\lambda_{ij}}{\partial x_j} = 0, \quad \delta\lambda_{ij} = \delta\tau_{ij} + L^0_{ijkl}\delta\varepsilon'_{kl}. \tag{10.46}$$

A useful identity is the following:

$$\int_{\mathcal{B}} \varepsilon'_{ij}\lambda_{ij}\,\mathrm{d}V = \int_{\mathcal{B}} \frac{\partial u'_i}{\partial x_j}\lambda_{ij}\,\mathrm{d}V = \int_{\partial\mathcal{B}} u'_i\lambda_{ij}n_j\,\mathrm{d}S - \int_{\mathcal{B}} u'_i\frac{\partial \lambda_{ij}}{\partial x_j}\,\mathrm{d}V = 0 \tag{10.47}$$

due to the equilibrium (10.45) and the boundary condition (10.42). In a similar manner,

$$\int_{\mathcal{B}} \varepsilon'_{ij}\lambda_{ij}\,\mathrm{d}V = \int_{\mathcal{B}} \varepsilon'_{ij}\delta\lambda_{ij}\,\mathrm{d}V = \int_{\mathcal{B}} \delta\varepsilon'_{ij}\lambda_{ij}\,\mathrm{d}V = \int_{\mathcal{B}} \delta\varepsilon'_{ij}\delta\lambda_{ij}\,\mathrm{d}V = 0. \tag{10.48}$$

Using relations (10.46) and identities (10.48), we have

$$\begin{aligned}\int_{\mathcal{B}} \varepsilon'_{ij}\delta\tau_{ij}\,\mathrm{d}V &= \int_{\mathcal{B}} \varepsilon'_{ij}\delta\lambda_{ij}\,\mathrm{d}V - \int_{\mathcal{B}} \varepsilon'_{ij}L^0_{ijkl}\delta\varepsilon'_{kl}\,\mathrm{d}V = -\int_{\mathcal{B}} \varepsilon'_{ij}L^0_{ijkl}\delta\varepsilon'_{kl}\,\mathrm{d}V \\ &= -\int_{\mathcal{B}} \lambda_{ij}\delta\varepsilon'_{kl}\,\mathrm{d}V + \int_{\mathcal{B}} \tau_{ij}\delta\varepsilon'_{kl}\,\mathrm{d}V = \int_{\mathcal{B}} \tau_{ij}\delta\varepsilon'_{kl}\,\mathrm{d}V.\end{aligned} \tag{10.49}$$

Considering the last identity and the symmetries of H_{ijkl}, the variation of U^* given by Eq. (10.43) is written

$$\delta U^* = -\frac{1}{2V}\int_{\mathcal{B}} \left[2\tau_{kl}H_{ijkl}\delta\tau_{ij} - \varepsilon'_{ij}\delta\tau_{ij} - \tau_{ij}\delta\varepsilon'_{ij} - 2\varepsilon^0_{ij}\delta\tau_{ij}\right] \mathrm{d}V$$

$$= -\frac{1}{2V}\int_{\mathcal{B}} \left[2\tau_{kl} H_{ijkl}\delta\tau_{ij} - 2\varepsilon'_{ij}\delta\tau_{ij} - 2\varepsilon^0_{ij}\delta\tau_{ij}\right] \mathrm{d}V$$

$$= -\frac{1}{V}\int_{\mathcal{B}} \left[H_{ijkl}\tau_{kl} - \varepsilon'_{ij} - \varepsilon^0_{ij}\right] \delta\tau_{ij}\, \mathrm{d}V. \tag{10.50}$$

The stationary condition $\delta U^* = 0$ yields

$$H_{ijkl}\tau_{kl} = \varepsilon^0_{ij} + \varepsilon'_{ij} = \varepsilon_{ij},$$

which concludes the proof of the Hashin–Shtrikman variational principle. □

We can notice that the stationarity condition gives

$$U^*_S = W^0 + \frac{1}{2V}\int_{\mathcal{B}} \tau_{ij}\varepsilon^0_{ij}\, \mathrm{d}V = W \tag{10.51}$$

due to (10.40). Thus the stationary point of the functional U^* is equal to the elastic energy of the heterogeneous medium. We further discuss the conditions under which the stationary point is maximum or minimum.

BOX 10.2 Maximum/minimum of Hashin–Shtrikman functional

- If $L_{ijkl} - L^0_{ijkl}$ is positive definite, then the elastic energy W of the heterogeneous medium is a maximum of the Hashin–Shtrikman functional U^*.
- If $L_{ijkl} - L^0_{ijmn}$ is negative definite, then the elastic energy W of the heterogeneous medium is a minimum of the Hashin–Shtrikman functional U^*.

Proof. The second variation of the functional U^* is given by the expression (see Subsection 10.1.2)

$$\delta^2 U^* = -\frac{1}{2V}\int_{\mathcal{B}} \left[\delta\tau_{ij} H_{ijkl}\delta\tau_{kl} - \delta\varepsilon'_{ij}\delta\tau_{ij}\right] \mathrm{d}V. \tag{10.52}$$

Substituting $(10.46)_2$ into the second term of the integrand and accounting for the identities (10.48) yield

$$\delta^2 U^* = -\frac{1}{2V}\int_{\mathcal{B}} \left[\delta\tau_{ij} H_{ijkl}\delta\tau_{kl} - \delta\varepsilon'_{ij}\delta\lambda_{ij} + \delta\varepsilon'_{ij} L^0_{ijkl}\delta\varepsilon'_{kl}\right] \mathrm{d}V$$

$$= -\frac{1}{2V}\int_{\mathcal{B}} \left[\delta\tau_{ij} H_{ijkl}\delta\tau_{kl} + \delta\varepsilon'_{ij} L^0_{ijkl}\delta\varepsilon'_{kl}\right] \mathrm{d}V. \tag{10.53}$$

<u>Maximum:</u> The elasticity tensor L^0_{ijkl} is always positive definite. If H_{ijkl} is also positive definite (and, consequently, $L_{ijkl} - L^0_{ijkl}$ is positive definite), then $\delta^2 U^* < 0$ for all possible choices of $\delta\varepsilon'_{ij}$ and $\delta\tau_{ij}$, leading to a maximum for U^*.

<u>Minimum:</u> Consider the functional

$$I = \int_{\mathcal{B}} \delta\tau_{ij} M^0_{ijkl} \delta\tau_{kl} \, \mathrm{d}V, \tag{10.54}$$

where M^0_{ijkl} is the positive definite inverse of L^0_{ijkl}. Substituting (10.46)$_2$ into the above integral and accounting for the identities (10.48) yield

$$\begin{aligned} I &= \int_{\mathcal{B}} [\delta\lambda_{ij} - L^0_{ijmn}\delta\varepsilon'_{mn}] M^0_{ijkl} [\delta\lambda_{kl} - L^0_{klpq}\delta\varepsilon'_{pq}] \, \mathrm{d}V \\ &= \int_{\mathcal{B}} [\delta\lambda_{ij} M^0_{ijkl} \delta\lambda_{kl} + \delta\varepsilon'_{ij} L^0_{ijkl} \delta\varepsilon'_{kl} - 2\delta\varepsilon'_{ij}\delta\lambda_{ij}] \, \mathrm{d}V \\ &= \int_{\mathcal{B}} [\delta\lambda_{ij} M^0_{ijkl} \delta\lambda_{kl} + \delta\varepsilon'_{ij} L^0_{ijkl} \delta\varepsilon'_{kl}] \, \mathrm{d}V. \end{aligned} \tag{10.55}$$

Comparing (10.54) and (10.55) and accounting for the fact that both L^0_{ijkl} and M^0_{ijkl} are positive definite, it becomes clear that

$$\int_{\mathcal{B}} \delta\tau_{ij} M^0_{ijkl} \delta\tau_{kl} \, \mathrm{d}V \geqslant \int_{\mathcal{B}} \delta\varepsilon'_{ij} L^0_{ijkl} \delta\varepsilon'_{kl} \, \mathrm{d}V. \tag{10.56}$$

Substituting the latter into (10.53) yields

$$\delta^2 U^* \geqslant -\frac{1}{2V} \int_{\mathcal{B}} \delta\tau_{ij} [H_{ijkl} + M^0_{ijkl}] \delta\tau_{kl} \, \mathrm{d}V. \tag{10.57}$$

If $H_{ijkl} + M^0_{ijkl}$ is negative definite, then $\delta^2 U^* > 0$, and U^* obtains its minimum. According to identity 2 of Subsection 10.1.1, if $H_{ijkl} + M^0_{ijkl}$ is negative definite, then $L_{ijkl} - L^0_{ijmn}$ is also negative definite and vice versa. □

Remark:

If $L_{ijkl} = L^0_{ijkl}$ in a subspace $\mathcal{B}^0$ of $\mathcal{B}$, then the maximum and minimum conditions still hold. In every position of $\mathcal{B}^0$, Eq. (10.36) gives $\tau_{ij} = 0$, and the terms that vanish in the integrals inside the region $\mathcal{B}^0$ do not influence the behavior of δU^* and $\delta^2 U^*$.

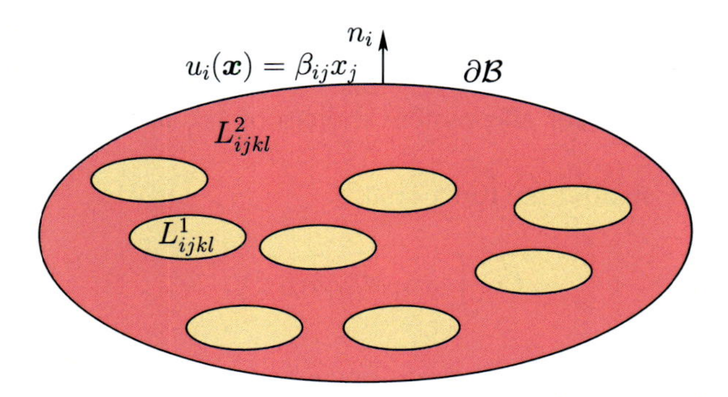

FIGURE 10.2 Bi–phase medium (matrix–reinforcement) with constant elasticity tensors per phase, subjected to linear displacement field at the boundary. All the heterogeneities have the same shape.

10.3 Bounds in a bi–phase composite

Let us consider the case of a bi–phase composite. In this medium the first phase (noted as 1) is the reinforcement. It occupies the total space $\mathcal{B}^1$, has the elasticity tensor L^1_{ijkl}, and its volume fraction is c_1. All the heterogeneities have the same shape. The second phase (noted as 2) is the matrix. It occupies the space $\mathcal{B}^2$ ($\mathcal{B}^1 \cup \mathcal{B}^2 = \mathcal{B}$, $\mathcal{B}^1 \cap \mathcal{B}^2 = \emptyset$), has elasticity tensor L^2_{ijkl}, and its volume fraction is $c_2 = 1 - c_1$. The composite is subjected to the linear displacement $u_i = \beta_{ij}x_j$ at the boundary, where β_{ij} is a known second–order strain tensor (Fig. 10.2).

As the Hill–Mandel principle indicates, the linear displacement boundary condition allows us to write the elastic energy of the composite as

$$W = \frac{1}{2}\left[\frac{1}{V}\int_{\mathcal{B}} \sigma_{ij}\,\mathrm{d}V\right]\left[\frac{1}{V}\int_{\mathcal{B}} \varepsilon_{ij}\,\mathrm{d}V\right]. \tag{10.58}$$

If the medium is statistically homogeneous, then an equivalent homogenized elasticity tensor $\overline{L}_{ijkl}$ exists. Considering the average strain theorem, the above elastic energy can be expressed as

$$W = \frac{1}{2}\beta_{ij}\overline{L}_{ijkl}\beta_{kl}. \tag{10.59}$$

To identify bounds for this type of composite, we utilize the Hashin–Shtrikman variational principle. In this procedure the two phases are substituted by a new hypothetical homogeneous material with elasticity tensor L^0_{ijkl}. This new medium is subjected to a zero displacement field at the boundary. The polarization stress that arises is assumed to be piecewise constant, that is, a different constant tensor for the reinforcement (τ^1_{ij})

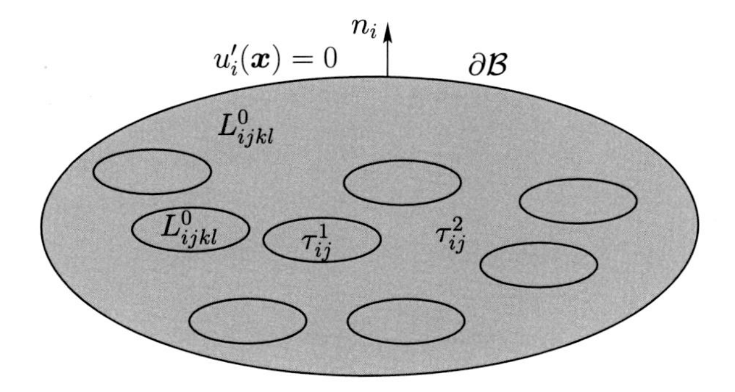

FIGURE 10.3 Hypothetical homogeneous medium for the Hashin–Shtrikman variational principle. Each material phase is under a constant polarization stress, and the body is subjected to zero displacement at the boundary.

and the matrix (τ^2_{ij}) (Fig. 10.3). This polarization stress can be written in the compact form

$$\tau_{ij}(\boldsymbol{x}) = \sum_{r=1}^{2} \tau^r_{ij}\chi_r(\boldsymbol{x}), \quad \chi_r(\boldsymbol{x}) = \begin{cases} 1, & \boldsymbol{x} \in \mathcal{B}^r, \\ 0, & \boldsymbol{x} \notin \mathcal{B}^r. \end{cases} \tag{10.60}$$

Assuming that the volume V of the space $\mathcal{B}$ is very large, the strain field for such a boundary value problem is given by (see Subsection 9.3.1)

$$\varepsilon'_{mn}(\boldsymbol{x}) = -\int_{\mathcal{B}} \Gamma^{\infty}_{mnij}(\boldsymbol{x}-\boldsymbol{y})[\tau_{ij}(\boldsymbol{y}) - \bar{\tau}_{ij}]\,\mathrm{d}V(\boldsymbol{y}), \tag{10.61}$$

where

$$\bar{\tau}_{ij} = \sum_{r=1}^{2} c_r \tau^r_{ij}, \tag{10.62}$$

is the average polarization stress tensor, and Γ^{∞}_{mnij} is the modified Green's tensor corresponding to the elastic material L^0_{ijkl}. Using this strain field and the integration order change rule, we can write the functional U^* of Eq. (10.43) for the bi–phase material as [2]

$$\begin{aligned} 2U^* &= 2W^0 - \frac{1}{V}\int_{\mathcal{B}} \left[\tau_{ij}H_{ijkl}\tau_{kl} - \tau_{ij}\varepsilon'_{ij} - 2\tau_{ij}\varepsilon^0_{ij}\right]\mathrm{d}V \\ &= \beta_{ij}L^0_{ijkl}\beta_{kl} - \sum_{r=1}^{2} c_r\tau^r_{ij}\left[L^r_{ijkl} - L^0_{ijkl}\right]^{-1}\tau^r_{kl} \\ &\quad - \sum_{r=1}^{2}\sum_{s=1}^{2} \tau^r_{ij}\Omega_{ijkl}\tau^s_{kl} + 2\sum_{r=1}^{2} c_r\tau^r_{ij}\beta_{ij}, \end{aligned} \tag{10.63}$$

where [3]

$$\Omega_{ijkl} = \int_{\mathcal{B}}\int_{\mathcal{B}} \chi_r(\boldsymbol{x})\Gamma^{\infty}_{ijkl}(\boldsymbol{x}-\boldsymbol{y})[\chi_s(\boldsymbol{y}) - c_s]\,\mathrm{d}V(\boldsymbol{y})\,\mathrm{d}V(\boldsymbol{x})$$
$$= \int_{\mathcal{B}} \Gamma^{\infty}_{ijkl}(\boldsymbol{x}-\boldsymbol{y})[\chi_{rs}(\boldsymbol{x}-\boldsymbol{y}) - c_r c_s]\,\mathrm{d}V(\boldsymbol{y}). \tag{10.64}$$

In these expressions, $\chi_{rs}(\boldsymbol{x}-\boldsymbol{y})$ is the two–point correlation function denoting the probability that phases r and s are found at the positions $\boldsymbol{x}$ and $\boldsymbol{y}$, respectively. The statistical homogeneity of the composite leads to the fact that χ_{rs} depends exclusively upon the difference $\boldsymbol{x}-\boldsymbol{y}$. Assuming that χ_{rs} is isotropic, we have [2]

$$\Omega_{ijkl} = P^0_{ijkl}[\chi_{rs}(\mathbf{0}) - c_r c_s], \tag{10.65}$$

where

$$P^0_{ijkl} = S^0_{ijmn}[L^0_{mnkl}]^{-1}, \quad \chi_{rs}(\mathbf{0}) = \begin{cases} c_r, & r = s, \\ 0, & r \neq s, \end{cases} \tag{10.66}$$

and S^0_{ijmn} is the Eshelby tensor that depends on the elasticity tensor L^0_{ijkl} and the shape of the inhomogeneities.[1]

For further computations, the tensorial notation allows us to simplify the expressions. Implementing (10.65) in (10.63) gives

$$2U^* = \boldsymbol{\beta}:\boldsymbol{L}_0:\boldsymbol{\beta} - \sum_{r=1}^{2} c_r\boldsymbol{\tau}_r : \left[[\boldsymbol{L}_r - \boldsymbol{L}_0]^{-1} + \boldsymbol{P}_0\right] : \boldsymbol{\tau}_r$$
$$+ \sum_{r=1}^{2}\sum_{s=1}^{2} c_r c_s \boldsymbol{\tau}_r : \boldsymbol{P}_0 : \boldsymbol{\tau}_s + 2\sum_{r=1}^{2} c_r\boldsymbol{\tau}_r : \boldsymbol{\beta}. \tag{10.67}$$

Clearly, U^* obtains its maximum or minimum when its derivatives with respect to each one of $\boldsymbol{\tau}_1$ and $\boldsymbol{\tau}_2$ vanish. This leads to the conditions

$$-\left[[\boldsymbol{L}_r - \boldsymbol{L}_0]^{-1} + \boldsymbol{P}_0\right] : \boldsymbol{\tau}_r + \boldsymbol{P}_0 : \overline{\boldsymbol{\tau}} + \boldsymbol{\beta} = \mathbf{0}, \quad r = 1, 2. \tag{10.68}$$

In the last equations, we took into account that $\boldsymbol{P}_0$ presents major symmetries since it is the volume integral of $\boldsymbol{\Gamma}^{\infty}$. From the last expressions we obtain

$$\boldsymbol{\tau}_r = \left[[\boldsymbol{L}_r - \boldsymbol{L}_0]^{-1} + \boldsymbol{P}_0\right]^{-1} : \boldsymbol{P}_0 : \overline{\boldsymbol{\tau}} + \left[[\boldsymbol{L}_r - \boldsymbol{L}_0]^{-1} + \boldsymbol{P}_0\right]^{-1} \boldsymbol{\beta} \tag{10.69}$$

[1] Whereas for the matrix material an Eshelby tensor cannot be established, this proof assumes that it has the same shape with the heterogeneities. The final form of the boundary conditions eliminates the presence of the Eshelby tensor for the matrix. A more rigorous approach about this issue is discussed by [3], leading to the same result for the bounds.

for $r = 1, 2$. Multiplying by the volume fractions and adding the expressions for $r = 1$ and $r = 2$ yield

$$\bar{\boldsymbol{\tau}} = \left[\mathcal{I} - \sum_{r=1}^{2} c_r \left[[\boldsymbol{L}_r - \boldsymbol{L}_0]^{-1} + \boldsymbol{P}_0\right]^{-1} : \boldsymbol{P}_0\right]^{-1} : \left[\sum_{s=1}^{2} c_s \left[[\boldsymbol{L}_s - \boldsymbol{L}_0]^{-1} + \boldsymbol{P}_0\right]^{-1}\right] : \boldsymbol{\beta}. \quad (10.70)$$

Using condition (10.68), the stationary value of U^* is given by

$$2U_S^* = \boldsymbol{\beta} : \boldsymbol{L}_0 : \boldsymbol{\beta} + \sum_{r=1}^{2} c_r \boldsymbol{\tau}_r : \boldsymbol{\beta} = \boldsymbol{\beta} : \boldsymbol{L}_0 : \boldsymbol{\beta} + \bar{\boldsymbol{\tau}} : \boldsymbol{\beta} = \boldsymbol{\beta} : [\boldsymbol{L}_0 + \bar{\boldsymbol{Q}}] : \boldsymbol{\beta} \quad (10.71)$$

with

$$\bar{\boldsymbol{Q}} = [\mathcal{I} - \boldsymbol{R} : \boldsymbol{P}_0]^{-1} : \boldsymbol{R}, \quad \boldsymbol{R} = \sum_{r=1}^{2} c_r \left[[\boldsymbol{L}_r - \boldsymbol{L}_0]^{-1} + \boldsymbol{P}_0\right]^{-1}. \quad (10.72)$$

Using matrix algebra, we obtain

$$\boldsymbol{R} = \sum_{r=1}^{2} c_r \left[[\boldsymbol{L}_r - \boldsymbol{L}_0]^{-1} + \boldsymbol{P}_0\right]^{-1} = \sum_{r=1}^{2} c_r [\boldsymbol{L}_r - \boldsymbol{L}_0] : \boldsymbol{T}_r \quad (10.73)$$

with

$$\boldsymbol{T}_r = [\mathcal{I} + \boldsymbol{P}_0 : [\boldsymbol{L}_r - \boldsymbol{L}_0]]^{-1}. \quad (10.74)$$

In addition,

$$\bar{\boldsymbol{Q}} = [\mathcal{I} - \boldsymbol{R} : \boldsymbol{P}_0]^{-1} : \boldsymbol{R} = \left[\boldsymbol{R}^{-1} - \boldsymbol{P}_0\right]^{-1} = \boldsymbol{R} : [\mathcal{I} - \boldsymbol{P}_0 : \boldsymbol{R}]^{-1}. \quad (10.75)$$

However,

$$\mathcal{I} - \boldsymbol{P}_0 : \boldsymbol{R} = \sum_{r=1}^{2} c_r [\mathcal{I} + \boldsymbol{P}_0 : [\boldsymbol{L}_r - \boldsymbol{L}_0] : \boldsymbol{T}_r] \quad (10.76)$$

and

$$\mathcal{I} + \boldsymbol{P}_0 : [\boldsymbol{L}_r - \boldsymbol{L}_0] : \boldsymbol{T}_r = \left[[\boldsymbol{T}_r]^{-1} + \boldsymbol{P}_0 : [\boldsymbol{L}_r - \boldsymbol{L}_0]\right] : \boldsymbol{T}_r = \boldsymbol{T}_r. \quad (10.77)$$

Regrouping (10.75), (10.76), and (10.77) yields

$$\bar{\boldsymbol{Q}} = \sum_{r=1}^{2} c_r [\boldsymbol{L}_r - \boldsymbol{L}_0] : \boldsymbol{A}_r,$$

$$
\boldsymbol{A}_r = \boldsymbol{T}_r : \left[\sum_{s=1}^{2} c_s \boldsymbol{T}_s \right]^{-1},
$$

$$
\boldsymbol{T}_r = \left[\boldsymbol{\mathcal{I}} + \boldsymbol{P}_0 : \left[\boldsymbol{L}_r - \boldsymbol{L}_0 \right] \right]^{-1}. \tag{10.78}
$$

Returning back to (10.71), we have the final expression

$$
2U_S^* = \boldsymbol{\beta} : \left[\boldsymbol{L}_0 + \sum_{r=1}^{2} c_r \left[\boldsymbol{L}_r - \boldsymbol{L}_0 \right] : \boldsymbol{A}_r \right] : \boldsymbol{\beta}. \tag{10.79}
$$

Assuming that the hypothetical material 0 has the same properties with the matrix phase (i.e., $\boldsymbol{L}_0 = \boldsymbol{L}_2$ and $\boldsymbol{P}_0 = \boldsymbol{P}_2$), we obtain the formula

$$
2U_S^* = \boldsymbol{\beta} : \left[\boldsymbol{L}_2 + c_1 \left[\boldsymbol{L}_1 - \boldsymbol{L}_2 \right] : \boldsymbol{T} : \left[c_2 \boldsymbol{\mathcal{I}} + c_1 \boldsymbol{T} \right]^{-1} \right] : \boldsymbol{\beta}, \tag{10.80}
$$

where

$$
\boldsymbol{T} = \left[\boldsymbol{\mathcal{I}} + \boldsymbol{S}_2 : \left[\boldsymbol{L}_2 \right]^{-1} : \left[\boldsymbol{L}_1 - \boldsymbol{L}_2 \right] \right]^{-1}. \tag{10.81}
$$

The Eshelby tensor $\boldsymbol{S}_2$ depends on the shape of reinforcement and the properties of the matrix. The fourth–order tensor of Eq. (10.80) is similar to the Mori–Tanaka estimate of a bi–phase composite material.

As discussed in the previous section, the stationary point of U^* is equal to the elastic energy of the composite (10.59). According to the positive or negative definiteness of $\boldsymbol{L}_1 - \boldsymbol{L}_2$, W is maximum or minimum, respectively. All the derived formulas are extendable to account for more than one type of reinforcement [2].

Summarizing the results of the above analysis, we can draw the following conclusion:

BOX 10.3 Hashin–Shtrikman bounds for a bi–phase composite

Consider a bi–phase composite consisting of a matrix with elasticity tensor $\boldsymbol{L}_2$ and volume fraction c_2 and inhomogeneities (reinforcement) with elasticity tensor $\boldsymbol{L}_1$ and total volume fraction c_1. If all the inhomogeneities have the same ellipsoidal shape, then the Eshelby tensor $\boldsymbol{S}_2$ is constant and depends on this shape and properties $\boldsymbol{L}_2$. Let us define the tensor

$$
\boldsymbol{L}_{HS} = \boldsymbol{L}_2 + c_1 \left[\boldsymbol{L}_1 - \boldsymbol{L}_2 \right] : \boldsymbol{T} : \left[c_2 \boldsymbol{\mathcal{I}} + c_1 \boldsymbol{T} \right]^{-1} \tag{10.82}
$$

with

$$T = \left[I + S_2 : [L_2]^{-1} : [L_1 - L_2] \right]^{-1}. \tag{10.83}$$

Denoting by $\bar{L}$ the macroscopic elasticity tensor of this composite, the Hashin–Shtrikman bounds state the following:

- If $L_1 - L_2$ is positive definite, then

$$\beta : L_{HS} : \beta \leqslant \beta : \bar{L} : \beta \tag{10.84}$$

  for every strain tensor β, that is, L_{HS} provides a lower bound for the composite elastic energy.

- If $L_1 - L_2$ is negative definite, then

$$\beta : L_{HS} : \beta \geqslant \beta : \bar{L} : \beta \tag{10.85}$$

  for every strain tensor β, that is, L_{HS} provides an upper bound for the composite elastic energy.

Quite often composites exhibit microstructure that is unknown or very complicated. Also, in many occasions the notion of matrix is not clear (for instance, in polycrystals). In such cases the Hashin–Shtrikman formulas can be directly used to compute both upper and lower bounds in the composite energy. If in addition the reinforcement is of unknown shape or consists of randomly oriented fibers, then the simplest hypothesis is to consider spherical shape. For isotropic material phases, the latter hypothesis leads to isotropic composite response and the original Hashin–Shtrikman bounds are obtained (Eqs. (3.40) and (3.41) of Chapter 3).

10.4 Examples

Example 1

Assume a composite with transversely isotropic behavior, made of two isotropic material phases with elasticity tensors L_1 and L_2. Identify the Hashin–Shtrikman bounds of the transverse shear modulus for this composite.

Solution:

Due to the isotropic response of the material phases, their elasticity tensors can be expressed in terms of the bulk and shear moduli K_1, μ_1, K_2, and μ_2 according to formula (2.24). On the other hand, the elasticity tensor of the transversely isotropic composite material is written with the help of (2.19) in terms of the transverse bulk modulus $\bar{K}^{tr}$, the transverse shear modulus $\bar{\mu}^{tr}$, the axial shear modulus $\bar{\mu}^{ax}$, and the stiffness terms $\bar{l}$ and $\bar{n}$.

Since the composite presents transversely isotropic behavior, the simplest hypothesis is that it behaves similarly to a cylindrical long fiber composite. For long cylindrical inclusions into an isotropic matrix, the Eshelby tensor is given by expression (5.8).

With the help of the above data and hypotheses, formulas (10.82) and (10.83) provide the Hashin–Shtrikman fourth–order tensor that bounds the energy of the composite medium. After some algebra, the transverse shear term of this tensor is obtained as

$$L_{44}^{HS} = \mu_2 + \frac{c_1}{\dfrac{1}{\mu_1 - \mu_2} + \dfrac{c_2[3K_2 + 7\mu_2]}{2\mu_2[3K_2 + 4\mu_2]}}, \qquad (10.86)$$

where c_1 and c_2 are the volume fractions of the two materials ($c_1 + c_2 = 1$). By setting on the Hashin–Shtrikman functional the strain tensor

$$\boldsymbol{\beta} = [0\;0\;0\;b\;0\;0]^T,$$

where b is an arbitrary constant, the Hashin–Shtrikman bounds on the energy are defined as

$$b\bar{\mu}^{\text{tr}} b \;(\leqslant \text{ or } \geqslant)\; bL_{44}^{HS} b. \qquad (10.87)$$

The direction of the inequality depends on the sign of $\mu_1 - \mu_2$. Since b is arbitrary, the bounds on the macroscopic transverse modulus are defined as

$$\bar{\mu}^{\text{tr}} \;(\leqslant \text{ or } \geqslant)\; L_{44}^{HS}. \qquad (10.88)$$

Let $\mu_1 > \mu_2$. Then the lower bound is given by

$$\mu_2 + \frac{c_1}{\dfrac{1}{\mu_1 - \mu_2} + \dfrac{c_2[3K_2 + 7\mu_2]}{2\mu_2[3K_2 + 4\mu_2]}} \leqslant \bar{\mu}^{\text{tr}}. \qquad (10.89)$$

By interchanging indices 1 and 2 in (10.86) the upper bound can be defined as

$$\bar{\mu}^{\text{tr}} \leqslant \mu_1 + \frac{c_2}{\dfrac{1}{\mu_2 - \mu_1} + \dfrac{c_1[3K_1 + 7\mu_1]}{2\mu_1[3K_1 + 4\mu_1]}}. \qquad (10.90)$$

Example 2

Consider the fiber composite of Example 2 of Chapter 8. Demonstrate that the generalized self–consistent composite cylinder assemblage estimate of the transverse shear modulus is within the Hashin–Shtrikman bounds.

Solution:

Fig. 10.4 illustrates the Hashin–Shtrikman bounds computed from formulas (10.89) and (10.90), as well as the solution obtained from the GSC-CCA method (Example 2 of Chapter 8). The lower bound coincides with the Mori–Tanaka estimate of the composite response.

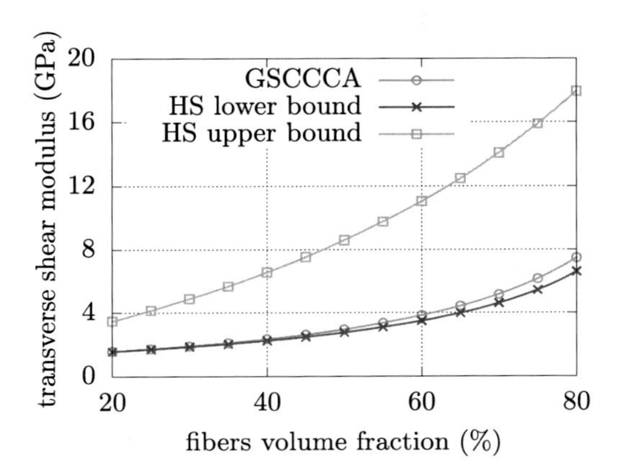

FIGURE 10.4 Macroscopic transverse shear modulus versus fiber volume fraction using GSCCCA and comparison with Hashin–Shtrikman bounds.

References

[1] Z. Hashin, S. Shtrikman, On some variational principles in anisotropic and nonhomogeneous elasticity, Journal of the Mechanics and Physics of Solids 10 (1962) 335–342.

[2] J.R. Willis, Bounds and self–consistent estimates for the overall properties of anisotropic composites, Journal of the Mechanics and Physics of Solids 25 (1977) 185–202.

[3] J.R. Willis, Variational and related methods for the overall properties of composites, Advances in Applied Mechanics 21 (1981) 1–78.

CHAPTER

11

Mathematical homogenization theory

OUTLINE

11.1 Preliminaries

The mathematical treatment of homogenization for composite media has led to the development of powerful theories, such as the compensated compactness, the H–convergence, the G–convergence, and the two–scale convergence. A very popular approach in the engineering community is the asymptotic expansion homogenization method [1,2].

Multiscale Modeling Approaches for Composites
https://doi.org/10.1016/B978-0-12-823143-2.00022-9

In this chapter, we denote all the variables that correspond to the general composite (without scale separation) with a superscript ϵ. Every function ϕ^ϵ is assumed as a two–scale function, periodic at the microscopic level:

$$\phi^\epsilon(\bar{\boldsymbol{x}}) = \phi(\bar{\boldsymbol{x}}, \boldsymbol{x}), \quad \text{periodic with respect to } \boldsymbol{x} = \frac{\bar{\boldsymbol{x}}}{\epsilon}.$$

When considering the composite in a global sense, the gradient operator $\left(\frac{\partial}{\partial \bar{x}_i}\right)^\epsilon$ is connected with the gradient operators of the macroscale, $\frac{\partial}{\partial \bar{x}_i}$, and microscale, $\frac{\partial}{\partial x_i}$, through the scale decomposition rule [2]

$$\left(\frac{\partial\{\bullet\}}{\partial \bar{x}_i}\right)^\epsilon = \frac{\partial\{\bullet\}}{\partial \bar{x}_i} + \frac{1}{\epsilon}\frac{\partial\{\bullet\}}{\partial x_i}. \tag{11.1}$$

The boundary value problem in quasi–static elasticity, including body forces b_i^ϵ and ignoring inertia effects, takes the form (for fixed ϵ)

$$\begin{aligned}
\left(\frac{\partial \sigma_{ij}^\epsilon}{\partial \bar{x}_j}\right)^\epsilon + b_i^\epsilon &= 0 && \text{in } \mathcal{B}^\epsilon, \\
\sigma_{ij}^\epsilon &= L_{ijkl}^\epsilon \varepsilon_{kl}^\epsilon && \text{in } \mathcal{B}^\epsilon, \\
\varepsilon_{ij}^\epsilon = \frac{1}{2}\left[\left(\frac{\partial u_i^\epsilon}{\partial \bar{x}_j}\right)^\epsilon + \left(\frac{\partial u_j^\epsilon}{\partial \bar{x}_i}\right)^\epsilon\right] & && \text{in } \mathcal{B}^\epsilon, \\
\sigma_{ij}^\epsilon n_j^\epsilon &= t_i^\epsilon && \text{on } \partial\mathcal{B}^{t\epsilon}, \\
u_i^\epsilon &= 0 && \text{on } \partial\mathcal{B}^{u\epsilon},
\end{aligned} \tag{11.2}$$

where $\partial\mathcal{B}^{t\epsilon}$ and $\partial\mathcal{B}^{u\epsilon}$ are the surfaces of the essential and the natural boundary conditions, respectively (Fig. 11.1). We assume that

$$\partial\mathcal{B}^{t\epsilon} \cup \partial\mathcal{B}^{u\epsilon} = \partial\mathcal{B}^\epsilon \quad \text{and} \quad \partial\mathcal{B}^{t\epsilon} \cap \partial\mathcal{B}^{u\epsilon} = \varnothing. \tag{11.3}$$

Moreover, $L_{ijkl}^\epsilon \equiv L_{ijkl}(\boldsymbol{x})$ is periodic.

System (11.2) in its differential form is solvable as long as $\boldsymbol{L}^\epsilon$ is a continuous and differentiable function of the position. This assumption though does not hold for composites, which generally consist of different phases attached close to each other with very dissimilar properties [6]. For this reason, a proper numerical treatment of system (11.2) requires expressing it in its variational form.

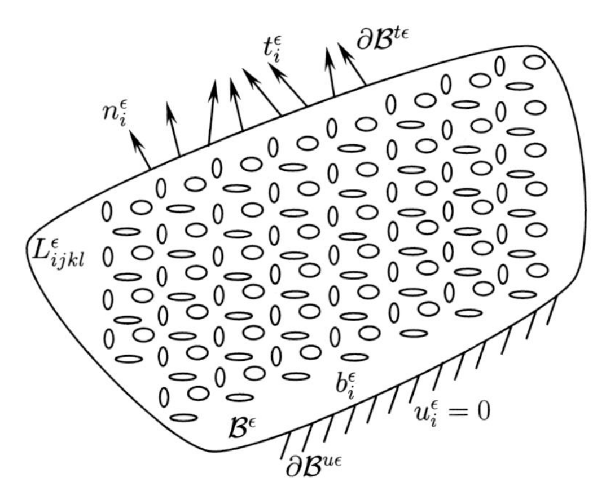

FIGURE 11.1 Composite structure under various boundary conditions. Schematic considering no scale separation.

11.2 Variational formulation

In mechanics the equations of equilibrium can appear in two forms, differential and variational. The latter is very useful when the stresses are discontinuous or non–smooth. This is often the case for composite materials, where interfaces between material phases introduce stress jumps. In this section, we provide a short discussion about the variational formulation in mechanics and its numerical implementation into the finite element method.

11.2.1 Functional spaces

Let $\mathcal{B}$ be a body with boundary surface $\partial\mathcal{B}$. To establish the variational formulation of the mechanical problem for this body, we need to introduce special spaces for functions.

The space $C(\mathcal{B})$ of continuous functions and the space $C^\infty(\mathcal{B})$ of continuous functions with continuous derivatives of all orders are the simplest spaces to work with, especially when analytical solutions of the problems can be identified. However, the latter is very seldom in engineering applications. For that reason, approximate techniques have been established. These techniques require the use of functions that are not always continuous.

Hilbert spaces are the common spaces used to establish approximate solutions in mechanics. Some characteristic examples of Hilbert spaces are the following:

- The Lebesgue spaces $L^p(\mathcal{B})$ of general order p with finite norm

$$\psi \in L^p(\mathcal{B}) \quad \rightarrow \quad \|\psi\|_{L^p(\mathcal{B})} = \left[\int_{\mathcal{B}} \psi^p \, \mathrm{d}V\right]^{1/p} \quad \text{finite.} \tag{11.4}$$

A frequently used Lebesgue space is the $L^2(\mathcal{B})$ of square–integrable functions ψ with finite norm

$$\psi \in L^2(\mathcal{B}) \quad \rightarrow \quad \|\psi\|_{L^2(\mathcal{B})} = \left[\int_{\mathcal{B}} \psi^2 \, \mathrm{d}V\right]^{1/2} \quad \text{finite.} \tag{11.5}$$

An important property of the space $L^2(\mathcal{B})$ is that it allows us to define the weak derivative in the sense that if there exist functions $\dfrac{\partial q}{\partial x_j} \in L^2(\mathcal{B})$, then

$$\begin{gathered} \int_{\mathcal{B}} \frac{\partial q}{\partial x_j} \phi \, \mathrm{d}V = -\int_{\mathcal{B}} q \frac{\partial \phi}{\partial x_j} \mathrm{d}V, \quad \forall \phi \in \mathcal{D}(\mathcal{B}), \\ \mathcal{D}(\mathcal{B}) = \{\phi \in C^{\infty}(\mathcal{B}), \ \text{with } \phi = 0 \text{ on } \partial\mathcal{B}\}. \end{gathered} \tag{11.6}$$

The weak derivation extends the integration by parts to L^2–functions. Moreover, a continuous function on the closure of $\mathcal{B}$ with piecewise continuous derivatives is weakly differentiable in $L^2(\mathcal{B})$. It is immediate to conclude that the above definition of weak derivative can be generalized to differential operators as the divergence of a function.

- The Sobolev spaces W_p^m of functions $\psi \in \mathcal{D}'(\mathcal{B})$ in which ψ and all of its distributional partial derivatives of order $\leqslant m$ belong to $L^p(\mathcal{B})$, $p \geqslant 1$. Note that $\mathcal{D}'(\mathcal{B})$ denotes the space of distributions in which a linear functional q possesses distributional partial derivatives in the sense of expression (11.6) for $j = 1, 2, 3$. For more details on distributions and Sobolev spaces, the reader is invited to consult [7].
Sobolev spaces are important for developing the variational approach of elliptic problems such as the elasticity problem and the Stokes equations. The Lax–Milgram theory and the energy minimization principle can be employed in functions belonging to Sobolev spaces to prove the existence and uniqueness of solutions in elasticity. The reader is invited to read [8] for a consistent study of elliptic problems.
A characteristic Sobolev space is the space $H^1(\mathcal{B})$ of functions whose L^2–norm and the L^2–norm of their first derivatives are bounded, that is,

$$H^1(\mathcal{B}) = \left\{\psi \in L^2(\mathcal{B}), \ \frac{\partial \psi}{\partial x_j} \in L^2(\mathcal{B}) \text{ for } j = 1, 2, 3\right\}. \tag{11.7}$$

Generally, functions belonging to a Sobolev space do not need to be differentiable at every point. For instance, the piecewise continuous functions with piecewise continuous derivatives and with compact support

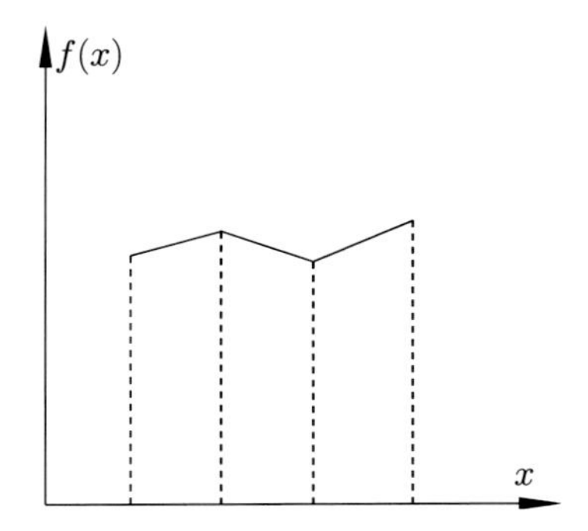

FIGURE 11.2 One–dimensional continuous function with discontinuities in the derivatives at certain points.

in the closure of $\mathcal{B}$ belong to $H^1(\mathcal{B})$ (see Fig. 11.2). Piecewise continuity appears frequently in the displacement field of heterogeneous elastic materials, where the constituents are perfectly bonded: at interfaces the displacements are continuous, but the strains are discontinuous. Thus problems involving heterogeneous bodies cannot exhibit classical (smooth) solutions, and natural spaces the displacements belong to are Sobolev spaces.

Two key characteristics of H^1–functions are the following.

1. The Green formula is applicable to H^1–functions, that is,

$$\int_{\mathcal{B}} v \frac{\partial u_i}{\partial x_i} \, \mathrm{d}V + \int_{\mathcal{B}} u_i \frac{\partial v}{\partial x_i} \, \mathrm{d}V = \int_{\partial \mathcal{B}} v u_i n_i \, \mathrm{d}S, \quad \forall \, u_i, v \in H^1(\mathcal{B}), \tag{11.8}$$

   where n_i is the normal vector on the boundary $\partial\mathcal{B}$ and $i = 1, 2, 3$. This expression can be extended to include higher–order tensor functions.
2. Every sequence simply bounded in $H^1(\mathcal{B})$ gives a sequence that converges in norm in $L^2(\mathcal{B})$. This property is used in the development of the finite element method.

11.2.2 Homogeneous body

Let us consider a body made of a homogeneous elastic material $\boldsymbol{L}$, occupying a space $\mathcal{B}$ with boundary surface $\partial\mathcal{B}$ and unit normal vector $\boldsymbol{n}$. The variational formulation in elasticity is an alternative way to express the boundary value problem

$$\frac{\partial \sigma_{ij}}{\partial x_j} + b_i = 0 \quad \text{in } \mathcal{B},$$

$$\sigma_{ij} = L_{ijkl} \varepsilon_{kl} \quad \text{in } \mathcal{B},$$

$$\varepsilon_{ij} = \frac{1}{2}\left[\frac{\partial u_i}{\partial x_j} + \frac{\partial u_j}{\partial x_i}\right] \quad \text{in } \mathcal{B},$$
$$\sigma_{ij} n_j = t_i \ \text{known on } \partial\mathcal{B}^t,$$
$$u_i = 0 \ \text{ on } \partial\mathcal{B}^u, \tag{11.9}$$

with

$$\partial\mathcal{B}^t \cup \partial\mathcal{B}^u = \partial\mathcal{B}, \quad \partial\mathcal{B}^t \cap \partial\mathcal{B}^u = \varnothing. \tag{11.10}$$

Multiplying the first of (11.9) by a smooth test function v_i equal to zero on $\partial\mathcal{B}^u$ and integrating over $\mathcal{B}$ yield

$$\int_{\mathcal{B}} v_i \frac{\partial \sigma_{ij}}{\partial x_j}\, \mathrm{d}V + \int_{\mathcal{B}} v_i b_i \, \mathrm{d}V = 0. \tag{11.11}$$

Integration by parts gives

$$\int_{\mathcal{B}} \sigma_{ij} \frac{\partial v_i}{\partial x_j}\, \mathrm{d}V = \int_{\mathcal{B}} v_i b_i \, \mathrm{d}V + \int_{\mathcal{B}} \frac{\partial}{\partial x_j}\left(v_i \sigma_{ij}\right) \mathrm{d}V. \tag{11.12}$$

The divergence theorem and the boundary conditions lead to

$$\int_{\mathcal{B}} \sigma_{ij} \frac{\partial v_i}{\partial x_j}\, \mathrm{d}V = \int_{\mathcal{B}} v_i b_i \, \mathrm{d}V + \int_{\partial\mathcal{B}^t} v_i t_i \, \mathrm{d}S. \tag{11.13}$$

Setting

$$\varepsilon_{ij}(\boldsymbol{v}) = \frac{1}{2}\left[\frac{\partial v_i}{\partial x_j} + \frac{\partial v_j}{\partial x_i}\right] \quad \text{in } \mathcal{B}, \tag{11.14}$$

in similarity with the third equation of (11.9), considering the symmetry of σ_{ij} and the fact that σ_{ij} is given by the second equation of (11.9), we obtain the following:

BOX 11.1 Variational form of the linear elasticity problem

For a linear elastic material, the variational form of the equilibrium equations is expressed as

$$\int_{\mathcal{B}} \sigma_{ij}(\boldsymbol{u})\varepsilon_{ij}(\boldsymbol{v})\, \mathrm{d}V = \int_{\mathcal{B}} v_i b_i \, \mathrm{d}V + \int_{\partial\mathcal{B}^t} v_i t_i \, \mathrm{d}S, \quad \forall v_i \text{ smooth, } v_i = 0 \ \text{ on } \partial\mathcal{B}^u. \tag{11.15}$$

11.2.3 Heterogeneous body with a surface of discontinuity

Let us now consider a body $\mathcal{B}$ with boundary surface $\partial\mathcal{B}$ composed of two parts $\mathcal{B}_1$ and $\mathcal{B}_2$, with elasticity tensors L^1_{ijkl} and L^2_{ijkl}, separated by a surface of discontinuity Γ (see Fig. 11.3). The body is subjected to body forces b_i in $\mathcal{B}$, tractions t_i on $\partial\mathcal{B}^t$, and it is fixed on $\partial\mathcal{B}^u$ ($\partial\mathcal{B}^t \cup \partial\mathcal{B}^u = \partial\mathcal{B}$, $\partial\mathcal{B}^t \cap \partial\mathcal{B}^u = \varnothing$). The elasticity tensors L^1_{ijkl} and L^2_{ijkl} can be constant or regular functions of the position vector $\boldsymbol{x}$.

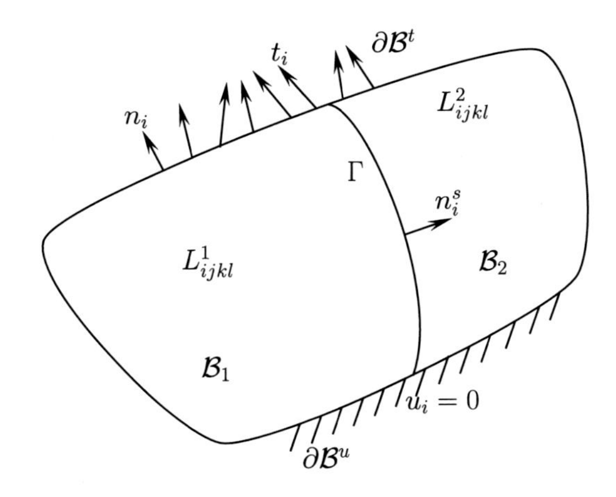

FIGURE 11.3 Heterogeneous body composed of two materials.

Taking into account the Cauchy hypothesis of displacement and traction continuity across the surface Γ, the elasticity problem for this body reads

$$
\begin{aligned}
\frac{\partial \sigma_{ij}}{\partial x_j} + b_i &= 0 && \text{in } \mathcal{B},\\
\sigma_{ij} &= L_{ijkl}\varepsilon_{kl} && \text{in } \mathcal{B},\\
\varepsilon_{ij} &= \frac{1}{2}\left[\frac{\partial u_i}{\partial x_j} + \frac{\partial u_j}{\partial x_i}\right] && \text{in } \mathcal{B},\\
[\![\sigma_{ij} n^s_j]\!] &= 0 && \text{on } \Gamma,\\
[\![u_i]\!] &= 0 && \text{on } \Gamma,\\
\sigma_{ij} n_j &= t_i && \text{on } \partial\mathcal{B}^t,\\
u_i &= 0 && \text{on } \partial\mathcal{B}^u,
\end{aligned}
\tag{11.16}
$$

where n_i and n^s_i are the unit vectors of the boundary $\partial\mathcal{B}$ and the surface Γ, respectively. For a variable ϕ, the symbol $[\![\phi]\!]$ denotes the difference in the values of this variable between the two edges of the discontinuity surface.

The presence of the discontinuity surface prohibits the use of continuous and differentiable functions for solving system (11.16). Instead, functions belonging to the Sobolev space $H^1(\mathcal{B})$ can be employed. By a procedure that, formally, seems to be analogous to that of a homogeneous material we obtain the following variational form:

BOX 11.2 Variational problem for heterogeneous body

Find

$$u_i \in \mathcal{V} := \{v_i \in H^1(\mathcal{B});\ v_i = 0 \text{ on } \partial\mathcal{B}^u\} \tag{11.17}$$

such that

$$\int_{\mathcal{B}} L_{ijkl}(\boldsymbol{x}) \frac{\partial u_k}{\partial x_l} \frac{\partial v_i}{\partial x_j} \, \mathrm{d}V = \int_{\mathcal{B}} v_i b_i \, \mathrm{d}V + \int_{\partial\mathcal{B}^t} v_i t_i \, \mathrm{d}S, \quad \forall v_i \in \mathcal{V}. \tag{11.18}$$

11.2.4 Approximating functions

The mechanical parameters and unknown fields are defined in the body $\mathcal{B}$ (homogeneous or heterogeneous) and are described by vector spaces of infinite dimension. Numerical analysis considers the above spaces as limits of spaces of finite dimension thanks to a discretization of $\mathcal{B}$, allowing for approximating solutions. A fundamental example is the approximation by a piecewise polynomial over a partition of $\mathcal{B}$ (Fig. 11.4). Another example is the approximation of periodic functions by Fourier series.

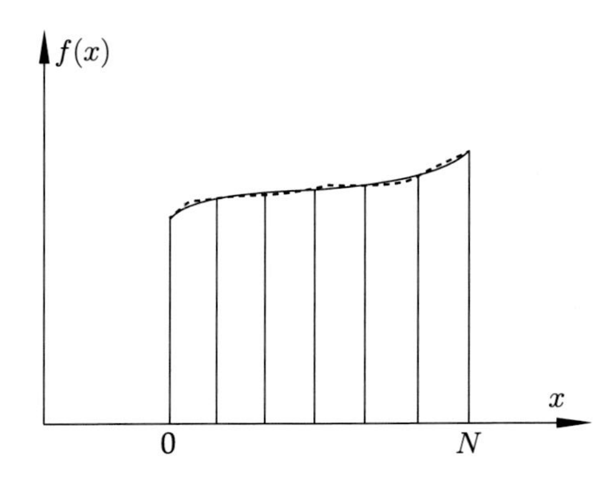

FIGURE 11.4 Approximation by a piecewise polynomial function in one dimension.

In the finite element (FE) method the approximating functions are piecewise polynomials on a partition of $\mathcal{B}$ adjusted smoothly between the pieces. This method is applied in particular to the numerical solution of the boundary value problem written in the variational form (11.15).

The identification of solutions of a system of partial differential equations in mechanics, such as the boundary value problem (11.9) or its variational expression (11.15), must rely on an appropriate functional space, that is, a space whose elements are not points but functions. The appropriate space for (11.15) is the space of square–integrable functions $\psi \in L^2(\mathcal{B})$ (see Eq. (11.5)).

An approximating solution for the displacement field u_i^n is defined in the approximating space $X_n \subset L^2(\mathcal{B})$, and its precision is compared with the error of the best approximation

$$\min_{u_i \in X_n} \|u_i - u_i^n\|_{L^2(\mathcal{B})}.$$

The approximation can be obtained by simple functions, as those belonging to the Lebesgue space $L^2(\mathcal{B})$, or to the space of continuous functions $C(\mathcal{B})$, or to the space of more regular functions. An important property in approximating theory is the density property that describes some simple functions g, for instance, piecewise constant, continuous with compact support, or well–known infinitely differentiable (used as regularizing functions by convolution). All these are dense in $L^2(\mathcal{B})$, that is,

$$\forall f \in L^2(\mathcal{B}) \text{ and } \epsilon \text{ small}, \qquad \|f - g\|_{L^2(\mathcal{B})} \leqslant \epsilon.$$

11.2.5 Finite element method

The two essential functional spaces used in elasticity, the spaces of L^2– and H^1–functions, are Hilbert spaces. In mechanics–oriented words, their practical interest consists in forming the context for application of the FE method, which comes directly from the variational approach of the problems. This method is based on the substitution of the Hilbert space ($L^2(\mathcal{B})$ or $H^1(\mathcal{B})$) by a subspace $\mathcal{B}_h$ of finite dimension. The concept of $\mathcal{B}_h$ responds to the necessity to have a good approximation of $\mathcal{B}$ by $\mathcal{B}_h$ and a solution u_h in $\mathcal{B}_h$, very close to the exact solution u in $\mathcal{B}$. The approximating problem on $\mathcal{B}_h$ consists in the numerical solution of a linear system. For the space $\mathcal{B}_h$, the conditions of applicability of the Lax–Milgram theory are considerably simplified.

The FE approach considers the following steps:

1. Discretize the domain $\mathcal{B}_h$ in several elements. The connection between the elements is through nodes (see Fig. 11.5).

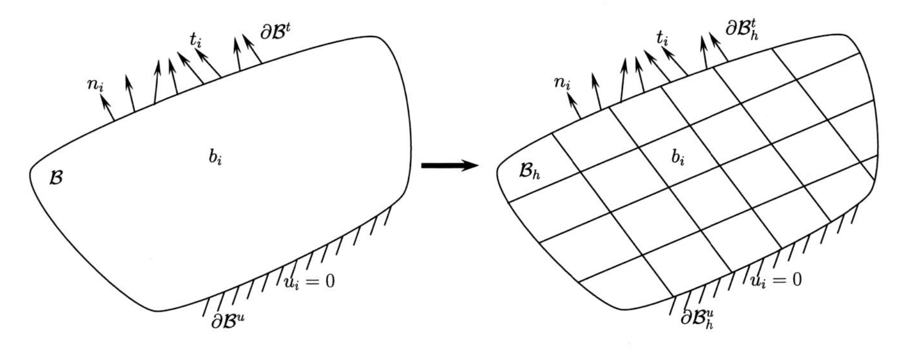

FIGURE 11.5 Discretization of a body in several elements for the use of the FE method.

2. Express the approximate solution of the displacement vector in terms of the basis functions ψ_i, that is,

$$\boldsymbol{u}(\boldsymbol{x}) \approx \sum_{i=1}^{N} \psi_i(\boldsymbol{x})\boldsymbol{c}_i,$$

where N is the number of nodes, and $\boldsymbol{c}_i$ corresponds to the nodal displacements.
3. Use the approximate solution to solve the variational problem.

11.3 Convergence of the heterogeneous problem

From a mathematical point of view, it would be useful to demonstrate how we can pass from the heterogeneous system of Eqs. (11.2) to those describing the equivalent homogeneous medium of Fig. 11.6$_a$, that is,

$$\begin{aligned}
\frac{\partial \bar{\sigma}_{ij}}{\partial \bar{x}_j} + \bar{b}_i &= 0 && \text{in } \bar{\mathcal{B}},\\
\bar{\sigma}_{ij} &= \bar{L}_{ijkl}\bar{\varepsilon}_{kl} && \text{in } \bar{\mathcal{B}},\\
\bar{\varepsilon}_{ij} = \frac{1}{2}\left[\frac{\partial \bar{u}_i}{\partial \bar{x}_j} \right.&\left.+ \frac{\partial \bar{u}_j}{\partial \bar{x}_i}\right] && \text{in } \bar{\mathcal{B}},\\
\bar{\sigma}_{ij} n_j &= \bar{t}_i && \text{on } \partial\bar{\mathcal{B}}^t,\\
\bar{u}_i &= 0 && \text{on } \partial\bar{\mathcal{B}}^u,
\end{aligned} \tag{11.19}$$

where $\partial\bar{\mathcal{B}}^t$ and $\partial\bar{\mathcal{B}}^u$ are the surfaces of the essential and natural boundary conditions, respectively, with

$$\partial\bar{\mathcal{B}}^t \cup \partial\bar{\mathcal{B}}^u = \partial\bar{\mathcal{B}} \quad \text{and} \quad \partial\bar{\mathcal{B}}^t \cap \partial\bar{\mathcal{B}}^u = \varnothing. \tag{11.20}$$

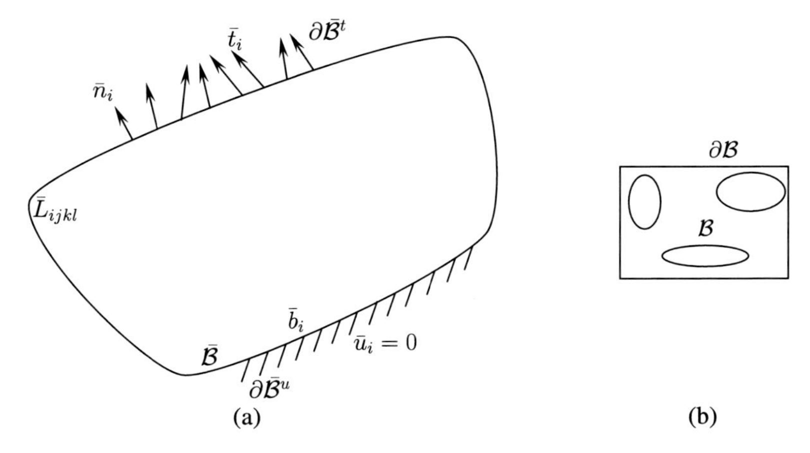

FIGURE 11.6 (a) Equivalent homogeneous medium for the composite of Fig. 11.1. (b) Unit cell of the same composite.

To simplify the demonstration, we assume that the body forces and the boundary traction conditions are independent[1] of the microscopic coordinates (i.e., $b_i^\epsilon = \bar{b}_i \equiv b_i$ and $t_i^\epsilon = \bar{t}_i \equiv t_i$). Recalling the variational formulation (11.18), the heterogeneous and the homogenized variational problems are written as follows.

- Heterogeneous: Find

$$u_i^\epsilon \in \mathcal{V}^\epsilon = \{u_i^\epsilon \in H^1(\mathcal{B}^\epsilon),\ u_i^\epsilon = 0 \text{ on } \partial\mathcal{B}^{u\epsilon}\} \tag{11.21}$$

such that

$$\int_{\mathcal{B}^\epsilon} L_{ijkl}^\epsilon \left(\frac{\partial u_k^\epsilon}{\partial \bar{x}_l}\right)^\epsilon \left(\frac{\partial v_i^\epsilon}{\partial \bar{x}_j}\right)^\epsilon \mathrm{d}V^\epsilon = \int_{\mathcal{B}^\epsilon} v_i^\epsilon b_i \,\mathrm{d}V^\epsilon + \int_{\partial\mathcal{B}^{t\epsilon}} v_i^\epsilon t_i \,\mathrm{d}S^\epsilon \tag{11.22}$$

for all $v_i^\epsilon \in \mathcal{V}^\epsilon$.

- Homogenized: Find

$$\bar{u}_i \in \bar{\mathcal{V}} = \{\bar{u}_i \in H^1(\bar{\mathcal{B}}),\ \bar{u}_i = 0 \text{ on } \partial\bar{\mathcal{B}}^u\} \tag{11.23}$$

such that

$$\int_{\bar{\mathcal{B}}} \bar{L}_{ijkl} \frac{\partial \bar{u}_k}{\partial \bar{x}_l} \frac{\partial \bar{v}_i}{\partial \bar{x}_j} \mathrm{d}\bar{V} = \int_{\bar{\mathcal{B}}} \bar{v}_i b_i \,\mathrm{d}\bar{V} + \int_{\partial\bar{\mathcal{B}}^t} \bar{v}_i t_i \,\mathrm{d}\bar{S} \tag{11.24}$$

[1] If b_i^ϵ and t_i^ϵ are periodically oscillating functions with respect to the microscopic coordinates, then certain slight modifications in the described procedure are required [9,10].

for all $\overline{v}_i \in \overline{\mathcal{V}}$, where the homogenized coefficients $\overline{L}_{ijkl}$ are given by (6.6) (Chapter 6) through proper volume integrations over the unit cell space $\mathcal{B}$ of volume V (Fig. 11.6$_b$).

The scope of the mathematical homogenization theory is proving that the variational equation (11.22) converges to the homogenized (11.24) when the length scale parameter ϵ tends to zero.

11.3.1 Weak convergence

In convergence studies for sequences, it is customary to look for strong limits. A sequence ϕ_n, $n \geqslant 1$, belonging to $L^2(\mathcal{B})$, is considered strongly convergent (in terms of norm) to a function $\phi \in L^2(\mathcal{B})$ when

$$\|\phi_n - \phi\|_{L^2(\mathcal{B})} \to 0 \quad \text{as} \quad n \to \infty. \tag{11.25}$$

Unfortunately, the study of (11.22) as $\epsilon \to 0$ is a complicated task due to the oscillating nature of the involved functions. Frequently, such expression does not converge for small values of ϵ and consequently a strong limit cannot be found. However, it can be demonstrated that (11.22) may converge weakly to (11.24). The definition of the weak convergence is as follows [11]:

BOX 11.3 Weak convergence

Let Ω be an open bounded smooth subset of $\Re^3$. Moreover, let $1 < p < \infty$, and let q be the conjugate exponent, $\dfrac{1}{p} + \dfrac{1}{q} = 1$.

A sequence $\{u_n\}_{n \geqslant 1} \subset L^p(\Omega)$ converges weakly to $u \in L^p(\Omega)$ if

$$\int_{\Omega} u_n v \,\mathrm{d}V \to \int_{\Omega} u v \,\mathrm{d}V \quad \text{as } n \to \infty, \quad \forall v \in L^q(\Omega).$$

If u_n is periodic with respect to $\epsilon = \dfrac{1}{n}$, then it can be denoted as u^{ϵ}, and its weak limit coincides with the mean value over a period.

This definition holds also for spaces Ω belonging to $\Re^N$, $N \neq 3$. A simple example illustrating the weak convergence concept is the case of the sequence

$$u^{\epsilon}(\overline{x}) = \sin\left(\frac{\overline{x}}{\epsilon}\right)$$

defined in $\Omega = [0, 2\pi]$. For $\epsilon > 0$, the function u^{ϵ} belongs to $L^2(\Omega)$ and is periodic with respect to the variable $\overline{x}$, with period depending of the value

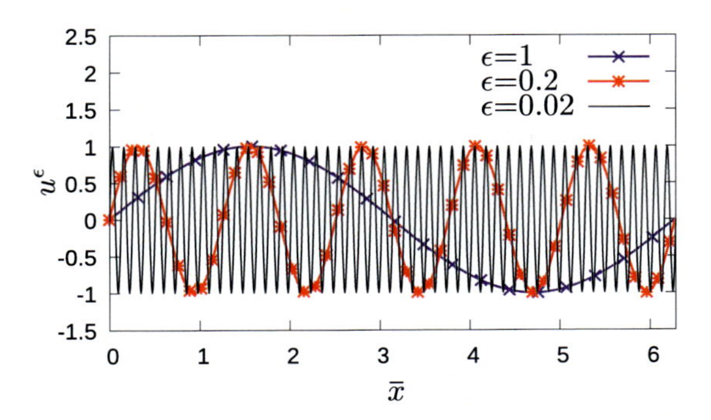

FIGURE 11.7 Spatial distribution of u^ϵ for various values of ϵ.

of ϵ. Whereas u^ϵ gas no limit as $\epsilon \to 0$ (Fig. 11.7), the integral

$$\int_0^{2\pi} u^\epsilon(\bar{x}) g(\bar{x}) \mathrm{d}\bar{x}, \qquad \forall g(\bar{x}) \in L^2(\Omega),$$

tends always to zero for any positive value of ϵ. Consequently, the expression

$$\frac{1}{2\pi} \int_0^{2\pi} u^\epsilon(\bar{x}) \mathrm{d}\bar{x} = 0,$$

is considered as the weak limit of the sequence u^ϵ.

11.3.2 Mathematical homogenization

Using the concept of the weak convergence, the two–scale convergence theory [12,4] demonstrates that (11.22) tends to the homogenized (11.24) as $\epsilon \to 0$.

As already mentioned previously, oscillating sequences belonging to $L^2(\bar{\mathcal{B}})$ are only weakly convergent sequences, in the sense that

$$\lim_{\epsilon \to 0} \int_{\mathcal{B}^\epsilon} \phi^\epsilon(\bar{\boldsymbol{x}}) \psi(\bar{\boldsymbol{x}}) \, \mathrm{d}V^\epsilon = \int_{\bar{\mathcal{B}}} \psi(\bar{\boldsymbol{x}}) \left[\frac{1}{V} \int_{\mathcal{B}} \phi(\bar{\boldsymbol{x}}, \boldsymbol{x}) \, \mathrm{d}V \right] \mathrm{d}\bar{V}, \;\; \forall \, \psi(\bar{\boldsymbol{x}}) \in L^2(\bar{\mathcal{B}}). \tag{11.26}$$

Note that the approximating domain of the body $\mathcal{B}^\epsilon$, illustrated in Fig. 11.1, tends to $\bar{\mathcal{B}}$ as $\epsilon \to 0$. Whereas this property is very useful for a single sequence, it cannot provide sufficient information for products of sequences. In the homogenization theory the mechanical work, which is the product of stress and strain, plays a very important role (see the Hill–Mandel theorem in Chapter 3). Unfortunately, the limit of the product of two weakly converging sequences is not generally equal to the product of the weak limits of the two sequences.

To overcome this difficult point, we need to seek for additional properties of weakly converging sequences to find the limit of their product. Such a task had led to the development of important mathematical theories, such as the compensated compactness theorem [3,5], applied to mechanical problems. The latter theorem provides a generalization of the Hill–Mandel theorem, independent from the type of boundary conditions.

The two–scale convergence theory [12,4] introduces two powerful mathematical tools:

1. A more detailed weak convergence, using oscillating test functions ψ^ϵ, utilizes the notion of two–scale convergence. According to this modified definition, a function $\phi^\epsilon(\bar{\boldsymbol{x}})$ two–scale converges to $\phi^{(0)}(\bar{\boldsymbol{x}}, \boldsymbol{x})$ if

$$\lim_{\epsilon \to 0} \int_{\mathcal{B}^\epsilon} \phi^\epsilon(\bar{\boldsymbol{x}}) \psi^\epsilon(\bar{\boldsymbol{x}}) \, \mathrm{d}V^\epsilon = \int_{\bar{\mathcal{B}}} \left[\frac{1}{V} \int_{\mathcal{B}} \phi^{(0)}(\bar{\boldsymbol{x}}, \boldsymbol{x}) \psi(\bar{\boldsymbol{x}}, \boldsymbol{x}) \, \mathrm{d}V \right] \mathrm{d}\bar{V},$$
$$\forall \psi^\epsilon(\bar{\boldsymbol{x}}) = \psi\left(\bar{\boldsymbol{x}}, \frac{\bar{\boldsymbol{x}}}{\epsilon}\right) \in L^2(\bar{\mathcal{B}} \times \mathcal{B}). \quad (11.27)$$

It should be pointed out that the two–scale limit $\phi^{(0)}$ carries more information than the weak limit

$$\langle \phi \rangle = \frac{1}{V} \int_{\mathcal{B}} \phi^{(0)}(\bar{\boldsymbol{x}}, \boldsymbol{x}) \, \mathrm{d}V. \quad (11.28)$$

2. An equally important lemma provided from this theory is the two–scale convergence of derivatives of H^1–functions: If the displacement field $u_i^\epsilon(\bar{\boldsymbol{x}})$ converges weakly in $H^1(\bar{\mathcal{B}})$ to $u_i^{(0)}(\bar{\boldsymbol{x}})$, then $u_i^\epsilon(\bar{\boldsymbol{x}})$ two–scale converges to $u_i^{(0)}$, and there exists a microscopic displacement field $u_i^{(1)}(\bar{\boldsymbol{x}}, \boldsymbol{x})$ such that

$$\lim_{\epsilon \to 0} \int_{\bar{\mathcal{B}}} \varepsilon_{ij}^\epsilon(\bar{\boldsymbol{x}}) \psi^\epsilon(\bar{\boldsymbol{x}}) \, \mathrm{d}V^\epsilon = \int_{\bar{\mathcal{B}}} \left[\frac{1}{V} \int_{\mathcal{B}} \left[\bar{\varepsilon}_{ij}(\bar{\boldsymbol{x}}) + \tilde{\varepsilon}_{ij}(\bar{\boldsymbol{x}}, \boldsymbol{x}) \right] \psi(\bar{\boldsymbol{x}}, \boldsymbol{x}) \, \mathrm{d}V \right] \mathrm{d}\bar{V},$$
$$\forall \psi^\epsilon(\bar{\boldsymbol{x}}) = \psi\left(\bar{\boldsymbol{x}}, \frac{\bar{\boldsymbol{x}}}{\epsilon}\right) \in L^2(\bar{\mathcal{B}} \times \mathcal{B}), \quad (11.29)$$

where

$$\varepsilon_{ij}^\epsilon = \frac{1}{2} \left[\left(\frac{\partial u_i^\epsilon}{\partial \bar{x}_j} \right)^\epsilon + \left(\frac{\partial u_j^\epsilon}{\partial \bar{x}_i} \right)^\epsilon \right],$$

$$\bar{\varepsilon}_{ij} = \frac{1}{2} \left[\frac{\partial u_i^{(0)}}{\partial \bar{x}_j} + \frac{\partial u_j^{(0)}}{\partial \bar{x}_i} \right],$$

$$\widetilde{\varepsilon}_{ij} = \frac{1}{2}\left[\frac{\partial u_i^{(1)}}{\partial x_j} + \frac{\partial u_j^{(1)}}{\partial x_i}\right]. \tag{11.30}$$

It is worth mentioning that in the case of linear elasticity the two–scale convergence of u_i^ϵ to $u_i^{(0)}$ can be proven rigorously [13].

These useful properties allow us to pass to the limit of (11.22) as $\epsilon \to 0$. Let us consider $v_i^\epsilon = u_i^\epsilon$. Since the body force and the traction boundary conditions are assumed independent of ϵ, the two–scale convergence of u_i^ϵ to $u_i^{(0)}$ allows us to pass to the limit on the right–hand side of (11.22):

$$\int_{\mathcal{B}^\epsilon} u_i^\epsilon b_i \,\mathrm{d}V^\epsilon + \int_{\partial\mathcal{B}^{t\epsilon}} u_i^\epsilon t_i \,\mathrm{d}S^\epsilon \to \int_{\bar{\mathcal{B}}} u_i^{(0)} b_i \mathrm{d}\bar{V} + \int_{\partial\bar{\mathcal{B}}^t} u_i^{(0)} t_i \mathrm{d}\bar{S}, \quad \text{as } \epsilon \to 0. \tag{11.31}$$

The left–hand side of (11.22) can be written in compact form as

$$\int_{\mathcal{B}^\epsilon} \sigma_{ij}^\epsilon \varepsilon_{ij}^\epsilon \,\mathrm{d}V^\epsilon.$$

Applying the two–scale convergence theory, we can pass to the limit:

$$\int_{\mathcal{B}^\epsilon} \sigma_{ij}^\epsilon \varepsilon_{ij}^\epsilon \,\mathrm{d}V^\epsilon \to \int_{\bar{\mathcal{B}}}\left[\frac{1}{V}\int_{\mathcal{B}} \sigma_{ij}\,\mathrm{d}V\right]\bar{\varepsilon}_{ij}\mathrm{d}\bar{V} + \int_{\bar{\mathcal{B}}}\left[\frac{1}{V}\int_{\mathcal{B}} \sigma_{ij}\widetilde{\varepsilon}_{ij}\,\mathrm{d}V\right]\mathrm{d}\bar{V}. \tag{11.32}$$

In this expression, we took into account that $\bar{\varepsilon}_{ij} \equiv \bar{\varepsilon}_{ij}(\bar{\boldsymbol{x}})$ is independent of the microscopic coordinates $\boldsymbol{x}$. Setting

$$\bar{\sigma}_{ij} = \frac{1}{V}\int_{\mathcal{B}} \sigma_{ij}\,\mathrm{d}V,$$

we provide the limit of (11.22) by combining (11.31) and (11.32). The latter leads to the equation

$$\int_{\bar{\mathcal{B}}} \bar{\sigma}_{ij}\bar{\varepsilon}_{ij}\mathrm{d}\bar{V} = \int_{\bar{\mathcal{B}}} u_i^{(0)} b_i \mathrm{d}\bar{V} + \int_{\partial\bar{\mathcal{B}}^t} u_i^{(0)} t_i \mathrm{d}\bar{S}, \tag{11.33}$$

which corresponds to the macroscopic equilibrium (11.24) and to the expression

$$\int_{\mathcal{B}} \sigma_{ij}\widetilde{\varepsilon}_{ij}\,\mathrm{d}V = 0 \tag{11.34}$$

representing the microscopic equilibrium.

11.4 Asymptotic expansion approach

The previously discussed powerful framework proves that the homogenization procedure leads asymptotically to an equivalent response to that of the actual composite as the characteristic length ϵ tends to zero. However, it does not give a direct answer on how to compute the macroscopic elasticity tensor $\bar{L}_{ijkl}$. This tensor can be obtained using additional mathematical tools. One such tool is the asymptotic expansion approach.

The main hypothesis of the asymptotic expansion homogenization method is that the displacement field is approximated with an asymptotic series expansion of the form

$$u_i^{\epsilon}(\bar{\boldsymbol{x}}) = u_i^{(0)}(\bar{\boldsymbol{x}}, \boldsymbol{x}) + \epsilon\, u_i^{(1)}(\bar{\boldsymbol{x}}, \boldsymbol{x}) + \epsilon^2 u_i^{(2)}(\bar{\boldsymbol{x}}, \boldsymbol{x}) + \cdots, \tag{11.35}$$

where the two–scale functions $\boldsymbol{u}^{(0)}$, $\boldsymbol{u}^{(1)}$, and so on are periodic in $\boldsymbol{x}$. This approximation is valid since such functions are the most typical example of two–scale converging functions [12,4]. Substituting (11.35) into $(11.2)_3$ and using the decomposition rule (11.1) yield

$$\begin{aligned}
\varepsilon_{ij}^{\epsilon} &= \frac{1}{\epsilon}\varepsilon_{ij}^{(-1)} + \varepsilon_{ij}^{(0)} + \epsilon\, \varepsilon_{ij}^{(1)} + \cdots, \\
\varepsilon_{ij}^{(-1)} &= \frac{1}{2}\left[\frac{\partial u_i^{(0)}}{\partial x_j} + \frac{\partial u_j^{(0)}}{\partial x_i}\right], \\
\varepsilon_{ij}^{(0)} &= \frac{1}{2}\left[\frac{\partial u_i^{(0)}}{\partial \bar{x}_j} + \frac{\partial u_j^{(0)}}{\partial \bar{x}_i}\right] + \frac{1}{2}\left[\frac{\partial u_i^{(1)}}{\partial x_j} + \frac{\partial u_j^{(1)}}{\partial x_i}\right], \\
\varepsilon_{ij}^{(1)} &= \frac{1}{2}\left[\frac{\partial u_i^{(1)}}{\partial \bar{x}_j} + \frac{\partial u_j^{(1)}}{\partial \bar{x}_i}\right] + \frac{1}{2}\left[\frac{\partial u_i^{(2)}}{\partial x_j} + \frac{\partial u_j^{(2)}}{\partial x_i}\right], \quad \text{and so on.}
\end{aligned} \tag{11.36}$$

The constitutive law $(11.2)_2$ allows us to write the stress in the asymptotic form

$$\sigma_{ij}^{\epsilon} = \frac{1}{\epsilon}\sigma_{ij}^{(-1)} + \sigma_{ij}^{(0)} + \epsilon\, \sigma_{ij}^{(1)} + \cdots, \qquad \sigma_{ij}^{(q)} = L_{ijkl}\varepsilon_{kl}^{(q)}. \tag{11.37}$$

Combining the last expression with $(11.2)_1$ and (11.1) leads to

$$\begin{aligned}
\frac{1}{\epsilon^2}\frac{\partial \sigma_{ij}^{(-1)}}{\partial x_j} + \frac{1}{\epsilon}\left[\frac{\partial \sigma_{ij}^{(-1)}}{\partial \bar{x}_j} + \frac{\partial \sigma_{ij}^{(0)}}{\partial x_j}\right] + \left[\frac{\partial \sigma_{ij}^{(0)}}{\partial \bar{x}_j} + \frac{\partial \sigma_{ij}^{(1)}}{\partial x_j}\right] + b_i \\
+ \epsilon\left[\frac{\partial \sigma_{ij}^{(1)}}{\partial \bar{x}_j} + \frac{\partial \sigma_{ij}^{(2)}}{\partial x_j}\right] + \cdots = 0.
\end{aligned} \tag{11.38}$$

Since the mathematical homogenization examines the behavior of (11.38) as $\epsilon \to 0$, the terms that are multiplied with ϵ^q, $q \geqslant 1$, vanish, and for the remaining three terms, we obtain the following results:

- The term ϵ^{-2} should be exactly equal to 0. Thus the displacement $\boldsymbol{u}^{(0)}$ must satisfy the homogeneous equation

$$\frac{\partial}{\partial x_j}\left(L_{ijkl}\frac{\partial u_k^{(0)}}{\partial x_l}\right) = 0. \tag{11.39}$$

  Since the elasticity tensor is always positive definite, (11.39) eventually states that $\boldsymbol{u}^{(0)}$ is independent of $\boldsymbol{x}$. So $\boldsymbol{\varepsilon}^{(-1)} = \boldsymbol{\sigma}^{(-1)} = \boldsymbol{0}$.
- The term ϵ^{-1} should be exactly equal to 0. Combining with the previous result, we obtain the microscopic equilibrium

$$\frac{\partial \sigma_{ij}^{(0)}}{\partial x_j} = 0. \tag{11.40}$$

  This equation is generally referred to as the unit cell problem.
- The remaining part of the equilibrium,

$$\frac{\partial \sigma_{ij}^{(0)}}{\partial \bar{x}_j} + \frac{\partial \sigma_{ij}^{(1)}}{\partial x_j} + b_i = 0,$$

  has an oscillating response, and its limit cannot be directly obtained. Instead, using the concept of weak convergence, its volume average is the weak limit of the equilibrium. Due to the periodicity of $\boldsymbol{\sigma}^{(1)}$ in $\mathcal{B}$, its volume average vanishes,

$$\left\langle \frac{\partial \sigma_{ij}^{(1)}}{\partial x_j} \right\rangle = \frac{1}{V}\int_{\mathcal{B}} \frac{\partial \sigma_{ij}^{(1)}}{\partial x_j}\,\mathrm{d}V = \frac{1}{V}\int_{\partial\mathcal{B}} \sigma_{ij}^{(1)} n_j\,\mathrm{d}S = 0,$$

  leading to the final expression for the macroscopic equilibrium

$$\frac{\partial\left\langle \sigma_{ij}^{(0)} \right\rangle}{\partial \bar{x}_j} + b_i = 0. \tag{11.41}$$

Comparing these results with the expressions of Table 6.1, the asymptotic expansion analysis shows that the stress $\boldsymbol{\sigma}^{(0)}$ represents the microscopic stress and its volume average is the macroscopic stress $\bar{\boldsymbol{\sigma}}$. Moreover, the microscopic strain is equal to $\boldsymbol{\varepsilon}^{(0)}$, whose volume average is expressed as

$$\left\langle \varepsilon_{ij}^{(0)} \right\rangle = \frac{1}{2}\left[\frac{\partial u_i^{(0)}}{\partial \bar{x}_j} + \frac{\partial u_j^{(0)}}{\partial \bar{x}_i}\right] + \frac{1}{2}\left\langle \frac{\partial u_i^{(1)}}{\partial x_j} + \frac{\partial u_j^{(1)}}{\partial x_i} \right\rangle$$
$$= \frac{1}{2}\left[\frac{\partial u_i^{(0)}}{\partial \bar{x}_j} + \frac{\partial u_j^{(0)}}{\partial \bar{x}_i}\right] = \bar{\varepsilon}_{ij}. \tag{11.42}$$

Lastly, substituting $\boldsymbol{\varepsilon}^{(0)}$ into the unit cell problem (11.40) yields the same expression with (6.2) if the displacement $\boldsymbol{u}^{\text{per}}$ is replaced by $\boldsymbol{u}^{(1)}$ (both are periodic in $\mathcal{B}$). This means that the asymptotic expansion homogenization and the engineering approach provide the same micro–equilibrium equation and lead to the same macroscopic elasticity tensor (6.6).

Table 11.1 summarizes the qualitative differences between the engineering approach of periodic homogenization, presented in Section 6.2 of Chapter 6, and the asymptotic expansion homogenization theory.

TABLE 11.1 Engineering vs asymptotic expansion homogenization (AEH) for a two–scale composite.

Engineering	AEH
• The two scales are separate from the beginning	• The two scales are related with the characteristic length ϵ
• Macro– and micro–equations are considered separate	• Macro– and micro–equations are obtained by requiring $\epsilon \to 0$
• Micro–displacement is split in linear and periodic part	• Micro–displacement is written in asymptotic series form
• Connection between the scales through Hill–Mandel principle	• Connection between the scales through weak convergence

As a final note, it is worth noting that in the so–called "higher–order homogenization theories" the general expression (11.38) is solved by keeping additional terms of the expansion. These approaches tend to be more reliable when the real characteristic size ϵ of the microstructure is not sufficiently small. However, the price that one pays for considering these terms is that strain gradients need to be taken into account in the overall behavior of the composite [14,15,16].

11.5 Examples

Example 1: FE solution of a 1–D bar

Consider the one–dimensional problem of Fig. 11.8. The elasticity prob-

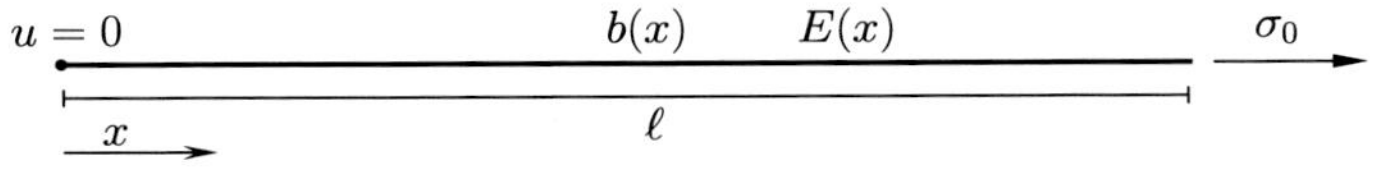

$u = 0$ $b(x)$ $E(x)$ σ_0

x ℓ

FIGURE 11.8 1–D bar under body forces b. Zero displacement at one end and applied stress σ_0 at the other end. The Young modulus of the bar may vary with the position.

lem in differential form for this 1–D bar is expressed as

$$\begin{aligned} \frac{\mathrm{d}}{\mathrm{d}x}\left(E\frac{\mathrm{d}u}{\mathrm{d}x}\right) + b = 0 & \quad \forall x, \\ u = 0 & \quad \text{at } x = 0, \\ \sigma = E\frac{\mathrm{d}u}{\mathrm{d}x} = \sigma_0 & \quad \text{at } x = \ell, \end{aligned} \tag{11.43}$$

where u is the displacement, σ is the stress, b represents the body forces, and σ_0 is the boundary stress condition at one edge of the bar. Moreover, ℓ is the length, and E is the Young modulus of the bar. Use the finite element method to solve this elasticity problem.

Solution:

As long as the Young modulus is constant or differentiable function of the position x, the system of Eqs. (11.43) can be solved with classical tools. If though E is not differentiable, or even discontinuous (as it is the case of composites), then (11.43) is not appropriate for describing the elastic problem of the bar. Instead, the variational form is required. Following similar steps with those discussed in Subsection 11.2.2, the variational form of (11.43) is given by

$$\int_0^\ell E\frac{\mathrm{d}u}{\mathrm{d}x}\frac{\mathrm{d}v}{\mathrm{d}x}\mathrm{d}x' = \sigma_0\, v(\ell) + \int_0^\ell bv\mathrm{d}x', \tag{11.44}$$

where

$$v \in \mathcal{V} = \{v \in H^1(0,\ell);\, v = 0 \text{ at } x = 0\} \tag{11.45}$$

is a test function. The solution u of the problem belongs also to the same space $\mathcal{V}$. To proceed to the numerical solution by the finite element method, the space $\mathcal{B} = (0,\ell)$ is approximated by $\mathcal{B}_h$: the continuous bar is subdivided into N smaller bars (elements) of constant length $h = \ell/N$, as shown in Fig. 11.9. The position of a point i in the discretized bar is denoted as x_i.

To apply the FE method, the introduction of the basis functions

$$\phi_i(x), \quad 0 \leqslant i \leqslant N,$$

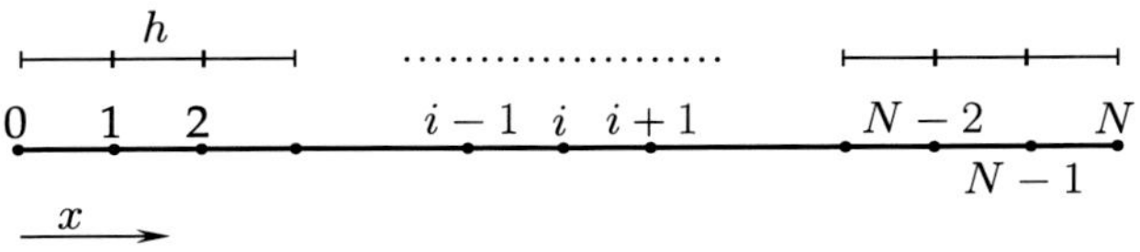

h
0
1
2
i − 1
i
i + 1
N − 2
N
N − 1
x

FIGURE 11.9 Discretization of 1–D bar in N elements of constant size h.

is required. These functions are usually polynomials. The simplest choice is to consider them as piecewise continuous and linear in every interval. In the 1–D case examined here, they can take the form (see Fig. 11.10)

$$\phi_0 = \begin{cases} \dfrac{x_1 - x}{h} & \text{if } x_0 \leqslant x \leqslant x_1, \\ 0 & \text{otherwise,} \end{cases}$$

$$\phi_i = \begin{cases} \dfrac{x - x_{i-1}}{h} & \text{if } x_{i-1} \leqslant x < x_i, \\ \dfrac{x_{i+1} - x}{h} & \text{if } x_i \leqslant x \leqslant x_{i+1}, \\ 0 & \text{otherwise,} \end{cases} \qquad \text{for } 1 \leqslant i \leqslant N-1,$$

$$\phi_N = \begin{cases} \dfrac{x - x_{N-1}}{h} & \text{if } x_{N-1} \leqslant x \leqslant x_N, \\ 0 & \text{otherwise.} \end{cases} \tag{11.46}$$

ϕ0
ϕ1
ϕ2
ϕi
ϕN−1
ϕN
1
0
1
2
3
i-1
i
i+1
N-2
N-1
N

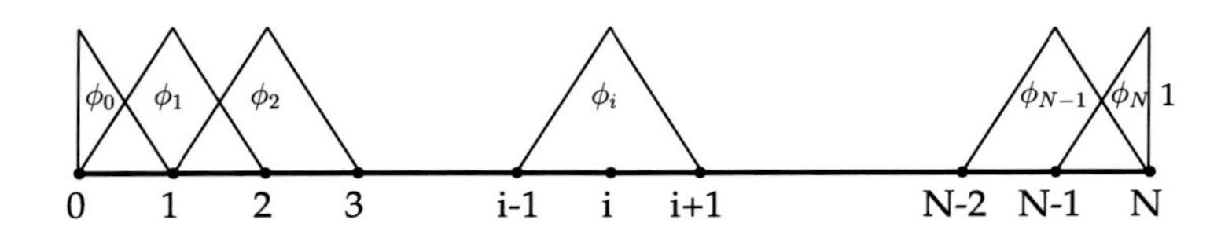

FIGURE 11.10 Basis functions for the finite element computations in 1–D bar.

Then the unknown displacement field u is expressed with the help of the functions ϕ_i through the expression

$$u(x) = \sum_{j=0}^{N} c_j \phi_j(x), \tag{11.47}$$

where c_i denotes the nodal displacement at the point i. Substituting (11.47) into (11.44) gives

$$\sum_{j=0}^{N} \left[\int_0^{\ell} E \frac{\mathrm{d}\phi_j}{\mathrm{d}x} \frac{\mathrm{d}v}{\mathrm{d}x} \mathrm{d}x' \right] c_j = \sigma_0 \, v(\ell) + \int_0^{\ell} bv\mathrm{d}x'. \tag{11.48}$$

In the last stage of the FE method a proper choice of the test function v is required. Setting $v = \phi_i$ for $i = 1, 2, ..., N$, we obtain the following linear system of equations:

$$\sum_{j=0}^{N}\left[\int_0^\ell E\frac{\mathrm{d}\phi_j}{\mathrm{d}x}\frac{\mathrm{d}\phi_i}{\mathrm{d}x}\mathrm{d}x'\right]c_j = \int_0^\ell b\phi_i\mathrm{d}x', \quad 1 \leqslant i \leqslant N-1,$$

$$\sum_{j=0}^{N}\left[\int_0^\ell E\frac{\mathrm{d}\phi_j}{\mathrm{d}x}\frac{\mathrm{d}\phi_N}{\mathrm{d}x}\mathrm{d}x'\right]c_j = \sigma_0 + \int_0^\ell b\phi_N\mathrm{d}x', \quad i = N. \tag{11.49}$$

From the forms (11.46) of the functions ϕ_i it becomes clear that the integrals appearing in (11.49) are nonzero only in the region $[x_{i-1}, x_{i+1}]$. Thus we can express (11.49) in the form

$$\left[\int_{x_{i-1}}^{x_i} E\frac{\mathrm{d}\phi_{i-1}}{\mathrm{d}x}\frac{\mathrm{d}\phi_i}{\mathrm{d}x}\mathrm{d}x'\right]c_{i-1} + \left[\int_{x_{i-1}}^{x_{i+1}} E\frac{\mathrm{d}\phi_i}{\mathrm{d}x}\frac{\mathrm{d}\phi_i}{\mathrm{d}x}\mathrm{d}x'\right]c_i + \left[\int_{x_i}^{x_{i+1}} E\frac{\mathrm{d}\phi_{i+1}}{\mathrm{d}x}\frac{\mathrm{d}\phi_i}{\mathrm{d}x}\mathrm{d}x'\right]c_{i+1} = \int_{x_{i-1}}^{x_{i+1}} b\phi_i\mathrm{d}x',$$

or

$$-\frac{1}{h^2}\left[\int_{x_{i-1}}^{x_i} E\mathrm{d}x'\right]c_{i-1} + \frac{1}{h^2}\left[\int_{x_{i-1}}^{x_{i+1}} E\mathrm{d}x'\right]c_i - \frac{1}{h^2}\left[\int_{x_i}^{x_{i+1}} E\mathrm{d}x'\right]c_{i+1} = \int_{x_{i-1}}^{x_{i+1}} b\phi_i\mathrm{d}x' \tag{11.50}$$

for $1 \leqslant i \leqslant N-1$, and

$$\left[\int_{x_{N-1}}^{x_N} E\frac{\mathrm{d}\phi_{N-1}}{\mathrm{d}x}\frac{\mathrm{d}\phi_N}{\mathrm{d}x}\mathrm{d}x'\right]c_{N-1} + \left[\int_{x_{N-1}}^{x_N} E\frac{\mathrm{d}\phi_N}{\mathrm{d}x}\frac{\mathrm{d}\phi_N}{\mathrm{d}x}\mathrm{d}x'\right]c_N = \sigma_0 + \int_{x_{N-1}}^{x_N} b\phi_N\mathrm{d}x',$$

or

$$-\frac{1}{h^2}\left[\int_{x_{N-1}}^{x_N} E\mathrm{d}x'\right]c_{N-1} + \frac{1}{h^2}\left[\int_{x_{N-1}}^{x_N} E\mathrm{d}x'\right]c_N = \sigma_0 + \int_{x_{N-1}}^{x_N} b\phi_N\mathrm{d}x' \tag{11.51}$$

for $i = N$. At the position $x = 0$, (11.47) yields the displacement boundary condition

$$0 = u(0) = u^{(0)} = c_0. \tag{11.52}$$

We can write Eqs. (11.50), (11.51), and (11.52) in the compact matrix form

$$\boldsymbol{K} \cdot \boldsymbol{U} = \boldsymbol{F}, \tag{11.53}$$

where $\boldsymbol{U}$ is the $N+1$ vector of the nodal displacements c_i, $\boldsymbol{F}$ is the $N+1$ vector of nodal forces (accounting also for the effect of body forces at the nodes), and $\boldsymbol{K}$ is the $N+1 \times N+1$ rigidity matrix.

Example 2: simple shearing of a periodic multilayered elastic material

Consider a multilayered composite structure whose microstructure consists of two phases with equal volume fractions and elastic shear coefficients G_α and G_β. The structure is fixed at $x=0$ and sheared by a steady force S at $x=h$ (Fig. 11.11). Prove that this medium has a homogenized behavior as $\epsilon \to 0$ and obtain the overall shear modulus $\bar{G}$ of the composite using the mathematical homogenization theory.

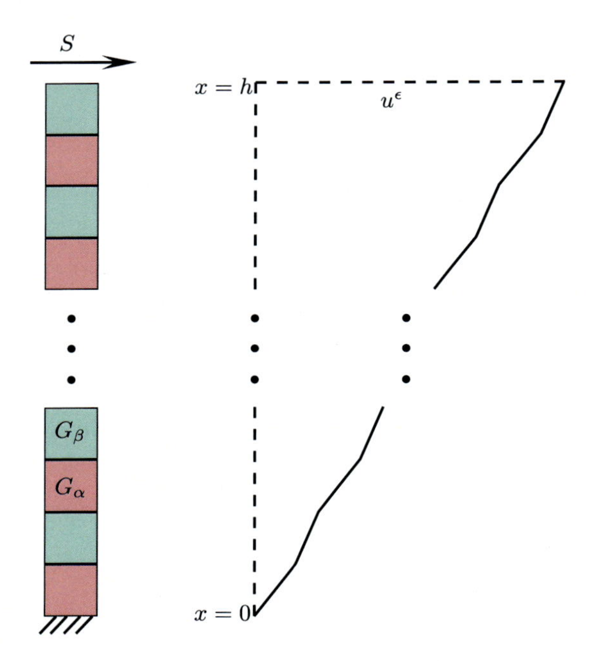

FIGURE 11.11 Multilayered material subjected to shear and displacement profile normal to the direction of the layers.

Solution:

The heterogeneous medium under simple shear is described by the following set of 1–D equations:

$$\frac{\partial \tau^{\epsilon}}{\partial x} = 0, \tag{11.54}$$

$$\tau^{\epsilon} = G^{\epsilon} \frac{\partial u^{\epsilon}}{\partial x}, \tag{11.55}$$

under the boundary conditions

$$\tau^{\epsilon}(h) = S, \quad u^{\epsilon}(0) = 0, \tag{11.56}$$

where τ^{ϵ} denotes the shear stress, u^{ϵ} is the displacement, and G^{ϵ} is the shear modulus, which takes the value of G_{α} or G_{β} according to the position that it refers to.

The first step is defining an appropriate functional setting for the displacement. This can be done by using standard techniques of functional analysis. Then it can be proved that, for fixed ϵ, u^{ϵ} belongs to

$$H_0^1(0,h) = \left\{ u^{\epsilon} \in H^1(0,h); u^{\epsilon}(0) = 0 \right\}.$$

As a consequence, a subsequence can be extracted such that, as $\epsilon \to 0$, $u^{\epsilon} \to \bar{u}$ in $H_0^1(0,h)$ weakly. The question now that arises is the following: What can we expect for this $\bar{u}$? More specifically, what equation is satisfied by $\bar{u}$? Is the constitutive law of the same type as the equation satisfied by u^{ϵ}?

To answer this question, it is important to notice that $\dfrac{\partial u^{\epsilon}}{\partial x}$ is only bounded in $L^2(0,h)$, thus we cannot pass directly to the limit in the expression

$$\frac{\partial}{\partial x}\left(G^{\epsilon} \frac{\partial u^{\epsilon}}{\partial x} \right) = 0,$$

since G^{ϵ} is only bounded in $L^2(0,h)$. Using the boundary condition, we can easily identify that the shear stress is everywhere constant, that is, $\tau^{\epsilon}(x) \equiv S$. Thus

$$\frac{\partial u^{\epsilon}}{\partial x} = \frac{1}{G^{\epsilon}} S.$$

Fig. 11.11 illustrates schematically the distribution of the displacement in the direction normal to the layers. We can pass to the limit in the last equality, since on the right–hand side, there is the product of the weakly converging sequence $\dfrac{1}{G^{\epsilon}}$ and a constant. Therefore

$$\frac{\partial \bar{u}}{\partial x} = S \int_0^h \frac{1}{G^{\epsilon}} \mathrm{d}x = \frac{1}{2}\left[\frac{1}{G_{\alpha}} + \frac{1}{G_{\beta}} \right] S.$$

Hence

$$S = \bar{G}\frac{\partial \bar{u}}{\partial x},$$

where the effective shear modulus is equal to the Reuss bound,

$$\bar{G} = 2\frac{G_\alpha G_\beta}{G_\alpha + G_\beta}.$$

Remark 1:

In the above problem, the facts that S is constant (more generally, strongly converging sequence, for instance, a function of the macroscopic position, independent of ϵ) and that the body force is zero (more generally, strongly converging sequence) play a crucial role: body forces, boundary conditions, and (in dynamical problems) initial conditions are potential agents of compactness.

Remark 2:

For $u^\epsilon \in H^1(0, h)$, the functional analysis theory allows us to prove that there exists a unique solution. It is worth noting that, exceptionally in one dimension, functions belonging to H^1 are continuous. This is not necessarily true in dimensions 2 or 3. In these cases, the continuity of the displacement and the traction vectors must be imposed at the interfaces. However, discontinuities of strain that belongs to L^2 are allowed at the interfaces.

Example 3: tension of a periodic multilayered elastic material

Consider the multilayered composite of Fig. 11.12. The unit cell of this composite consists of two layers that have a common surface normal to the x_1 axis. Both layers are assumed to be elastic isotropic materials with

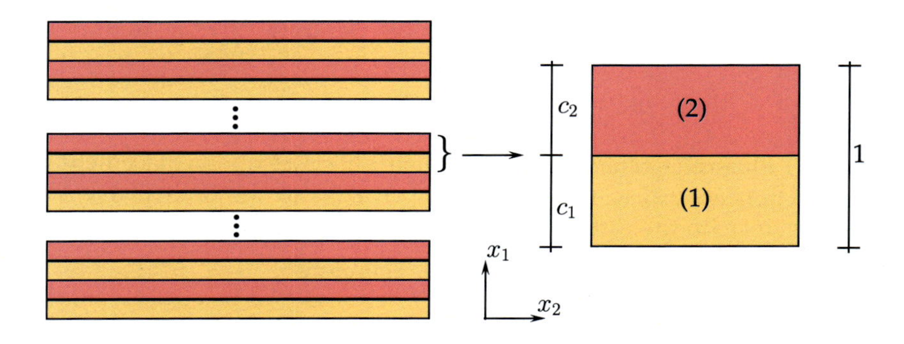

FIGURE 11.12 Schematic representation of multilayered composite and its unit cell.

Young moduli $E_1 = 3$ GPa for layer 1 and $E_2 = 55$ GPa for layer 2. The two layers have equal volume fractions ($c_1 = c_2 = 50\%$). Considering that in simple traction the elasticity problem is reduced to 1–D, use the finite element method to compute the macroscopic Young modulus $\bar{E}$.

Solution:

The FE discretization of the unit cell is illustrated in Fig. 11.13. Two elements are considered of equal length $h = 0.5$. The first element has Young modulus E_1, and the second E_2. There are three nodes in total with coordinates $x_0 = 0$, $x_1 = 0.5$, and $x_2 = 1$, which are assigned with the nodal displacements u_0, u_1, and u_2 respectively. Following the same method-

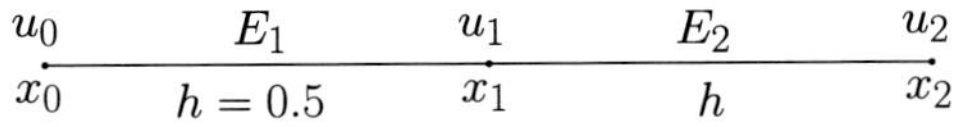

u_0 E_1 u_1 E_2 u_2

x_0 $h = 0.5$ x_1 h x_2

FIGURE 11.13 FE discretization of the unit cell of Fig. 11.12.

ology with that of Example 1, the middle node has to obey an equation similar to (11.50), that is,

$$-\frac{1}{h^2}\left[\int_{x_0}^{x_1} E\mathrm{d}x'\right]u_0 + \frac{1}{h^2}\left[\int_{x_0}^{x_2} E\mathrm{d}x'\right]u_1 - \frac{1}{h^2}\left[\int_{x_1}^{x_2} E\mathrm{d}x'\right]u_2 = 0. \quad (11.57)$$

Since

$$E = \begin{cases} E_1, & x_0 \leqslant x \leqslant x_1, \\ E_2, & x_1 \leqslant x \leqslant x_2, \end{cases}$$

Eq. (11.57) is reduced to

$$-E_1 u_0 + [E_1 + E_2] u_1 - E_2 u_2 = 0. \quad (11.58)$$

At the end nodes of the unit cell the periodicity condition should be imposed (see Section 6.3 of Chapter 6). This implies that

$$u_0 = \bar{\varepsilon} \cdot 0 + u^{\text{per}}, \quad u_2 = \bar{\varepsilon} \cdot 1 + u^{\text{per}}. \quad (11.59)$$

Setting a unit value for $\bar{\varepsilon}$ the average stress in the unit cell should be equal to the macroscopic Young modulus $\bar{E}$. The linear system (11.58) and (11.59) is not solvable unless one of the nodes displacements is assigned with an arbitrary value. For simplicity reasons, we can impose $u_0 = 0$, leading to

$$u_2 = 1 \quad \text{and} \quad u_1 = \frac{E_2}{E_1 + E_2}.$$

Since the two chosen finite elements are linear, their strains are computed by the simple expressions

$$\varepsilon_1 = \frac{u_1 - u_0}{h} = \frac{1}{h}\frac{E_2}{E_1 + E_2} \quad \text{and} \quad \varepsilon_2 = \frac{u_2 - u_1}{h} = \frac{1}{h}\frac{E_1}{E_1 + E_2}.$$

Consequently, the stress at both elements are equal:

$$\sigma_1 = E_1\varepsilon_1 = \frac{1}{h}\frac{E_1E_2}{E_1 + E_2} \quad \text{and} \quad \sigma_2 = E_2\varepsilon_2 = \frac{1}{h}\frac{E_1E_2}{E_1 + E_2},$$

and the average stress, which is equal to the macroscopic Young modulus, is given by

$$\bar{\sigma} = \bar{E} = \frac{1}{h}\frac{E_1E_2}{E_1 + E_2} = 5.69\ \text{GPa}.$$

From the last expression it becomes obvious that $\bar{E}$ coincides with the Reuss bound.

References

[1] A. Bensoussan, J. Lions, G. Papanicolaou, Asymptotic Methods for Periodic Structures, North–Holland, Amsterdam, 1978.

[2] E. Sanchez–Palencia, Non–homogeneous Media and Vibration Theory, Lecture Notes in Physics, vol. 127, Springer–Verlag, Berlin, 1978, pp. 1–398.

[3] L. Tartar, Nonlinear constitutive relations and homogenization, in: Contemporary Developments in Continuum Mechanics and Partial Differential Equations, North–Holland, Amsterdam, 1978, pp. 472–484.

[4] G. Allaire, Homogenization and two–scale convergence, SIAM Journal on Mathematical Analysis 23 (1992) 1482–1518.

[5] F. Murat, L. Tartar, H–convergence, in: A. Cherkaev, R.V. Kohn (Eds.), Topics in the Mathematical Modelling of Composite Materials, in: Progress in Nonlinear Differential Equations and Their Applications, vol. 31, Birkhäuser, Boston, 1997, pp. 21–43.

[6] N. Charalambakis, G. Chatzigeorgiou, Y. Chemisky, F. Meraghni, Mathematical homogenization of inelastic dissipative materials: a survey and recent progress, Continuum Mechanics and Thermodynamics 30 (1) (2018) 1–51.

[7] P.A. Raviart, J.M. Thomas, Introduction à l'analyse numérique des équations aux dérivées partielles, Masson, Paris, 1983.

[8] G. Allaire, Analyse Numérique et Optimisation, Les Éditions de l'École Polytechnique, Paris, 2006.

[9] J.M. Guedes, N. Kikuchi, Preprocessing and postprocessing for materials based on the homogenization method with adaptive finite element methods, Computer Methods in Applied Mechanics and Engineering 83 (1990) 143–198.

[10] K. Terada, N. Kikuchi, A class of general algorithms for multi–scale analyses of heterogeneous media, Computer Methods in Applied Mechanics and Engineering 190 (2001) 5427–5464.

[11] G. Chatzigeorgiou, N. Charalambakis, Y. Chemisky, F. Meraghni, Thermomechanical Behavior of Dissipative Composite Materials, ISTE Press – Elsevier, London, 2018.

[12] G. Nguetseng, A general convergence result for a functional related to the theory of homogenization, SIAM Journal on Mathematical Analysis 20 (1989) 608–623.

[13] A.L. Kalamkarov, A.G. Kolpakov, Analysis, Design and Optimization of Composite Structures, Wiley, West Sussex, 1997.
[14] J. Fish, W. Chen, Higher–order homogenization of initial/boundary–value problem, Journal of Engineering Mechanics 127 (2001) 1223–1230.
[15] S. Forest, Homogenization methods and the mechanics of generalized continua – Part 2, Theoretical and Applied Mechanics 28–29 (2002) 113–143.
[16] Z. He, M.J. Pindera, Locally exact asymptotic homogenization of periodic materials under anti–plane shear loading, European Journal of Mechanics. A, Solids 81 (2020) 103972.

12

Nonlinear composites

O U T L I N E

12.1 Introduction

As already discussed in the previous chapters, the macroscopic behavior of a composite with elastic constituents is also an elastic medium, usually more anisotropic than that of its constituents. Recall that the composite elasticity tensor is estimated directly from RVE computations (using mean–field or full–field methods), and it depends on the elasticity tensors, the geometric characteristics and the volume fractions of the constituents. In addition, the boundary conditions at the level of the heterogeneous structure do not influence the $\bar{\boldsymbol{L}}$ tensor.

In contrast to the elastic case, an inelastic composite is characterized by nonlinear behavior, and its response cannot be identified exclusively by RVE computations. Indeed, there is a strong dependency on the boundary conditions at the level of the structure. Consequently, any approach toward developing multiscale models for nonlinear composites requires simultaneous treatment of both microscale and macroscale equilibrium

Multiscale Modeling Approaches for Composites
https://doi.org/10.1016/B978-0-12-823143-2.00023-0

equations. When dealing with dynamic and/or fully coupled thermomechanical processes, initial conditions also play a vital role [1,2]. In the following sections, we discuss general computational strategies for the quasi–static response of nonlinear heterogeneous media using full–field (i.e., periodic homogenization) or mean–field approaches.

12.2 Inelastic mechanisms in periodic homogenization

From a theoretical point of view, the main issue with the periodic homogenization of nonlinear composites is that we cannot pass to the limit in the system of partial differential equations describing the problem as it happens in the elastic case. Thus, strictly mathematically speaking, the homogenized system and the homogenized solution cannot be found explicitly due to the nonlinear terms describing the constitutive behavior. There are a few exceptions in one–dimensional problems, as in viscoplastic shearing, in which the effective coefficients can be defined explicitly by a procedure of passing to the limit, but the general approach for nonlinear composites is based on some additional assumptions concerning, for instance, internal variables like the plastic strain. To overcome this difficulty, fully computational procedures are often adopted.

Incorporating inelastic mechanisms in full–field approaches like periodic homogenization is nowadays a well–studied field in the micromechanics community. Various researchers have studied the behavior of viscoelastic, elastoplastic, viscoplastic, damageable, and shape memory alloy–based composites using the periodic homogenization framework. Due to the nonlinear character of the response in these cases, the majority of the developed methods propose incremental techniques for applying the loading conditions, and the overall behavior is obtained iteratively through appropriate schemes, like the return mapping algorithms [3].

Viscoelastic composites present time–dependent response and a proper homogenization framework needs to account for these effects [4,5]. In the usual time domain, identifying the viscoelastic behavior of the composite requires the development of incremental iterative schemes. However, the correspondence principle allows us to solve the problem in the Laplace transform domain [6] using classical tools from homogenization of elastic media.

Homogenization of periodic elastoplastic composites is the most commonly studied case in the literature of inelastic micromechanics. Frequently, the numerical strategies utilize finite element computations [7,8] and adaptive return mapping algorithm schemes [3]. However, the computational cost of the finite element implementation can be reduced by adopting alternative computational tools, like the fast Fourier transform [9] or the finite volume–based homogenization [10]. The study of composites with viscoplastic response follows closely the methodologies used for elastoplastic composites [11,12]. Similar computational strategies can also be adopted for composites with damageable [13,14] or shape memory alloy [15,16] material phases.

A common strategy for treating composites with inelastic phases considers that the weak form of the unit cell problem is written (similar to expression (11.34) of Chapter 11 for elastic media) in the linearized form

$$\int_{\mathcal{B}} [\sigma_{ij} + \delta\sigma_{ij}] \frac{\partial \widehat{u}_i}{\partial x_j} \, \mathrm{d}V = 0, \tag{12.1}$$

where $\widehat{u}_i$ is a periodic displacement test function, and σ_{ij} and $\delta\sigma_{ij}$ are the microscopic stress and its iteration increment, respectively. The latter is expressed in terms of the iteration increment of the microscopic strains $\delta\varepsilon_{ij}$ according to the formula

$$\delta\sigma_{ij} = L^{\tan}_{ijkl} \delta\varepsilon_{kl}. \tag{12.2}$$

The tensor $L^{\tan}_{ijkl}$ represents the instantaneous tangent modulus tensor and is a function of the state of the material at the current loading conditions. Solving (12.1) iteratively allows us to pass to the macroscopic scale. In the latter a similar relation between the iteration increments of the stress and the strain hold, that is,

$$\delta\bar{\sigma}_{ij} = \bar{L}^{\tan}_{ijkl} \delta\bar{\varepsilon}_{kl}. \tag{12.3}$$

The $\bar{L}^{\tan}_{ijkl}$ is the homogenized instantaneous tangent operator, which can be computed as a function of its microscopic counterpart through classical homogenization scheme. The macroscopic problem is also written in a linearized form, following similar structure with expression (11.33) of Chapter 11 for elastic media.

In nonlinear composite structures, solving simultaneously the microscopic and macroscopic equilibrium is often a task with high computational cost. A typical algorithmic scheme for the complete homogenization problem is illustrated in Fig. 12.1. According to this strategy, each macroscopic point with position $\bar{\boldsymbol{x}}_\beta$ is linked to its own unit cell β. The macroscale analysis computes at every macroscopic point the macroscopic strain tensor $\bar{\boldsymbol{\varepsilon}}_\beta$. The latter is transferred as information to the corresponding unit cell to compute a) the microscopic strain $\boldsymbol{\varepsilon}_\beta$ and stress $\boldsymbol{\sigma}_\beta$ fields, as well as the macroscopic stress tensor $\bar{\boldsymbol{\sigma}}_\beta$ and b) the macroscopic instantaneous tangent modulus tensor $\bar{\boldsymbol{L}}^{\tan}_\beta$. Both $\bar{\boldsymbol{\sigma}}_\beta$ and $\bar{\boldsymbol{L}}^{\tan}_\beta$ pass as information to the macroscale analysis (i.e., the level of the structure). This procedure is repeated until the chosen local and global convergence criteria are reached, in which case the solution at the current step is assumed to be achieved, and the analysis passes to the new time step. When the finite element method is used for both scales, the adopted strategy is called FE^2 [17,14].

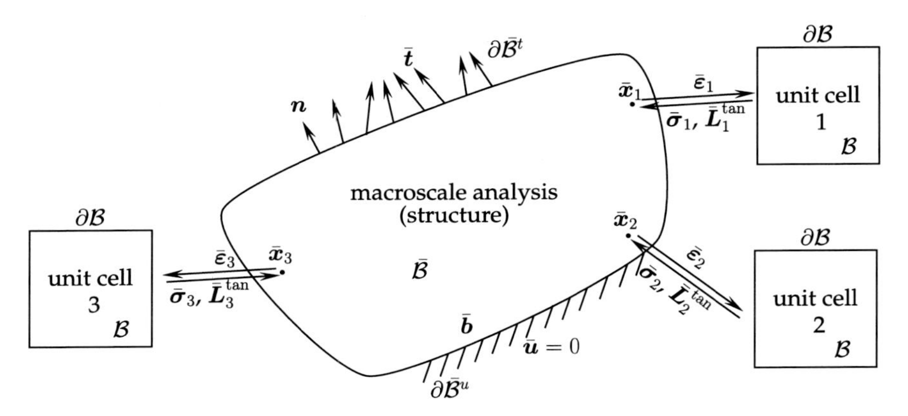

FIGURE 12.1 Simultaneous solution of macroscale analysis (structure) and unit cell problem. The structure may be subjected to macroscopic body forces $\bar{b}$, surface tractions $\bar{t}$, and zero surface displacements ($\bar{u} = \mathbf{0}$) conditions. In each macroscopic point with position $\bar{x}_\beta$, the unit cell β is attached. The various symbols of this figure have been presented in Chapter 11.

12.3 Inelastic mechanisms in mean–field theories

The mean–field approaches presented in Chapter 5 have been originally designed to study elastic composites. Trying to integrate the same techniques in the case of inelastic media is not trivial. One of the main difficulties lies on the fact that the Eshelby solutions have been derived exclusively for elastic media. Expanding the Eshelby inclusion and inhomogeneity problems to materials undergoing nonlinear phases is not obvious, as analytical solutions are hard to be obtained.

For viscoelastic composite media, the dynamic correspondence principle has been implemented successfully in mean–field approaches [18,19]. This principle takes advantage of the half–sided Fourier or Laplace–Carson transform and allows us to treat viscoelasticity in the frequency domain with micromechanics tools of elasticity. However, such nice property cannot be extended to more complicated nonlinear behaviors, like plasticity or damage.

Nowadays there is a plethora of developed methods to tackle the homogenization of elastoplastic composite media. A popular one follows a similar scheme with the periodic homogenization strategy of nonlinear composites: substituting the elasticity tensor with the instantaneous, time–dependent, tangent modulus tensor [20] and proceeding with an incremental iterative scheme [2]. However, such approach usually leads to very stiff behavior of the composite, especially when the matrix phase is strongly nonlinear [21]. Several theories have been developed to overcome this issue. Among the first ones is the so–called "isotropization" of the tangent or secant modulus [22,21], which has also been implemented in viscoelastic and viscoplastic behaviors [23]. Other solution strategies consider the description of a linear comparison composite, whose macroscopic properties can be defined in a simpler manner compared to the actual composite [24,25,26]. Alternative techniques account for increased inelastic strains by introducing a special interphase layer between the matrix and reinforcement [27].

The integration of damage mechanisms in mean–field homogenization schemes is also a challenging task. The main difficulty is that after a certain level of damage, the material enters into the softening regime, and the continuum damage theories fail to predict such unstable behavior. When dealing with composite media, an alternative way to integrate damage is by considering micro–crack density that appears as void creation in the form of spheroids or ellipsoids. This additional void occupies its own volume fraction by extracting it from the damaged material phase volume fraction [28,29,30,31] (see Example 3 of this chapter). Other techniques consider special load transfer conditions through shear lag models at the interface between the matrix and reinforcement [32,33]. More sophisticated approaches integrate non–local damage models into the mean–field micromechanics schemes [34].

Further in the section, we present a solution strategy for nonlinear composites based on the combination of the Mori–Tanaka theory and the well–known theory of transformation field analysis [35,36]. This methodology should be utilized with caution, especially when the nonlinearity appears mainly in the matrix phase.

12.3.1 Inhomogeneity problem with two eigenstrains

Consider a small inhomogeneity Ω with elasticity $\boldsymbol{L}_r$ in an infinite elastic body $\mathcal{B}$ of elasticity $\boldsymbol{L}'$ with stress $\boldsymbol{\sigma}'$ and strain $\boldsymbol{\varepsilon}'$ in the far field. Both the body and the inhomogeneity have their own eigenstrains $\boldsymbol{\varepsilon}'^{p}$ and $\boldsymbol{\varepsilon}_r^{p}$, respectively (Fig. 12.2).

The eigenstrains taken into account in this problem can represent inelastic deformation mechanisms (thermal, plastic, viscoelastic strains, etc.). In the present approach, they are assumed constant; thus the overall problem in incremental schemes remains elastic, and the superposition principle holds. Strictly speaking, inelastic strains are usually non–uniform. Since mean–field theories consider only average fields and spatial variation of strains cannot be taken into account, a crucial assumption is required: the inelastic strains or stresses are introduced per phase only with their average value [20], thus allowing us to utilize the elastic Eshelby problems. The accuracy of this hypothesis and the range of validity of the mean–field approaches in inelastic problems have been extensively discussed in the literature [35].

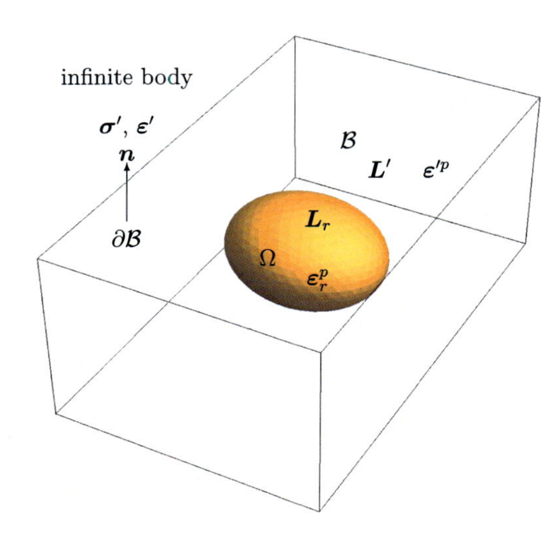

FIGURE 12.2 Inhomogeneity problem with two eigenstrains.

In mathematical formalism the following set of equations needs to be solved:

$$\begin{aligned}
&\operatorname{div}\boldsymbol{\sigma} = \mathbf{0} \quad \text{in } \mathcal{B},\\
&\boldsymbol{\sigma} = \begin{cases} \boldsymbol{L}' : \left[\boldsymbol{\varepsilon} - \boldsymbol{\varepsilon}'^{p}\right] & \text{in } \mathcal{B} - \Omega,\\ \boldsymbol{L}_r : \left[\boldsymbol{\varepsilon} - \boldsymbol{\varepsilon}_r^{p}\right] & \text{in } \Omega,\end{cases}\\
&\boldsymbol{\sigma} = \boldsymbol{\sigma}' \quad \text{and} \quad \boldsymbol{\varepsilon} = \boldsymbol{\varepsilon}' \quad \text{at far distance.}
\end{aligned} \tag{12.4}$$

This problem can be split into two simpler ones, shown in Fig. 12.3:

1. A problem with constant stresses $\boldsymbol{\sigma}'$ and strains $\boldsymbol{\varepsilon}'$ at far distance, where the inhomogeneity has the same properties and the same eigenstrain with the rest of the material (Fig. 12.3_a). The body's eigenstrain is also accounted. In mathematical formalism, this is expressed as

$$\operatorname{div}\boldsymbol{\sigma}^a = \operatorname{div}\left(\boldsymbol{L}' : \left[\boldsymbol{\varepsilon}^a - \boldsymbol{\varepsilon}'^{p}\right]\right) = \mathbf{0} \text{ in } \mathcal{B}, \quad \boldsymbol{\sigma}^a = \boldsymbol{\sigma}' \text{ at far distance.} \tag{12.5}$$

   The solution of this problem is trivial and is expressed as uniform stress $\boldsymbol{\sigma}^a = \boldsymbol{\sigma}'$ and uniform strain $\boldsymbol{\varepsilon}^a = \boldsymbol{\varepsilon}'$ inside the entire body $\mathcal{B}$.
2. A boundary value problem with zero surface tractions, where the inhomogeneity has been substituted by an equivalent inclusion problem. The difference between the eigenstrains of the inhomogeneity and the body are introduced as eigenstrain in the inclusion. Under these conditions, we assume that the region Ω has the same properties with the rest of the body and the disturbance due to the inhomogeneity is recovered

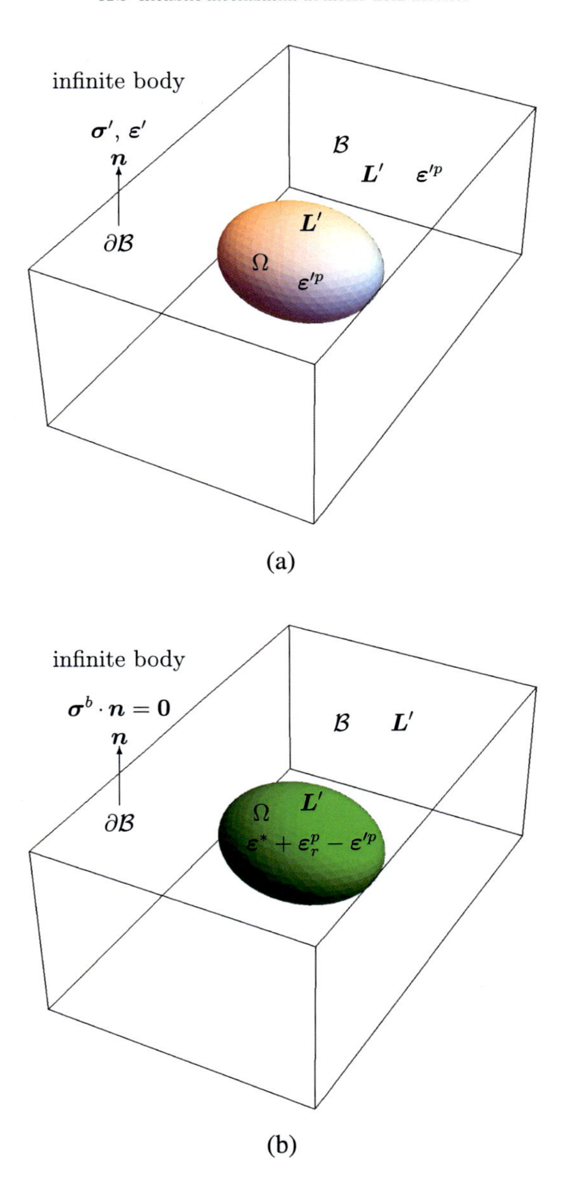

FIGURE 12.3 Inhomogeneity problem with two eigenstrains split into two problems.

by an imaginary additional eigenstrain $\boldsymbol{\varepsilon}^*$ (Fig. 12.3$_b$). In mathematical formalism, the second problem can be expressed as

$$\mathrm{div}\boldsymbol{\sigma}^b = \mathbf{0} \quad \text{in } \mathcal{B},$$

$$\begin{aligned}\boldsymbol{\sigma}^b &= \boldsymbol{L}': [\boldsymbol{\varepsilon}^b - \boldsymbol{\varepsilon}^* - \boldsymbol{\varepsilon}_r^p + \boldsymbol{\varepsilon}'^p] \quad \text{in } \mathcal{B},\\ \boldsymbol{\sigma}^b \cdot \boldsymbol{n} &= \mathbf{0} \quad \text{in } \partial\mathcal{B},\\ \boldsymbol{\varepsilon}^* + \boldsymbol{\varepsilon}_r^p - \boldsymbol{\varepsilon}'^p &= \begin{cases} \text{constant} \neq \mathbf{0} & \text{in } \Omega,\\ \mathbf{0} & \text{in } \mathcal{B} - \Omega.\end{cases}\end{aligned} \tag{12.6}$$

This is the classical Eshelby problem, which means that inside the inclusion,

$$\boldsymbol{\varepsilon}^b = \text{constant} = \boldsymbol{S}: \left[\boldsymbol{\varepsilon}^* + \boldsymbol{\varepsilon}_r^p - \boldsymbol{\varepsilon}'^p\right] \quad \text{in } \Omega. \tag{12.7}$$

All the problems are elastic, and the principle of superposition holds. Thus the solution of the initial problem is the sum of the solutions of two simpler problems. Focusing only in the region inside the inhomogeneity, we can write

$$\boldsymbol{\sigma} = \boldsymbol{\sigma}^a + \boldsymbol{\sigma}^b = \boldsymbol{\sigma}' + \boldsymbol{\sigma}^b \quad \text{and} \quad \boldsymbol{\varepsilon} = \boldsymbol{\varepsilon}^a + \boldsymbol{\varepsilon}^b = \boldsymbol{\varepsilon}' + \boldsymbol{\varepsilon}^b \quad \text{in } \Omega. \tag{12.8}$$

Eq. $(12.8)_1$, with the help of $(12.4)_2$, $(12.5)_2$, $(12.6)_2$, and $(12.8)_2$, gives the modified Eshelby equivalence principle

$$\boldsymbol{L}_r: [\boldsymbol{\varepsilon}' + \boldsymbol{\varepsilon}^b - \boldsymbol{\varepsilon}_r^p] = \boldsymbol{L}': [\boldsymbol{\varepsilon}' - \boldsymbol{\varepsilon}'^p + \boldsymbol{\varepsilon}^b - \boldsymbol{\varepsilon}^* - \boldsymbol{\varepsilon}_r^p + \boldsymbol{\varepsilon}'^p] \quad \text{in } \Omega \tag{12.9}$$

with

$$\boldsymbol{\varepsilon}^b = \boldsymbol{S}: \left[\boldsymbol{\varepsilon}^* + \boldsymbol{\varepsilon}_r^p - \boldsymbol{\varepsilon}'^p\right] = \boldsymbol{S}: \boldsymbol{\varepsilon}^{**} \quad \text{in } \Omega. \tag{12.10}$$

Substituting (12.10) into (12.9) yields

$$\boldsymbol{\varepsilon}^{**} = -\left[[\boldsymbol{L}_r - \boldsymbol{L}'] : \boldsymbol{S} + \boldsymbol{L}'\right]^{-1} : \left[[\boldsymbol{L}_r - \boldsymbol{L}'] : \boldsymbol{\varepsilon}' + \boldsymbol{L}': \boldsymbol{\varepsilon}'^p - \boldsymbol{L}_r : \boldsymbol{\varepsilon}_r^p\right]. \tag{12.11}$$

Substituting (12.11) inyo (12.10) gives

$$\boldsymbol{\varepsilon}^b = [\boldsymbol{T} - \boldsymbol{\mathcal{I}}] : \boldsymbol{\varepsilon}' + \boldsymbol{T}'^p : \boldsymbol{\varepsilon}'^p + \boldsymbol{T}_r^p : \boldsymbol{\varepsilon}_r^p \quad \text{in } \Omega \tag{12.12}$$

with

$$\begin{aligned}\boldsymbol{T} &= \left[\boldsymbol{\mathcal{I}} + \boldsymbol{S}: [\boldsymbol{L}']^{-1}: [\boldsymbol{L}_r - \boldsymbol{L}']\right]^{-1},\\ \boldsymbol{T}'^p &= [\boldsymbol{T} - \boldsymbol{\mathcal{I}}]: [\boldsymbol{L}_r - \boldsymbol{L}']^{-1}: \boldsymbol{L}' = -\boldsymbol{T}: \boldsymbol{S},\\ \boldsymbol{T}_r^p &= -[\boldsymbol{T} - \boldsymbol{\mathcal{I}}]: [\boldsymbol{L}_r - \boldsymbol{L}']^{-1}: \boldsymbol{L}_r = \boldsymbol{T}: \boldsymbol{S}: [\boldsymbol{L}']^{-1}: \boldsymbol{L}_r.\end{aligned} \tag{12.13}$$

Finally, substitution of (12.12) into $(12.8)_2$ provides the strain in the inhomogeneity as a function of the total strain at far distance and the two eigenstrains,

$$\boldsymbol{\varepsilon} = \boldsymbol{T}: \boldsymbol{\varepsilon}' + \boldsymbol{T}'^p : \boldsymbol{\varepsilon}'^p + \boldsymbol{T}_r^p : \boldsymbol{\varepsilon}_r^p \quad \text{in } \Omega. \tag{12.14}$$

The last expression is very useful for the development of mean–field homogenization strategies for inelastic composites, since it allows us to compute concentration tensors per phase for elastic and inelastic fields.

As a remark, it is worth mentioning that in composites with strongly nonlinear response and complex microstructure, computational techniques based on the Mori–Tanaka method or the self–consistent method may require multiple recalculations of the interaction tensors. Although the Eshelby tensor has a clear physical meaning, as indicated in Box 5.1 of Chapter 5, direct use of the polarization tensor $\boldsymbol{P} = \boldsymbol{S} : [\boldsymbol{L}']^{-1}$ is numerically more appropriate and requires less computational cost.

12.3.2 Mori–Tanaka/TFA method for composites with inelastic strains

Identifying elastic and inelastic concentration tensors to study inelastic composites is an established approach in the literature, known as transformation field analysis (TFA) [35]. The interaction tensors defined in the previous subsection allow us to combine the TFA approach with classical mean–field homogenization schemes, like Eshelby dilute, Mori–Tanaka, or self–consistent.

As a characteristic example, we can consider a composite with unidirectional particles of ellipsoidal shape. The volume fraction of the particles and the matrix are equal to c_1 and c_0, respectively, with $c_0 + c_1 = 1$. According to the Mori–Tanaka method, for this composite, each individual particle is assumed to lie inside the matrix material with elasticity modulus $\boldsymbol{L}_0$ and is subjected to i) the eigenstrain $\boldsymbol{\varepsilon}_0^p$ and ii) at far distance to the average strain $\boldsymbol{\varepsilon}_0$ of the matrix. Thus the underline hypothesis is that in the modified inhomogeneity problem of the previous subsection we should consider

$$\boldsymbol{L}' = \boldsymbol{L}_0, \quad \boldsymbol{\varepsilon}' = \boldsymbol{\varepsilon}_0, \quad \boldsymbol{\varepsilon}'^p = \boldsymbol{\varepsilon}_0^p. \tag{12.15}$$

For convenience, we adopt the following notation:

$$\boldsymbol{T} \equiv \boldsymbol{T}_1, \quad \boldsymbol{T}'^p \equiv \boldsymbol{T}_0^p,$$

allowing us to express the average strain in the particle using expression (12.14) as

$$\boldsymbol{\varepsilon}_1 = \boldsymbol{T}_1 : \boldsymbol{\varepsilon}_0 + \boldsymbol{T}_0^p : \boldsymbol{\varepsilon}_0^p + \boldsymbol{T}_1^p : \boldsymbol{\varepsilon}_1^p.$$

Taking into account the general relation between the macroscopic and the average microscopic strains,

$$\bar{\boldsymbol{\varepsilon}} = c_0 \boldsymbol{\varepsilon}_0 + c_1 \boldsymbol{\varepsilon}_1,$$

after some algebra we can obtain

$$\boldsymbol{\varepsilon}_q = \boldsymbol{A}_q : \bar{\boldsymbol{\varepsilon}} + \boldsymbol{A}^p_{0,q} : \boldsymbol{\varepsilon}^p_0 + \boldsymbol{A}^p_{1,q} : \boldsymbol{\varepsilon}^p_1, \quad q = 0, 1, \tag{12.16}$$

with

$$\begin{aligned} \boldsymbol{A}_0 &= \left[c_0 \boldsymbol{\mathcal{I}} + c_1 \boldsymbol{T}_1\right]^{-1}, & \boldsymbol{A}_1 &= \boldsymbol{T}_1 : \boldsymbol{A}_0, \\ \boldsymbol{A}^p_{0,0} &= -c_1 \boldsymbol{A}_0 : \boldsymbol{T}^p_0, & \boldsymbol{A}^p_{1,0} &= -c_1 \boldsymbol{A}_0 : \boldsymbol{T}^p_1, \\ \boldsymbol{A}^p_{0,1} &= \boldsymbol{T}^p_0 - c_1 \boldsymbol{A}_1 : \boldsymbol{T}^p_0, & \boldsymbol{A}^p_{1,1} &= \boldsymbol{T}^p_1 - c_1 \boldsymbol{A}_1 : \boldsymbol{T}^p_1. \end{aligned} \tag{12.17}$$

Expressions (12.16) represent a special case of the general context of the transformation field analysis [35,36]. Finally, considering the relation between the macroscopic and the average microscopic stresses,

$$\bar{\boldsymbol{\sigma}} = c_0 \boldsymbol{\sigma}_0 + c_1 \boldsymbol{\sigma}_1,$$

and the constitutive laws

$$\boldsymbol{\sigma}_q = \boldsymbol{L}_q : \left[\boldsymbol{\varepsilon}_q - \boldsymbol{\varepsilon}^p_q\right], \quad q = 0, 1, \tag{12.18}$$

the macroscopic constitutive law is obtained in the form

$$\bar{\boldsymbol{\sigma}} = \bar{\boldsymbol{L}} : \left[\bar{\boldsymbol{\varepsilon}} - \bar{\boldsymbol{\varepsilon}}^p\right], \tag{12.19}$$

where

$$\begin{aligned} &\bar{\boldsymbol{L}} = c_0 \boldsymbol{L}_0 : \boldsymbol{A}_0 + c_1 \boldsymbol{L}_1 : \boldsymbol{A}_1, \quad \bar{\boldsymbol{\varepsilon}}^p = \boldsymbol{\mathcal{A}}^p_0 : \boldsymbol{\varepsilon}^p_0 + \boldsymbol{\mathcal{A}}^p_1 : \boldsymbol{\varepsilon}^p_1, \\ &\boldsymbol{\mathcal{A}}^p_q = \bar{\boldsymbol{L}}^{-1} : \left[c_q \boldsymbol{L}_q - c_0 \boldsymbol{L}_0 : \boldsymbol{A}^p_{q,0} - c_1 \boldsymbol{L}_1 : \boldsymbol{A}^p_{q,1}\right], \quad q = 0, 1. \end{aligned} \tag{12.20}$$

12.4 Examples

Example 1: Mori–Tanaka to compute the thermal expansion tensor for a particulate composite

Consider a bi–phase elastic material in which an isotropic matrix (phase 0) is reinforced with isotropic spherical particles (phase 1). The bulk modulus K, the shear modulus μ, and the thermal expansion coefficient α of the two phases are given below:

matrix	glass particles
$K_0 = 2.5$ GPa	$K_1 = 44$ GPa
$\mu_0 = 1.15$ GPa	$\mu_1 = 30$ GPa
$\alpha_0 = 5 \cdot 10^{-5}$ 1/K	$\alpha_1 = 1 \cdot 10^{-5}$ 1/K

The volume fraction of the spherical particles is $c_1 = 30\%$. Compute the macroscopic thermal expansion coefficient of the composite.

Solution:

The presence of the temperature difference $\Delta\theta$ in a material point of an isotropic elastic medium causes thermal expansion strains, which are expressed as

$$\boldsymbol{\varepsilon}^{\theta} = \boldsymbol{\alpha}\Delta\theta \quad \text{with } \boldsymbol{\alpha} = \begin{bmatrix} \alpha & 0 & 0 \\ 0 & \alpha & 0 \\ 0 & 0 & \alpha \end{bmatrix}.$$

Inside the RVE of a composite material the temperature difference is considered to be uniform in all phases [37,38,39], whereas the developed thermal strains represent the eigenstrains in the Eshelby approach.

Using the Hill notation (Example 2 in Section 1.4 of Chapter 1), we express the elasticity tensors of both phases as

$$\boldsymbol{L}_0 = (3K_0, 2\mu_0) = (7.5, 2.3) \text{ GPa},$$
$$\boldsymbol{L}_1 = (3K_1, 2\mu_1) = (132, 60) \text{ GPa}.$$

Since the particles are spherical, the Eshelby tensor is given by

$$\boldsymbol{S}_1 = (3\gamma_0, 2\delta_0) = \left(\frac{3K_0}{3K_0 + 4\mu_0}, \frac{6K_0 + 12\mu_0}{15K_0 + 20\mu_0}\right) = (0.6198, 0.476).$$

Using formulas (12.13), the elastic and inelastic interaction tensors are given by

$$\begin{aligned}
\boldsymbol{T}_1 &= \left(3T_1^b, 2T_1^s\right) = \left[\boldsymbol{I} + \boldsymbol{S}_1 : \boldsymbol{L}_0^{-1} : [\boldsymbol{L}_1 - \boldsymbol{L}_0]\right]^{-1} \\
&= \left[(1,1) + (3\gamma_0, 2\delta_0) \cdot \left(\frac{1}{3K_0}, \frac{1}{2\mu_0}\right) \cdot (3K_1 - 3K_0, 2\mu_1 - 2\mu_0)\right]^{-1} \\
&= \left[\left(1 + \frac{3\gamma_0[K_1 - K_0]}{K_0}, 1 + \frac{2\delta_0[\mu_1 - \mu_0]}{\mu_0}\right)\right]^{-1} \\
&= \left(\frac{K_0}{K_0 + 3\gamma_0[K_1 - K_0]}, \frac{\mu_0}{\mu_0 + 2\delta_0[\mu_1 - \mu_0]}\right) = (0.0886, 0.0773),
\end{aligned}$$

$$\begin{aligned}
\boldsymbol{T}_0^p &= \left(3T_0^{pb}, 2T_0^{ps}\right) = -\boldsymbol{T}_1 : \boldsymbol{S}_1 = -\left(3T_1^b, 2T_1^s\right) \cdot (3\gamma_0, 2\delta_0) \\
&= (-0.0549, -0.0368), \\
\boldsymbol{T}_1^p &= \left(3T_1^{pb}, 2T_1^{ps}\right) = \boldsymbol{T}_1 : \boldsymbol{S}_1 : \boldsymbol{L}_0^{-1} : \boldsymbol{L}_1 = -\boldsymbol{T}_0^p : \boldsymbol{L}_0^{-1} : \boldsymbol{L}_1 \\
&= -\left(3T_0^{pb}, 2T_0^{ps}\right) \cdot \left(\frac{1}{3K_0}, \frac{1}{2\mu_0}\right) \cdot (3K_1, 2\mu_1) = (0.9663, 0.9595).
\end{aligned}$$

According to the Mori–Tanaka approach, the elastic and inelastic concentration tensors are given by formulas (12.17):

$$\begin{aligned}
\boldsymbol{A}_0 &= \left(3A_0^b, 2A_0^s\right) = [c_0\boldsymbol{I} + c_1\boldsymbol{T}_1]^{-1} = \left[0.7\cdot(1,1) + 0.3\cdot\left(3T_1^b, 2T_1^s\right)\right]^{-1} \\
&= (1.3763, 1.3828), \\
\boldsymbol{A}_1 &= \left(3A_1^b, 2A_1^s\right) = \boldsymbol{T}_1 : \boldsymbol{A}_0 = \left(3T_1^b, 2T_1^s\right)\cdot\left(3A_0^b, 2A_0^s\right) \\
&= (0.1219, 0.1068),
\end{aligned}$$

$$\begin{aligned}
\boldsymbol{A}_{0,0}^p &= \left(3A_{0,0}^{pb}, 2A_{0,0}^{ps}\right) = -c_1\boldsymbol{A}_0 : \boldsymbol{T}_0^p = -c_1\cdot\left(3A_0^b, 2A_0^s\right)\cdot\left(3T_0^{pb}, 2T_0^{ps}\right) \\
&= (0.0227, 0.0153), \\
\boldsymbol{A}_{1,0}^p &= \left(3A_{1,0}^{pb}, 2A_{1,0}^{ps}\right) = -c_1\boldsymbol{A}_0 : \boldsymbol{T}_1^p = -c_1\cdot\left(3A_0^b, 2A_0^s\right)\cdot\left(3T_1^{pb}, 2T_1^{ps}\right) \\
&= (-0.399, -0.398), \\
\boldsymbol{A}_{0,1}^p &= \left(3A_{0,1}^{pb}, 2A_{0,1}^{ps}\right) = \boldsymbol{T}_0^p - c_1\boldsymbol{A}_1 : \boldsymbol{T}_0^p \\
&= \left(3T_0^{pb}, 2T_0^{ps}\right) - c_1\cdot\left(3A_1^b, 2A_1^s\right)\cdot\left(3T_0^{pb}, 2T_0^{ps}\right) \\
&= (-0.0529, -0.0356), \\
\boldsymbol{A}_{1,1}^p &= \left(3A_{1,1}^{pb}, 2A_{1,1}^{ps}\right) = \boldsymbol{T}_1^p - c_1\boldsymbol{A}_1 : \boldsymbol{T}_1^p \\
&= \left(3T_0^{pb}, 2T_0^{ps}\right) - c_1\cdot\left(3A_1^b, 2A_1^s\right)\cdot\left(3T_0^{pb}, 2T_0^{ps}\right) \\
&= (0.9310, 0.9288).
\end{aligned}$$

In addition, Eqs. (12.20) yield

$$\begin{aligned}
\bar{\boldsymbol{L}} &= c_0\boldsymbol{L}_0 : \boldsymbol{A}_0 + c_1\boldsymbol{L}_1 : \boldsymbol{A}_1 = (12.0535, 4.1494)\ \text{GPa}, \\
\mathcal{A}_0^p &= \bar{\boldsymbol{L}}^{-1} : \left[c_0\boldsymbol{L}_0 - c_0\boldsymbol{L}_0 : \boldsymbol{A}_{0,0}^p - c_1\boldsymbol{L}_1 : \boldsymbol{A}_{0,1}^p\right] = (0.5995, 0.5365), \\
\mathcal{A}_1^p &= \bar{\boldsymbol{L}}^{-1} : \left[c_1\boldsymbol{L}_1 - c_0\boldsymbol{L}_0 : \boldsymbol{A}_{1,0}^p - c_1\boldsymbol{L}_1 : \boldsymbol{A}_{1,1}^p\right] = (0.4005, 0.4635),
\end{aligned}$$

and

$$\bar{\boldsymbol{\varepsilon}}^\theta = \bar{\boldsymbol{\alpha}}\Delta\theta = \mathcal{A}_0^p : \boldsymbol{\alpha}_0\Delta\theta + \mathcal{A}_1^p : \boldsymbol{\alpha}_1\Delta\theta \quad\Longrightarrow\quad \bar{\boldsymbol{\alpha}} = \mathcal{A}_0^p : \boldsymbol{\alpha}_0 + \mathcal{A}_1^p : \boldsymbol{\alpha}_1.$$

At this point, note that when a fourth–order isotropic tensor $\boldsymbol{C} = (3C^b, 2C^s)$ is multiplied with a second–order tensor of the form

$$\boldsymbol{c} = \begin{bmatrix} c & 0 & 0 \\ 0 & c & 0 \\ 0 & 0 & c \end{bmatrix},$$

then the tensor algebra yields

$$\boldsymbol{C}:\boldsymbol{c}=\begin{bmatrix}3C^bc & 0 & 0\\ 0 & 3C^bc & 0\\ 0 & 0 & 3C^bc\end{bmatrix}.$$

Thus

$$\bar{\boldsymbol{\alpha}}=\mathcal{A}_0^p:\boldsymbol{\alpha}_0+\mathcal{A}_1^p:\boldsymbol{\alpha}_1=\begin{bmatrix}3.398 & 0 & 0\\ 0 & 3.398 & 0\\ 0 & 0 & 3.398\end{bmatrix}\cdot 10^{-5}\ 1/\text{K}.$$

So the macroscopic thermal expansion coefficient is $3.398\cdot 10^{-5}$ 1/K.

Remark:

We can obtain the same result by using the Levin formula (Example 2 of Chapter 3).

Example 2: Mori–Tanaka for particulate composite undergoing plastic shear loading

A two–phase composite consisting of an elastic matrix reinforced with elastoplastic spherical particles is subjected to shear deformation. The elastoplastic particles are characterized by a J_2 type of plastic behavior with isotropic power law–type hardening. The shear stress vs shear angle response of the material is described by the following set of equations and inequalities:

$$\begin{aligned}\tau&=\mu[\gamma-\gamma^p],\\ \dot{\gamma}^p&=\sqrt{3}\frac{\tau}{|\tau|}\dot{p},\\ \Phi&=\sqrt{3}|\tau|-Hp^m-k,\\ \Phi\dot{p}&=0,\quad \Phi\leqslant 0,\quad \dot{p}\geqslant 0,\end{aligned}\tag{12.21}$$

where $\tau=\sigma_{12}$ and $\gamma=2\varepsilon_{12}$ denote the shear stress and shear angle, respectively, μ is the shear modulus, γ^p represents the plastic shear angle, p is the accumulated plastic strain, Φ is the elastoplastic yield criterion, k is the elastic limit, and H, m are plastic hardening–related parameters. In addition, $\dot{\{\bullet\}}$ corresponds to the rate of the quantity $\{\bullet\}$.

The volume fraction of the particles in the composite is $c=30\%$. The material properties of the two materials are summarized below:

property	matrix	particles
bulk modulus K (MPa)	833.333	8333.33
shear modulus μ (MPa)	500	5000
elastic limit k (MPa)	–	50
hardening multiplier H (MPa)	–	100
hardening exponent m	–	0.5

Using the Mori–Tanaka/TFA approach, identify the shear stress vs shear angle response of the composite, as well as for the matrix and the particles, when the macroscopic shear angle evolves from 0 to 0.07.

Solution:

For the mean–field strategy, the matrix phase is denoted with index 0 and the particles with index 1. With regard to the Mori–Tanaka scheme–related terms, we briefly present the calculations:

- The shear term of the Eshelby tensor for a composite with spherical particles is given by expression (5.5) as

$$S_1^G = \frac{6K_0 + 12\mu_0}{15K_0 + 20\mu_0}.$$

- The important elastic and inelastic interaction tensors for the computations ($\boldsymbol{T}_1$ and $\boldsymbol{T}_1^p$) are obtained from relation (12.13). With the help of the Hill notation (Example 2 in Section 1.4 of Chapter 1), we can obtain the shear terms of these tensors:

$$T_1^G = \frac{\mu_0}{\mu_0 + S_1^G[\mu_1 - \mu_0]}, \qquad T_1^{pG} = T_1^G S_1^G \frac{\mu_1}{\mu_0}.$$

- Expressions (12.17) provide the elastic and inelastic concentration tensors used in the computational procedure ($\boldsymbol{A}_1$ and $\boldsymbol{A}_{1,1}^p$). Their shear terms are given by

$$A_1^G = \frac{T_1^G}{1 - c + cT_1^G}, \qquad A_{1,1}^{pG} = T_1^{pG} - cA_1^G T_1^{pG}.$$

Due to the isotropic response of the particulate composite, macroscopic shear deformation activates only shear components in the constituents of the microstructure. In case of more complicated loading conditions, expressions (12.21) should be considered in their generalized 3D form [40].

The nonlinear character of the particles and the overall composite dictates the use of an incremental scheme. In incremental approaches the loading is provided gradually to the structure. In time–dependent responses the complete time of the process is split in several time steps. In elastoplastic problems, as that examined here, the material response is time–independent. In the latter case a virtual total time can be adopted, which is split into several time steps. At an arbitrary time step n, all the fields (stresses, strains, inelastic strains, etc.) are considered known, and the scope is to update these fields for the next time step $n+1$. In this computational approach, every field ϕ is expressed at the two time steps as $\phi^{(n)}$ and $\phi^{(n+1)}$. The time increment of this field ϕ is defined as

$$\Delta\phi = \phi^{(n+1)} - \phi^{(n)}.$$

An additional complexity that arises in nonlinear processes is that quite often the field ϕ at the time step $n+1$ is computed through an iterative procedure. If k and $k+1$ are two consecutive iteration steps, then $\phi^{(n+1)(k+1)}$ and $\phi^{(n+1)(k)}$ correspond to the values of $\phi^{(n+1)}$ at these steps.

For the micro–mechanical part, the numerical scheme summarized below is based on the well–known return mapping algorithm [3,2]:

1. The total applied macroscopic shear angle $\bar{\gamma} = 2\bar{\varepsilon}_{12}$ is split into N parts with constant size $\Delta\bar{\gamma}$.
2. At time step n the fields at all phases are known.
3. At time step $n+1$ the macroscopic shear angle takes the value

$$\bar{\gamma}^{(n+1)} = \bar{\gamma}^{(n)} + \Delta\bar{\gamma}.$$

4. Initial prediction: At the beginning of the iterative process, we assume that new plastic strains are not generated. Under this hypothesis, the increments of the shear angle at the particles are given by the expression

$$\Delta\gamma_1 = A_1^G \Delta\bar{\gamma}.$$

5. The plastic shear angle $\gamma_1^{p(n+1)}$ and the shear stress $\tau_1^{(n+1)}$ at the particles are computed from the *constitutive law algorithm*. If $\gamma_1^{p(n+1)} = \gamma_1^{p(n)}$, then the calculations pass to step 9; otherwise, the prediction is wrong, and an iterative procedure is required.
6. At the iteration k the plastic shear angle at the particles as $\gamma_1^{p(n+1)(k)}$, and its time increment is denoted as $\Delta\gamma_1^{p(k)} = \gamma_1^{p(n+1)(k)} - \gamma_1^{p(n)}$.
7. The increment of the particles shear angle at the new iteration $k+1$ is updated by the expression

$$\Delta\gamma_1^{(k+1)} = A_1^G \Delta\bar{\gamma} + A_{1,1}^{pG} \Delta\gamma_1^{p(k)}.$$

8. The plastic shear angle $\gamma_1^{p(n+1)(k+1)}$ and the shear stress $\tau_1^{(n+1)(k+1)}$ at the particles are recomputed from the *constitutive law algorithm*. If

$$\gamma_1^{p(n+1)(k+1)} - \gamma_1^{p(n+1)(k)} \leqslant \text{tolerance},$$

then the algorithm passes to the next step; otherwise, k is set equal to $k+1$, and the algorithm returns to step 6.

9. The shear angle and stress at the matrix are computed from the formulas

$$\gamma_0^{(n+1)} = \frac{1}{1-c}\left[\bar{\gamma}^{(n+1)} - c\gamma_1^{(n+1)}\right], \qquad \tau_0^{(n+1)} = \mu_0\gamma_0^{(n+1)},$$

and the macroscopic shear stress is provided by the expression

$$\bar{\tau}^{(n+1)} = [1-c]\tau_0^{(n+1)} + c\tau_1^{(n+1)}.$$

A new time step $n = n+1$ is assigned, and the algorithm returns to step 2.

The *constitutive law algorithm* is a special part of the complete analysis, in which the user provides the shear angle γ of the particles and receives as output the shear stress τ, the plastic shear angle γ^p, and the accumulated plastic strain p of the particles. This numerical process is based on a combination of the return mapping algorithm and the Newton–Raphson approach [3,2]. The calculation steps are the following:

1. At time step $n+1$ the shear angle $\gamma^{(n+1)}$ is known.
2. As initial guess, p and γ^p are assumed to not evolve, that is, they retain their values of time step n. With this hypothesis, the shear stress is computed by the expression

$$\tau^{(n+1)} = \mu[\gamma^{(n+1)} - \gamma^{p(n)}].$$

In addition, the yield criterion is computed from the formula

$$\Phi^{(n+1)} = \sqrt{3}|\tau^{(n+1)}| - H\left[p^{(n)}\right]^m - k.$$

3. If Φ is less than a tolerance, then the guess is correct, all the variables are updated, and the computations are completed. If Φ is greater than the tolerance, an iterative procedure starts.
4. Starting from known values at the iteration step k, the p, γ^p, and τ are updated to the iteration step $k+1$ according to the expressions

$$p^{(n+1)(k+1)} = p^{(n+1)(k)} + \frac{\Phi^{(n+1)(k)}}{3\mu + Hm\left[p^{(n+1)(k)}\right]^{m-1}},$$

$$\gamma^{p(n+1)(k+1)} = \gamma^{p(n)} + \sqrt{3}\frac{\tau^{(n+1)(k)}}{|\tau^{(n+1)(k)}|}\left[p^{(n+1)(k+1)} - p^{(n)}\right],$$

$$\tau^{(n+1)(k+1)} = \mu\left[\gamma^{(n+1)} - \gamma^{p(n+1)(k+1)}\right].$$

5. The criterion Φ is reevaluated for the iteration step $k+1$. If $|\Phi|$ is less than the tolerance, then the process has converged, and all the required variables are updated.

Combining the two numerical processes (the micro–mechanical part and the constitutive law algorithm), the shear stress vs shear angle curves for the overall composite and two phases are obtained and illustrated in Fig. 12.4. The corresponding Python script that performs the calculations

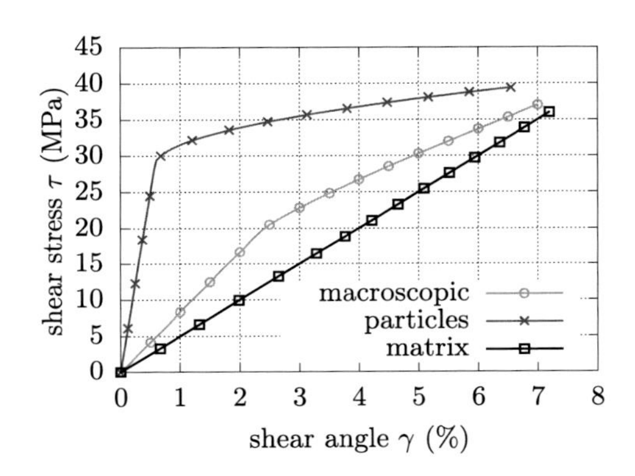

FIGURE 12.4 Shear stress vs shear angle curve for the particulate composite, the matrix phase, and the particles.

is given below:

```
import numpy as np

def plastic(mu,H,k,m,g,gp,p):
# initiation
    tol=1.E-10
    pold=p
    gpold=gp
# stress prediction and criterion check
    tau=mu*(g-gp)
    Phi=np.sqrt(3)*np.absolute(tau)-H*np.power(p+tol,m)-k
# loop when plastification occurs
    if Phi>tol:
      while np.absolute(Phi)>tol:
         p=p+Phi/(3*mu+H*m*np.power(p+tol,m-1))
         gp=gpold+np.sqrt(3)*(tau/np.absolute(tau))*(p-pold)
         tau=mu*(g-gp)
         Phi=np.sqrt(3)*np.absolute(tau)-H*np.power(p+tol,m)-k
    return tau,gp,p
```

```
# Data
K0=833.333
mu0=500.
K1=8333.33
mu1=5000.
H1=100.
k1=50.
m1=0.5
c=0.3

# shear concentration terms for particles
S1G=(6*K0+12*mu0)/(15*K0+20*mu0)
T1G=mu0/(mu0+S1G*(mu1-mu0))
T1PG=T1G*S1G*mu1/mu0
A1G=T1G/(1-c+c*T1G)
A11PG=T1PG-c*A1G*T1PG

# initiation of variables
g1p=0.
p1=0.
g1=0.
tau1=0.
g0=0.
tau0=0.
gbar=0.
taubar=0.

tol=1.E-10
deltagbar=0.001
f=open("macro.txt","w+")
f.write("%8.5f %5.2f\n" % (gbar, taubar))
f1=open("micro1.txt","w+")
f1.write("%8.5f %5.2f\n" % (g1, tau1))
f0=open("micro0.txt","w+")
f0.write("%8.5f %5.2f\n" % (g0, tau0))

# global procedure
for i in range(70):
  gbar=gbar+deltagbar
  g1old=g1
  p1old=p1
  g1pold=g1p

# initial guess
```

```
  Deltag1=A1G*deltagbar
  g1=g1old+Deltag1
  tau1,g1p,p1 = plastic(mu1,H1,k1,m1,g1,g1p,p1)
# iterative process
  if g1p-g1pold>tol:
    diff=1.
    while np.absolute(diff)>tol:
      g1ppr=g1p
      Deltag1p=g1p-g1pold
      Deltag1=A1G*deltagbar+A11PG*Deltag1p
      g1=g1old+Deltag1
      tau1,g1p,p1 = plastic(mu1,H1,k1,m1,g1,g1p,p1)
      diff=abs(g1p-g1ppr)
# end of iterations
  g0=(gbar-c*g1)/(1-c)
  tau0=mu0*g0
  taubar=(1-c)*tau0+c*tau1

  f.write("%10.7f %7.4f\n" % (gbar, taubar))
  f1.write("%10.7f %7.4f\n" % (g1, tau1))
  f0.write("%10.7f %7.4f\n" % (g0, tau0))
```

Example 3: particulate composite undergoing damage: Mori–Tanaka approach with void creation

Let us consider a two–phase composite made of an epoxy containing glass spherical particles with the following elastic constants for each phase:

matrix	glass particles
$E_0 = 3000$ MPa	$E_1 = 81000$ MPa
$\nu_0 = 0.3$	$\nu_1 = 0.25$

The volume fraction of the spherical particles is $c_1 = 40\%$. The matrix is assumed to undergo damage when the shear stress in the particles exceeds 30 MPa. The matrix damage is characterized by the creation of voids inside it. The volume fraction of voids is given as a function of the particles shear stress by the relation

$$c_v = 0.00125 \cdot \left\{\sigma^1_{12} - 30\right\}_+ ,$$

where $\{\{\bullet\}\}_+ = \dfrac{\{\bullet\} + |\{\bullet\}|}{2}$ denotes the Macaulay brackets. Identify the macroscopic shear stress vs macroscopic shear angle curve using the Mori–Tanaka method.

Solution:

Before passing to the solution of the problem, we need to clarify a few things:

- The choice of connecting the damage of the matrix with the stress developed on the particles is motivated by the expected evolution of damage in these types of composites. Usually, the damage initiation starts at the interface between the material phases. Due to the traction continuity, the stress vector at this interface is the same for both the matrix and the particle. Thus it is reasonable to consider that once the shear stress in the particle reaches the critical value ($\sigma_{12}^{d} = 30$ MPa), the matrix is also subjected to approximately the same shear stress at certain points close to the region.
- The micro–mechanics–driven damage considered in this example is described as follows: Below the critical shear stress level, the matrix is undamaged, and only two phases exist in the composite (the matrix and the particles). Once the shear stress surpasses its critical value, the damage appears in the region close to the particles in the form of micro–voids (Fig. 12.5). For simplicity, these voids are assumed to be of spherical shape (i.e., the damage mechanism is assumed to be of isotropic nature). A second simplification applied here is the hypothesis that in the Mori–Tanaka scheme the voids are considered uniformly distributed inside the RVE of the composite.[1]
- The damage mechanism described here is a simplified version that allows us to treat the problem quasi–analytically. A real damage mechanism follows a more complicate criterion and depends on the complete stress tensor.

Starting the computational procedure, the elasticity tensors for the matrix and the particles can be expressed in the Hill notation as

$$\boldsymbol{L}_0 = (3K_0, 2\mu_0) = \left(\frac{E_0}{1-2\nu_0}, \frac{E_0}{1+\nu_0}\right) = (7500, 2307.69),$$

$$\boldsymbol{L}_1 = (3K_1, 2\mu_1) = \left(\frac{E_1}{1-2\nu_1}, \frac{E_1}{1+\nu_1}\right) = (162000, 64800).$$

The properties of the tensors using the Hill notation are presented in Example 2 in Section 1.4 of Chapter 1. Since only the shear stress and strain are of interest, the bulk related terms can be omitted. The shear term of the Eshelby tensor for a particulate composite is given by expression (5.5) as

$$S_1^G = 2\delta_0 = \frac{6K_0 + 12\mu_0}{15K_0 + 20\mu_0} = 0.47619.$$

[1] A more rigorous approach would require defining an interphase layer in the proximity of the particles in which the matrix is damaged. However, such layer must be treated using a modified version of the Mori–Tanaka method [41].

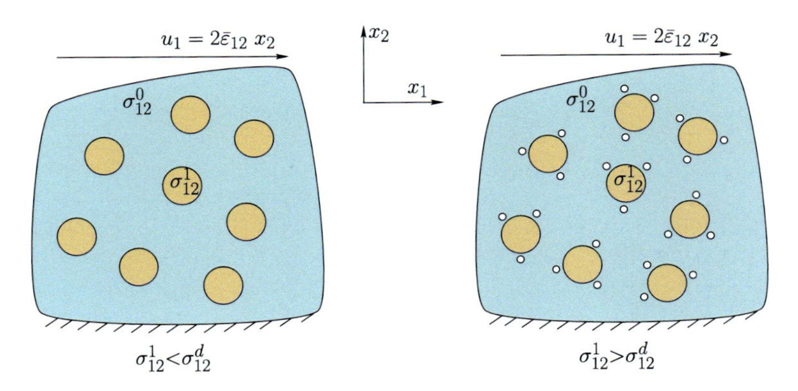

FIGURE 12.5 Schematic of typical damage evolution in particulate composite.

The interaction tensor for the particles utilized in the Mori–Tanaka technique is computed from relation (5.19),

$$\boldsymbol{T}_1 = \left[\boldsymbol{I} + \boldsymbol{S}_1 : \boldsymbol{L}_0^{-1} : [\boldsymbol{L}_1 - \boldsymbol{L}_0]\right]^{-1}.$$

Focusing exclusively on the shear term, this expression gives

$$T_1^G = \left[1 + \frac{0.47619}{2307.69} \cdot [64800 - 2307.69]\right]^{-1} = 0.071967.$$

Once the damage starts evolving, spherical voids appear in the RVE of the composite. These voids have also an interaction tensor given by the simple formula

$$\boldsymbol{T}_v = [\boldsymbol{I} - \boldsymbol{S}_v]^{-1}$$

due to the fact that the elasticity tensor of a void is null. Since the particles and the voids have the same shape, the corresponding Eshelby tensors are the same, that is, $\boldsymbol{S}_v = \boldsymbol{S}_1$. Thus the shear term of the tensor $\boldsymbol{T}_v$ can be computed:

$$T_v^G = [1 - 0.47619]^{-1} = 1.90909.$$

Due to the nonlinear nature of the problem, the macroscopic shear angle is imposed incrementally. The numerical procedure is the following:

1. For a given macroscopic shear angle $2\bar{\varepsilon}_{12}$, we consider an initial guess for the void volume fraction c_v. This guess value is the same with that of the previous shear angle value. At the beginning of loading, $c_v = 0$. With this guess, the volume fraction of the matrix becomes equal to $c_0 = 1 - c_1 - c_v$.

2. The concentration tensor for the particles and the matrix, according to the Mori–Tanaka scheme, are given by expressions (5.35) and (5.36),

$$
\begin{aligned}
\boldsymbol{A}_1 &= \boldsymbol{T}_1 : \left[c_0 \boldsymbol{I} + c_1 \boldsymbol{T}_1 + c_v \boldsymbol{T}_v\right]^{-1}, \\
\boldsymbol{A}_0 &= \left[c_0 \boldsymbol{I} + c_1 \boldsymbol{T}_1 + c_v \boldsymbol{T}_v\right]^{-1}.
\end{aligned}
$$

In addition, the average strain and stress at the matrix and at the particles are computed by the relations

$$
\begin{aligned}
\boldsymbol{\varepsilon}_1 &= \boldsymbol{A}_1 : \bar{\boldsymbol{\varepsilon}}, \boldsymbol{\sigma}_1 = \boldsymbol{L}_1 : \boldsymbol{\varepsilon}_1, \\
\boldsymbol{\varepsilon}_0 &= \boldsymbol{A}_0 : \bar{\boldsymbol{\varepsilon}}, \boldsymbol{\sigma}_0 = \boldsymbol{L}_0 : \boldsymbol{\varepsilon}_0.
\end{aligned}
$$

3. Using the void volume fraction relation

$$
c_v = 0.00125 \cdot \left\{\sigma_{12}^1 - 30\right\}_+,
$$

we obtain a new value of c_v. If it coincides with the previous value of c_v (the initial guess), then the computation is considered successful, and we can pass to the next value of the shear angle. In the opposite case (i.e., the new void volume fraction is higher than its guess value) the computations are repeated for the new value of c_v. A convergence criterion based on the difference between new and old values of c_v is imposed to allow identifying (within reasonable numerical accuracy) the final value of the c_v.
4. Once the iteration process is completed, we obtain the macroscopic stress by the averaging relation

$$
\bar{\boldsymbol{\sigma}} = c_0 \boldsymbol{\sigma}_0 + c_1 \boldsymbol{\sigma}_1.
$$

The shear stress vs shear angle curve for the composite in question is illustrated in Fig. 12.6. The corresponding Python script that performs the calculations is the following:

```
import numpy as np

# Data
E0=3000
v0=0.3
E1=81000
v1=0.25
c1=0.4
sigd=30
a=0.1/80

# Bulk, shear moduli, Eshelby term
```

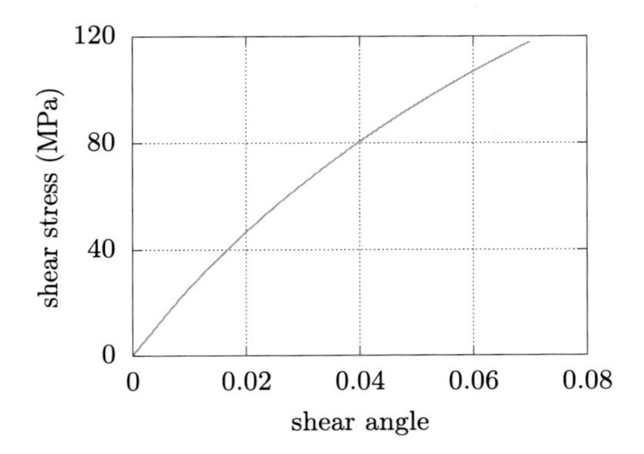

FIGURE 12.6 Macroscopic shear stress vs macroscopic shear angle curve for a particulate composite whose matrix phase is subjected to damage.

```
K0=E0/(3*(1-2*v0))
mu0=E0/(2*(1+v0))
delta0=(3*K0+6*mu0)/(15*K0+20*mu0)
mu1=E1/(2*(1+v1))

# Interaction tensors
L0G=2*mu0
L1G=2*mu1
S1G=2*delta0
T1G=1/(1+(S1G/L0G)*(L1G-L0G))
TvG=1/(1-S1G)

# Initial conditions
cv=0
Eps=0
tableA=np.zeros((71,2))
tableB=np.zeros((71,5))
tableC=np.zeros((71,1))

# macroscopic strain path loop
for i in range(71):
   error=1
   iter=0

   # loop until convergence
   while error>1.0E-5 and iter<100:
      c0=1-c1-cv
      iter=iter+1
      A0G=1/(c0+c1*T1G+cv*TvG)
```

```
        A1G=T1G*A0G
        AvG=TvG*A0G
        eps0=A0G*Eps
        eps1=A1G*Eps
        sig0=mu0*eps0
        sig1=mu1*eps1
        cv_new=a*(abs(sig1-sigd)+(sig1-sigd))/2
        error=abs(cv_new-cv)/(cv_new+1.0E-8)
        cv=cv_new

    # macroscopic stress and other info
    Sig=c0*sig0+c1*sig1
    tableA[i,:]=[Eps, Sig]
    tableB[i,:]=[eps0, sig0, eps1, sig1, cv]
    tableC[i,:]=[iter]
    Eps=Eps+0.001

np.savetxt('tableA.txt',tableA,fmt='%1.5f')
np.savetxt('tableB.txt',tableB,fmt='%1.5f')
np.savetxt('tableC.txt',tableC,fmt='%1.5f')
```

References

[1] N. Charalambakis, Homogenization techniques and micromechanics. A survey and perspectives, Applied Mechanics Reviews 63 (3) (2010) 030803.

[2] G. Chatzigeorgiou, N. Charalambakis, Y. Chemisky, F. Meraghni, Thermomechanical Behavior of Dissipative Composite Materials, ISTE Press – Elsevier, London, 2018.

[3] J.C. Simo, T.J.R. Hughes, Computational Inelasticity, Springer–Verlag, New York, 1998.

[4] P.W. Chung, K.K. Tamma, R.R. Namburu, A micro/macro homogenization approach for viscoelastic creep analysis with dissipative correctors for heterogeneous woven–fabric layered media, Composites Science and Technology 60 (2000) 2233–2253.

[5] Q. Yu, J. Fish, Multiscale asymptotic homogenization for multiphysics problems with multiple spatial and temporal scales: a coupled thermo–viscoelastic example problem, International Journal of Solids and Structures 39 (26) (2002) 6429–6452.

[6] G. Wang, M.J. Pindera, Locally–exact homogenization of viscoelastic unidirectional composites, Mechanics of Materials 103 (2016) 95–109.

[7] K. Terada, N. Kikuchi, A class of general algorithms for multi–scale analyses of heterogeneous media, Computer Methods in Applied Mechanics and Engineering 190 (2001) 5427–5464.

[8] T. Asada, N. Ohno, Fully implicit formulation of elastoplastic homogenization problem for two–scale analysis, International Journal of Solids and Structures 44 (22–23) (2007) 7261–7275.

[9] H. Moulinec, P. Suquet, A numerical method for computing the overall response of nonlinear composites with complex microstructure, Computer Methods in Applied Mechanics and Engineering 157 (1998) 69–94.

[10] M.A.A. Cavalcante, M.J. Pindera, Generalized FVDAM theory for elastic–plastic periodic materials, International Journal of Plasticity 77 (2016) 90–117.

[11] N. Ohno, T. Matsuda, X. Wu, A homogenization theory for elastic–viscoplastic composites with point symmetry of internal distributions, International Journal of Solids and Structures 38 (2001) 2867–2878.

[12] Q. Chen, X. Chen, Z. Zhai, Z. Yang, A new and general formulation of three-dimensional finite–volume micromechanics for particulate reinforced composites with viscoplastic phases, Composites. Part B, Engineering 85 (2016) 216–232.
[13] J. Fish, Q. Yu, K. Shek, Computational damage mechanics for composite materials based on mathematical homogenization, International Journal for Numerical Methods in Engineering 45 (1999) 1657–1679.
[14] E. Tikarrouchine, G. Chatzigeorgiou, Y. Chemisky, F. Meraghni, Non-linear FE^2 multiscale simulation of damage, micro and macroscopic strains in polyamide 66–woven composite structures: analysis and experimental validation, Composite Structures 255 (2021) 112926.
[15] H. Herzog, E. Jacquet, From a shape memory alloys model implementation to a composite behavior, Computational Materials Science 39 (2007) 365–375.
[16] G. Chatzigeorgiou, Y. Chemisky, F. Meraghni, Computational micro to macro transitions for shape memory alloy composites using periodic homogenization, Smart Materials and Structures 24 (2015) 035009.
[17] F. Feyel, Multiscale FE2 elastoviscoplastic analysis of composite structures, Computational Materials Science 16 (1) (1999) 344–354.
[18] F.T. Fisher, L.C. Brinson, Viscoelastic interphases in polymer–matrix composites: theoretical models and finite–element analysis, Composites Science and Technology 61 (2001) 731–748.
[19] D. Anagnostou, G. Chatzigeorgiou, Y. Chemisky, F. Meraghni, Hierarchical micromechanical modeling of the viscoelastic behavior coupled to damage in SMC and SMC–hybrid composites, Composites. Part B, Engineering 151 (2018) 8–24.
[20] D.C. Lagoudas, A.C. Gavazzi, H. Nigam, Elastoplastic behavior of metal matrix composites based on incremental plasticity and the Mori–Tanaka averaging scheme, Computational Mechanics 8 (1991) 193–203.
[21] J. Chaboche, P. Kanoute, A. Ross, On the capabilities of mean field approaches for the description of plasticity in metal matrix composites, International Journal of Plasticity 21 (2005) 1409–1434.
[22] I. Doghri, A. Ouaar, Homogenization of two–phase elasto–plastic composite materials and structures: study of tangent operators, cyclic plasticity and numerical algorithms, International Journal of Solids and Structures 40 (2003) 1681–1712.
[23] B. Miled, I. Doghri, L. Brassart, L. Delannay, Micromechanical modeling of coupled viscoelastic–viscoplastic composites based on an incrementally affine formulation, International Journal of Solids and Structures 50 (2013) 1755–1769.
[24] P. Ponte–Castañeda, Second–order homogenization estimates for nonlinear composites incorporating field fluctuations. I. Theory, Journal of the Mechanics and Physics of Solids 50 (2002) 737–757.
[25] N. Lahellec, P. Suquet, On the effective behavior of nonlinear inelastic composites: I. Incremental variational principles, Journal of the Mechanics and Physics of Solids 55 (9) (2007) 1932–1963.
[26] L. Wu, L. Adam, I. Doghri, L. Noels, An incremental–secant mean–field homogenization method with second statistical moments for elasto–visco–plastic composite materials, Mechanics of Materials 114 (2017) 180–200.
[27] M. Barral, G. Chatzigeorgiou, F. Meraghni, R. Léon, Homogenization using modified Mori–Tanaka and TFA framework for elastoplastic–viscoelastic–viscoplastic composites: theory and numerical validation, International Journal of Plasticity 127 (2020) 102632.
[28] F. Meraghni, M.L. Benzeggagh, Micromechanical modelling of matrix degradation in randomly oriented discontinuous–fibre composites, Composites Science and Technology 55 (2) (1995) 171–186.
[29] N. Despringre, Analyse et modélisation des mécanismes d'endommagement et de déformation en fatigue multiaxiale de matériaux composites: polyamide renforcé par des fibres courtes, Ph.D. thesis, Arts et Métiers ParisTech, Metz, 2016.

[30] F. Praud, G. Chatzigeorgiou, Y. Chemisky, F. Meraghni, Hybrid micromechanical–phenomenological modelling of anisotropic damage and anelasticity induced by micro–cracks in unidirectional composites, Composite Structures 182 (2017) 223–236.
[31] Q. Chen, G. Chatzigeorgiou, F. Meraghni, Extended mean–field homogenization of viscoelastic–viscoplastic polymer composites undergoing hybrid progressive degradation induced by interface debonding and matrix ductile damage, International Journal of Solids and Structures 210 (211) (2021) 1–17.
[32] J.A. Nairn, On the use of shear–lag methods for analysis of stress transfer in unidirectional composites, Mechanics of Materials 26 (2) (1997) 63–80.
[33] N. Despringre, Y. Chemisky, K. Bonnay, F. Meraghni, Micromechanical modeling of damage and load transfer in particulate composites with partially debonded interface, Composite Structures 155 (2016) 77–88.
[34] L. Wu, L. Noels, L. Adam, I. Doghri, An implicit–gradient–enhanced incremental–secant mean–field homogenization scheme for elasto–plastic composites with damage, International Journal of Solids and Structures 50 (2013) 3843–3860.
[35] G. Dvorak, Transformation field analysis of inelastic composite materials, Proceedings of the Royal Society of London. Series A 437 (1992) 311–327.
[36] G. Dvorak, Y. Benveniste, On transformation strains and uniform fields in multiphase elastic media, Proceedings of the Royal Society of London. Series A 437 (1992) 291–310.
[37] V.M. Levin, On the coefficients of thermal expansion of heterogeneous materials (in Russian), Mekhanika Tverdogo Tela 1 (1967) 88–94.
[38] B.W. Rosen, Z. Hashin, Effective thermal expansion coefficients and specific heats of composite materials, International Journal of Engineering Science 8 (1970) 157–173.
[39] H.I. Ene, On linear thermoelasticity of composite materials, International Journal of Engineering Science 21 (5) (1983) 443–448.
[40] J. Lemaitre, J.L. Chaboche, Mechanics of Solid Materials, Cambridge University Press, Cambridge, 2002.
[41] G. Chatzigeorgiou, F. Meraghni, Elastic and inelastic local strain fields in composites with coated fibers or particles: theory and validation, Mathematics and Mechanics of Solids 24 (9) (2019) 2858–2894.

APPENDIX

Fiber orientation in composites

A.1 Introduction

As already discussed previously, the macroscopic elasticity tensor $\bar{\boldsymbol{L}}$ of a composite consisting of N ellipsoidal phases (particles or fibers) and a matrix is computed from the formula

$$\bar{\boldsymbol{L}} = c_0 \boldsymbol{L}_0 : \boldsymbol{A}_0 + \sum_{r=1}^{N} c_r \boldsymbol{L}_r : \boldsymbol{A}_r, \tag{A.1}$$

where c_r, $\boldsymbol{L}_r$, and $\boldsymbol{A}_r$ are the volume fraction, the stiffness tensor, and the concentration tensor of a phase r, respectively. The index 0 denotes the matrix phase.

In certain types of composites, there is no specific number of inclusions that can be determined in an absolute manner. A composite with randomly oriented fibers has in theory fibers in every possible orientation. In addition to that, experimentally it is also possible to identify an orientation distribution function g for such composite, which will define a probability for each orientation to appear. In these cases, Eq. (A.1) should be transformed properly to account for the different orientations [1,2].

A.2 Reinforcement orientation in a plane

A common composite that appears frequently in the automotive industry consists of a matrix reinforced with fibers. Often, the structural components are produced in the form of thin plates, and the fibers are distributed in random orientation. In these structures the orientation distribution function can be considered independent of the axis normal to the plate.

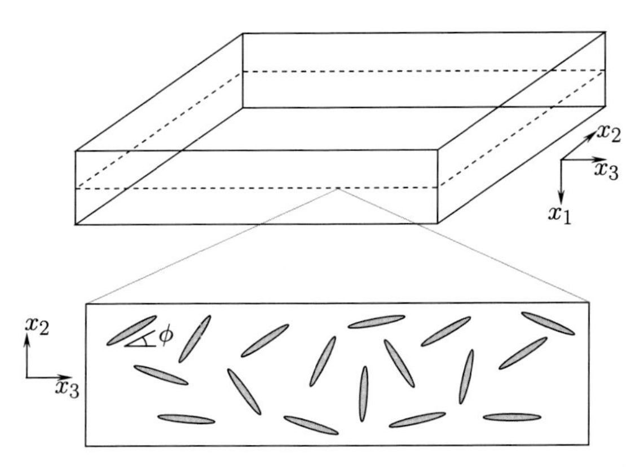

FIGURE A.1 Fiber reinforced composite plate. The fibers are oriented randomly in the middle plane.

Assume that ellipsoidal shape fibers of total volume fraction c have planar distribution and are oriented randomly inside a matrix phase (Fig. A.1). We denote the fiber elasticity tensor by $\boldsymbol{L}$. If ϕ represents the angle of planar orientation, then Eq. (A.1) is substituted by[1]

$$\bar{\boldsymbol{L}} = [1-c]\boldsymbol{L}_0 : \boldsymbol{A}_0 + c\int_0^{\pi} g(\phi)\boldsymbol{L}(\phi) : \boldsymbol{A}(\phi)\mathrm{d}\phi \Big/ \int_0^{\pi} g(\phi)\mathrm{d}\phi. \tag{A.2}$$

When $g(\phi)$ has an analytical expression, Eq. (A.2) can be computed exactly. An equal probability of appearance for all orientations yields g to be constant. In practice the value of g is experimentally determined at specific angles or angle ranges, not allowing us to establish proper analytical expressions. To resolve this issue, the distribution of g with respect to the angle ϕ is discretized using constant step h (Fig. A.2). Considering N orientations, the step h becomes π/N. With this approximation, Eq. (A.2) is expressed in the discrete form

$$\bar{\boldsymbol{L}} \approx [1-c]\boldsymbol{L}_0 : \boldsymbol{A}_0 + \frac{c}{g}\sum_{r=1}^{N} g_r \boldsymbol{L}_r : \boldsymbol{A}_r, \quad g = \sum_{r=1}^{N} g_r, \tag{A.3}$$

where g_r is the value of $g(\phi)$ at the r_{th} orientation. The index r is also used for the reinforcement elasticity tensor $\boldsymbol{L}$ and concentration tensor $\boldsymbol{A}$ due to their dependence on the orientation angle. Recall from Chapter 1 that a fourth–order tensor $\boldsymbol{A}$ is transformed from one coordinate system $\boldsymbol{x}$ to another $\boldsymbol{x}'$ through the general formulas

[1]Note that similar integrals appear in the formulas concerning the computation of concentration tensors.

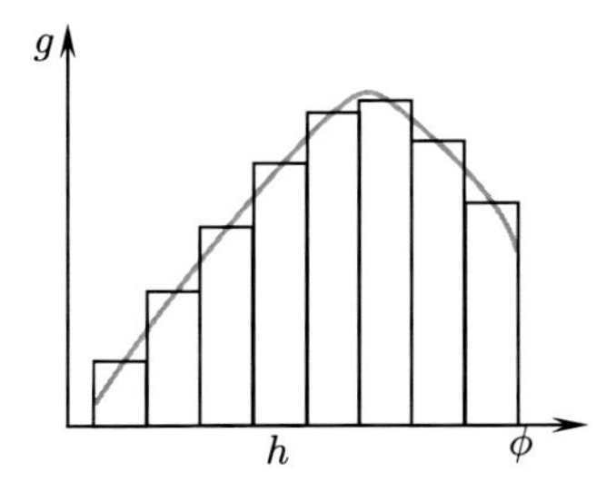

FIGURE A.2 Approximation of integrals involving the orientation distribution function $g(\phi)$ using rectangles of constant width h.

$$A'_{ijkl} = R_{mi} R_{nj} R_{pk} R_{ql} A_{mnpq} \quad \text{and} \quad A_{ijkl} = R_{im} R_{jn} R_{kp} R_{lq} A'_{mnpq},$$

where $\boldsymbol{R}$ is the rotator tensor between the two coordinate systems. If the fibers are lying on the $x_2 - x_3$ plane and ϕ represents a rotation around the x_1 axis, then the rotator is written as the 3×3 non–Voigt form matrix

$$\boldsymbol{R} = \begin{bmatrix} 1 & 0 & 0 \\ 0 & \cos\phi & -\sin\phi \\ 0 & \sin\phi & \cos\phi \end{bmatrix}.$$

Comparing (A.3) with (A.1), we conclude that each orientation can be "seen" as a separate phase with volume fraction

$$c_r = \frac{g_r}{g} c. \tag{A.4}$$

A.3 Reinforcement orientation in 3–D space

Let us assume that ellipsoidal shape fibers of total volume fraction c are oriented randomly with respect to a Cartesian coordinate system (x_1, x_2, x_3) inside a matrix phase (Fig. A.3).

If ϑ is the angle of orientation around the x_3 axis and ϕ is the angle of orientation around the x_1 axis (Fig. A.4), then Eq. (A.1) is substituted by[2] [2]

$$\bar{\boldsymbol{L}} = [1 - c]\boldsymbol{L}_0 : \boldsymbol{A}_0 + c \int_0^{2\pi} \int_0^{\pi} g(\vartheta, \phi) \boldsymbol{L}(\vartheta, \phi) : \boldsymbol{A}(\vartheta, \phi) \sin\phi \, \mathrm{d}\phi \mathrm{d}\vartheta \Bigg/ \int_0^{2\pi} \int_0^{\pi} g(\vartheta, \phi) \sin\phi \, \mathrm{d}\phi \mathrm{d}\vartheta, \tag{A.5}$$

[2]Note that similar integrals appear in the formulas concerning the computation of concentration tensors.

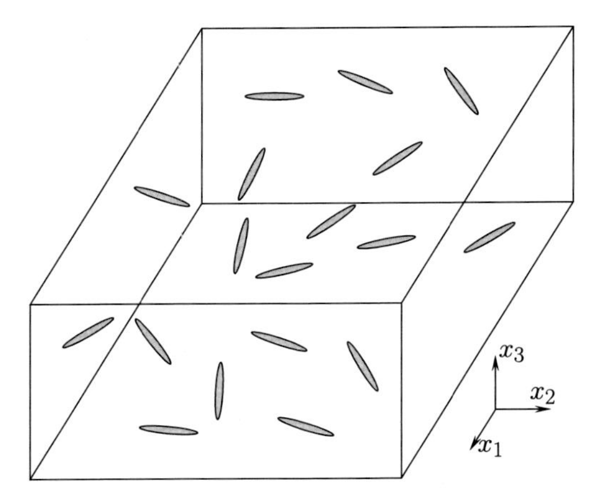

FIGURE A.3 Fiber reinforced composite. The fibers are oriented randomly in all possible directions.

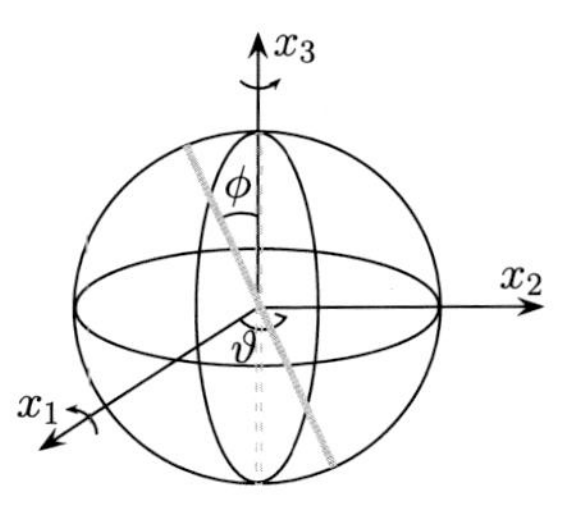

FIGURE A.4 The relative position of a fiber in space is defined through the rotation of the x_3 and x_1 axes. The dashed line shows the initial position of the fiber, and the solid rectangle shows the final position of the fiber after the rotations.

where $\boldsymbol{L}_0$ and $\boldsymbol{L}$ are the matrix and fiber elasticity tensors, respectively. Similarly to the 2–D case examined previously, an equal probability of appearance for all orientations yields g to be constant. If the orientation distribution function is known at specific angles or angle ranges, then g appears in a discretized form with respect to the angles ϕ and ϑ. Considering N_1 equidistant orientations ϕ and N_2 equidistant orientations ϑ, the total steps are $N = N_1 N_2$, and Eq. (A.5) is expressed in the discrete form

$$\bar{\boldsymbol{L}} \approx [1-c]\boldsymbol{L}_0 : \boldsymbol{A}_0 + \frac{c}{f}\sum_{r=1}^{N} f_r \boldsymbol{L}_r : \boldsymbol{A}_r,$$

$$f_r = g_r \sin\phi_r, \quad f = \sum_{r=1}^{N} f_r, \tag{A.6}$$

where g_r is the value of $g(\vartheta, \phi)$ at the r_{th} orientation. The index r is also used for the reinforcement elasticity tensor $\boldsymbol{L}$ and concentration tensor $\boldsymbol{A}$ due to their dependence on the orientation angles. Recall from Chapter 1 that a fourth–order tensor $\boldsymbol{A}$ is transformed from one coordinate system $\boldsymbol{x}$ to another $\boldsymbol{x}'$ through the general formulas

$$A'_{ijkl} = R_{mi} R_{nj} R_{pk} R_{ql} A_{mnpq} \quad \text{and} \quad A_{ijkl} = R_{im} R_{jn} R_{kp} R_{lq} A'_{mnpq},$$

where $\boldsymbol{R}$ is the rotator tensor between the two coordinate systems, which here represents a rotation first around the x_3 axis and then around the x_1 axis. The rotator tensors related to these rotations are written as the 3×3 non–Voigt form matrices

$$\boldsymbol{R}^3 = \begin{bmatrix} \cos\vartheta & -\sin\vartheta & 0 \\ \sin\vartheta & \cos\vartheta & 0 \\ 0 & 0 & 1 \end{bmatrix} \quad \text{and} \quad \boldsymbol{R}^1 = \begin{bmatrix} 1 & 0 & 0 \\ 0 & \cos\phi & -\sin\phi \\ 0 & \sin\phi & \cos\phi \end{bmatrix},$$

respectively. The first rotation transforms an arbitrary vector $\boldsymbol{v}$ from the coordinate system $\boldsymbol{x}$ to the coordinate system $\boldsymbol{x}''$,

$$v''_k = R^3_{ik} v_i,$$

whereas the second rotation transforms the latter to the coordinate system $\boldsymbol{x}'$,

$$v'_j = R^1_{kj} v''_k, = R^1_{kj} R^3_{ik} v_i = R_{ij} v_i.$$

Thus the rotator $\boldsymbol{R}$ is given by

$$R_{ij} = R^3_{ik} R^1_{kj},$$

or it can be expressed as the 3×3 non–Voigt matrix

$$\boldsymbol{R} = \boldsymbol{R}^3 \cdot \boldsymbol{R}^1 = \begin{bmatrix} \cos\vartheta & -\cos\phi\sin\vartheta & \sin\phi\sin\vartheta \\ \sin\vartheta & \cos\phi\cos\vartheta & -\sin\phi\cos\vartheta \\ 0 & \sin\phi & \cos\phi \end{bmatrix}.$$

Comparing (A.6) with (A.1), we conclude that each orientation can be "seen" as a separate phase with volume fraction

$$c_r = \frac{f_r}{f} c. \tag{A.7}$$

Before closing the discussion, it is worth mentioning the following:

- Expressions (A.3) and (A.6) should be applied with certain caution. Several researchers [3,4] have pointed out that for general ellipsoidal fiber shape, the obtained macroscopic tensor from mean–field homogenization schemes may loose symmetry when multiple orientations are taken into account. To correct this issue, proper regularization of $\bar{\boldsymbol{L}}$ is required to symmetrize it.

- Identifying an orientation distribution function in short fiber composites requires a complicated experimental procedure. When the processing conditions for the design of a composite are known, it is possible to identify the so–called orientation tensors that represent surface or volume integrals of the orientation distribution function [5]. These tensors simplify expressions (A.2) and (A.5) and allow estimation of $\bar{\boldsymbol{L}}$.

A.4 Examples

Example 1: Mori–Tanaka for 3–D randomly oriented long fiber composite

Let us consider a fiber reinforced composite in which the fibers are infinitely long and randomly oriented in three directions inside the matrix. We assume the uniform orientation distribution function. We assume the matrix to be a polymer (isotropic material) with Young modulus $E_0 = 3$ GPa and Poisson ratio $\nu_0 = 0.3$, whereas we consider the fibers to be made of glass (isotropic material) with Young modulus $E_1 = 72$ GPa and Poisson ratio $\nu_1 = 0.2$. The overall fiber volume fraction is equal to $c = 20\%$.

To identify the macroscopic response through the Mori–Tanaka scheme, we need to start by considering a discrete number of orientations. According to the discussion on Section A.3, the angle ϕ varies between 0 and 180° (π), whereas ϑ varies between 0 and 360° (2π). A reasonable discretization could consider angles every 2.5° for both ϕ and ϑ. This leads to $N_1 = 72$ and $N_2 = 144$ orientations. Moreover, g is constant due to the uniform distribution.

For $\phi = \vartheta = 0$ (fiber axis parallel to x_3 direction), the interaction tensor is given in the Voigt notation by the expression (Example 2 in Section 1.8 of Chapter 1)

$$\widetilde{\boldsymbol{T}}_1 = \left[\widetilde{\boldsymbol{I}} + \widetilde{\boldsymbol{S}}_1 \cdot \boldsymbol{L}_0^{-1} \cdot [\boldsymbol{L}_1 - \boldsymbol{L}_0]\right]^{-1},$$

where $\widetilde{\boldsymbol{S}}_1$ is evaluated from (5.8) of Chapter 5. In every other orientation the interaction tensor $\widetilde{\boldsymbol{T}}_r$ is connected with $\widetilde{\boldsymbol{T}}_1$ through a proper "local to global" rotation

$$\widetilde{\boldsymbol{T}}_r = \breve{\boldsymbol{Q}}^T \cdot \widetilde{\boldsymbol{T}}_1 \cdot \widetilde{\boldsymbol{Q}},$$

where $\breve{\boldsymbol{Q}}$ and $\widetilde{\boldsymbol{Q}}$ are the Voigt forms of the fourth–order tensor obtained by the usual second–order rotator $\boldsymbol{R}$ (for details, see Section 1.7 of Chapter 1). The strain concentration tensor of the polymer, according to the Mori–Tanaka scheme and the discussion of Section A.3, is given in the Voigt notation by

$$\widetilde{\boldsymbol{A}}_0 = \left[[1-c]\widetilde{\boldsymbol{I}} + \frac{c}{f} \sum_{r=1}^{N} f_r \widetilde{\boldsymbol{T}}_r \right]^{-1}$$

with

$$N = N_1 N_2, \qquad f_r = \sin\phi_r, \qquad f = \sum_{r=1}^{N} f_r.$$

The strain concentration tensor of the fibers with r_{th} orientation is given by

$$\widetilde{\boldsymbol{A}}_r = \widetilde{\boldsymbol{T}}_r \cdot \widetilde{\boldsymbol{A}}_0.$$

Finally, the Mori–Tanaka macroscopic elasticity tensor $\bar{\boldsymbol{L}}$ is given by Eq. (A.3), which, after small algebraic manipulation, is expressed in the Voigt notation as

$$\bar{\boldsymbol{L}} = [1-c]\boldsymbol{L}_0 \cdot \widetilde{\boldsymbol{A}}_0 + \frac{c}{f}\left[\sum_{r=1}^{N} f_r \boldsymbol{L}_r \cdot \widetilde{\boldsymbol{T}}_r \right] \cdot \widetilde{\boldsymbol{A}}_0, \qquad \boldsymbol{L}_r = \widetilde{\boldsymbol{Q}}^T \cdot \boldsymbol{L}_1 \cdot \widetilde{\boldsymbol{Q}}.$$

Note that the elasticity tensor $\boldsymbol{L}_r$ is always equal to $\boldsymbol{L}_1$ due to the isotropic response of the fibers. Thus the above expression is simplified to

$$\bar{\boldsymbol{L}} = [1-c]\boldsymbol{L}_0 \cdot \widetilde{\boldsymbol{A}}_0 + \frac{c}{f}\boldsymbol{L}_1 \cdot \left[\sum_{r=1}^{N} f_r \widetilde{\boldsymbol{T}}_r \right] \cdot \widetilde{\boldsymbol{A}}_0.$$

The obtained result for the specific numerical example, considering two significant digits, is

$$\bar{\boldsymbol{L}} = \begin{bmatrix} 8.35 & 3.09 & 3.09 & 0 & 0 & 0 \\ 3.09 & 8.35 & 3.09 & 0 & 0 & 0 \\ 3.09 & 3.09 & 8.35 & 0 & 0 & 0 \\ 0 & 0 & 0 & 2.63 & 0 & 0 \\ 0 & 0 & 0 & 0 & 2.63 & 0 \\ 0 & 0 & 0 & 0 & 0 & 2.63 \end{bmatrix} \text{GPa}.$$

The final composite presents isotropic behavior. This result agrees with the geometrical characteristics of the microstructure: since every fiber orientation appears with equal probability, there is no preferable orientation, and thus the composite material is "geometrically" isotropic.

The corresponding python script for obtaining the $\bar{L}$ is the following:

```
import numpy as np
# Data
E0=3
v0=0.3
E1=72.0
v1=0.2
c=0.2

K0=E0/(3*(1-2*v0))
mu0=E0/(2*(1+v0))
K1=E1/(3*(1-2*v1))
mu1=E1/(2*(1+v1))

I=np.eye(6)

# Elasticity tensors
L0=np.matrix([[K0+4*mu0/3, K0-2*mu0/3, K0-2*mu0/3,\
               0., 0., 0.],\
              [K0-2*mu0/3, K0+4*mu0/3, K0-2*mu0/3,\
               0., 0., 0.],\
              [K0-2*mu0/3, K0-2*mu0/3, K0+4*mu0/3,\
               0., 0., 0.],\
              [0., 0., 0., mu0, 0., 0.],\
              [0., 0., 0., 0., mu0, 0.],\
              [0., 0., 0., 0., 0., mu0]])
L1=np.matrix([[K1+4*mu1/3, K1-2*mu1/3, K1-2*mu1/3,\
               0., 0., 0.],\
              [K1-2*mu1/3, K1+4*mu1/3, K1-2*mu1/3,\
               0., 0., 0.],\
              [K1-2*mu1/3, K1-2*mu1/3, K1+4*mu1/3,\
               0., 0., 0.],\
              [0., 0., 0., mu1, 0., 0.],\
              [0., 0., 0., 0., mu1, 0.],\
              [0., 0., 0., 0., 0., mu1]])

# Eshelby tensor
S1=np.matrix([[(5-4*v0)/(8*(1-v0)), (4*v0-1)/(8*(1-v0)),\
                v0/(2*(1-v0)), 0., 0., 0.],\
              [(4*v0-1)/(8*(1-v0)), (5-4*v0)/(8*(1-v0)),\
                v0/(2*(1-v0)), 0., 0., 0.],\
              [0., 0., 0., 0., 0., 0.],\
```

```
                  [0., 0., 0., (3-4*v0)/(4*(1-v0)), 0., 0.],\
                  [0., 0., 0., 0., 0.5, 0.],\
                  [0., 0., 0., 0., 0., 0.5]])

# Reference interaction tensor
T1=(I+S1*(L0.I)*(L1-L0)).I

sumT=np.zeros((6,6))
f=0
N1=72
N2=144
for phi in range(N1):
  for theta in range(N2):
# Rotator
    R11=np.cos(theta*2*np.pi/N2)
    R12=-np.cos(phi*np.pi/N1)*np.sin(theta*2*np.pi/N2)
    R13=np.sin(phi*np.pi/N1)*np.sin(theta*2*np.pi/N2)
    R21=np.sin(theta*2*np.pi/N2)
    R22=np.cos(phi*np.pi/N1)*np.cos(theta*2*np.pi/N2)
    R23=-np.sin(phi*np.pi/N1)*np.cos(theta*2*np.pi/N2)
    R31=0.
    R32=np.sin(phi*np.pi/N1)
    R33=np.cos(phi*np.pi/N1)
    Qtilde=np.matrix([[R11*R11, R21*R21, R31*R31,\
                      R11*R21, R11*R31, R21*R31],\
                     [R12*R12, R22*R22, R32*R32,\
                      R12*R22, R12*R32, R22*R32],\
                     [R13*R13, R23*R23, R33*R33,\
                      R13*R23, R13*R33, R23*R33],\
                     [2*R11*R12, 2*R21*R22, 2*R31*R32,\
                      R12*R21+R11*R22, R12*R31+R11*R32,\
                      R22*R31+R21*R32],\
                     [2*R11*R13, 2*R21*R23, 2*R31*R33,\
                      R13*R21+R11*R23, R13*R31+R11*R33,\
                      R23*R31+R21*R33],\
                     [2*R12*R13, 2*R22*R23, 2*R32*R33,\
                      R13*R22+R12*R23, R13*R32+R12*R33,\
                      R23*R32+R22*R33]])
    Qchk=  np.matrix([[R11*R11, R21*R21, R31*R31,\
                      2*R11*R21, 2*R11*R31, 2*R21*R31],\
                     [R12*R12, R22*R22, R32*R32,\
                      2*R12*R22, 2*R12*R32, 2*R22*R32],\
                     [R13*R13, R23*R23, R33*R33,\
                      2*R13*R23, 2*R13*R33, 2*R23*R33],\
                     [R11*R12, R21*R22, R31*R32,\
```

```
                    R12*R21+R11*R22, R12*R31+R11*R32,\
                    R22*R31+R21*R32],\
                   [R11*R13, R21*R23, R31*R33,\
                    R13*R21+R11*R23, R13*R31+R11*R33,\
                    R23*R31+R21*R33],\
                   [R12*R13, R22*R23, R32*R33,\
                    R13*R22+R12*R23, R13*R32+R12*R33,\
                    R23*R32+R22*R33]])

# Sum of interaction tensors over all orientations
    Tr=(Qchk.T)*T1*Qtilde
    sumT=sumT+Tr*np.sin(phi*np.pi/N1)
# Divisor for fibers volume fraction
    f=f+np.sin(phi*np.pi/N1)

# Strain concentration tensor of matrix
A0=((1-c)*I+(c/f)*sumT).I

# Effective elastic properties
Lmt=(1-c)*L0*A0+(c/f)*L1*sumT*A0

np.savetxt('Lmt3D.txt',Lmt,fmt='%1.5f')
```

Example 2: Mori–Tanaka for 2–D randomly oriented long fiber composite

Let us consider again a long fiber reinforced composite, but this time the fibers are randomly oriented in the $x_2 - x_3$ plane inside the matrix. We assume the uniform orientation distribution function. The matrix and the fibers are the same as in the previous example. The overall fiber volume fraction is equal to $c = 20\%$.

The steps to solve this problem are similar to those of the previous example, and thus we omit the technical details. Inside the plane of the fibers, we selected 36 discrete orientations. The obtained result for the specific numerical example, considering three significant digits, is

$$\bar{\boldsymbol{L}} = \begin{bmatrix} 5.367 & 2.147 & 2.147 & 0 & 0 & 0 \\ 2.147 & 10.732 & 3.796 & 0 & 0 & 0 \\ 2.147 & 3.796 & 10.732 & 0 & 0 & 0 \\ 0 & 0 & 0 & 1.624 & 0 & 0 \\ 0 & 0 & 0 & 0 & 1.624 & 0 \\ 0 & 0 & 0 & 0 & 0 & 3.468 \end{bmatrix} \text{GPa}.$$

The final composite presents transversely isotropic behavior, in which $x_2 - x_3$ is the plane of isotropy. The corresponding Python script for obtaining the $\bar{\boldsymbol{L}}$ is the following:

```
import numpy as np
# Data
E0=3
v0=0.3
E1=72.0
v1=0.2
c=0.2

K0=E0/(3*(1-2*v0))
mu0=E0/(2*(1+v0))
K1=E1/(3*(1-2*v1))
mu1=E1/(2*(1+v1))

I=np.eye(6)

# Elasticity tensors
L0=np.matrix([[K0+4*mu0/3, K0-2*mu0/3, K0-2*mu0/3,\
               0., 0., 0.],\
              [K0-2*mu0/3, K0+4*mu0/3, K0-2*mu0/3,\
               0., 0., 0.],\
              [K0-2*mu0/3, K0-2*mu0/3, K0+4*mu0/3,\
               0., 0., 0.],\
              [0., 0., 0., mu0, 0., 0.],\
              [0., 0., 0., 0., mu0, 0.],\
              [0., 0., 0., 0., 0., mu0]])
L1=np.matrix([[K1+4*mu1/3, K1-2*mu1/3, K1-2*mu1/3,\
               0., 0., 0.],\
              [K1-2*mu1/3, K1+4*mu1/3, K1-2*mu1/3,\
               0., 0., 0.],\
              [K1-2*mu1/3, K1-2*mu1/3, K1+4*mu1/3,\
               0., 0., 0.],\
              [0., 0., 0., mu1, 0., 0.],\
              [0., 0., 0., 0., mu1, 0.],\
              [0., 0., 0., 0., 0., mu1]])

# Eshelby tensor
S1=np.matrix([[(5-4*v0)/(8*(1-v0)), (4*v0-1)/(8*(1-v0)),\
                v0/(2*(1-v0)), 0., 0., 0.],\
              [(4*v0-1)/(8*(1-v0)), (5-4*v0)/(8*(1-v0)),\
                v0/(2*(1-v0)), 0., 0., 0.],\
              [0., 0., 0., 0., 0., 0.],\
              [0., 0., 0., (3-4*v0)/(4*(1-v0)), 0., 0.],\
              [0., 0., 0., 0., 0.5, 0.],\
              [0., 0., 0., 0., 0., 0.5]])
```

```
# Reference interaction tensor
T1=(I+S1*(L0.I)*(L1-L0)).I

sumT=np.zeros((6,6))
N=36
for phi in range(N):
# Rotator
    R11=1.
    R12=0.
    R13=0.
    R21=0.
    R22=np.cos(phi*np.pi/N)
    R23=-np.sin(phi*np.pi/N)
    R31=0.
    R32=np.sin(phi*np.pi/N)
    R33=np.cos(phi*np.pi/N)
    Qtilde=np.matrix([[R11*R11, R21*R21, R31*R31,\
                      R11*R21, R11*R31, R21*R31],\
                     [R12*R12, R22*R22, R32*R32,\
                      R12*R22, R12*R32, R22*R32],\
                     [R13*R13, R23*R23, R33*R33,\
                      R13*R23, R13*R33, R23*R33],\
                     [2*R11*R12, 2*R21*R22, 2*R31*R32,\
                      R12*R21+R11*R22, R12*R31+R11*R32,\
                      R22*R31+R21*R32],\
                     [2*R11*R13, 2*R21*R23, 2*R31*R33,\
                      R13*R21+R11*R23, R13*R31+R11*R33,\
                      R23*R31+R21*R33],\
                     [2*R12*R13, 2*R22*R23, 2*R32*R33,\
                      R13*R22+R12*R23, R13*R32+R12*R33,\
                      R23*R32+R22*R33]])
    Qchk=  np.matrix([[R11*R11, R21*R21, R31*R31,\
                      2*R11*R21, 2*R11*R31, 2*R21*R31],\
                     [R12*R12, R22*R22, R32*R32,\
                      2*R12*R22, 2*R12*R32, 2*R22*R32],\
                     [R13*R13, R23*R23, R33*R33,\
                      2*R13*R23, 2*R13*R33, 2*R23*R33],\
                     [R11*R12, R21*R22, R31*R32,\
                      R12*R21+R11*R22, R12*R31+R11*R32,\
                      R22*R31+R21*R32],\
                     [R11*R13, R21*R23, R31*R33,\
                      R13*R21+R11*R23, R13*R31+R11*R33,\
                      R23*R31+R21*R33],\
                     [R12*R13, R22*R23, R32*R33,\
```

```
                    R13*R22+R12*R23, R13*R32+R12*R33,\
                    R23*R32+R22*R33]])

# Sum of interaction tensors over all orientations
    Tr=(Qchk.T)*T1*Qtilde
    sumT=sumT+Tr*(c/N)

# Strain concentration tensor of matrix
A0=((1-c)*I+sumT).I

# Effective elastic properties
Lmt=(1-c)*L0*A0+L1*sumT*A0

np.savetxt('Lmt2D.txt',Lmt,fmt='%1.5f')
```

References

[1] P.B. Entchev, D.C. Lagoudas, Modeling porous shape memory alloys using micromechanical averaging techniques, Mechanics of Materials 34 (2002) 1–24.

[2] G.D. Seidel, Micromechanics modeling of the multifunctional nature of carbon nanotube–polymer nanocomposites, Ph.D. thesis, Texas A&M University, College Station, 2007.

[3] Y. Benveniste, G.J. Dvorak, T. Chen, On diagonal and elastic symmetry of the approximate effective stiffness tensor oh heterogeneous media, Journal of the Mechanics and Physics of Solids 39 (7) (1991) 927–946.

[4] J. Schjødt–Thomsen, R. Pyrz, The Mori–Tanaka stiffness tensor: diagonal symmetry, complex fibre orientations and non–dilute volume fractions, Mechanics of Materials 33 (10) (2001) 531–544.

[5] S.G. Advani, C.L. Tucker, The use of tensors to describe and predict fiber orientation in short fiber composites, Journal of Rheology 31 (1987) 751–784.

Index

H

I

K

L

R

S

T

Printed in the United States
by Baker & Taylor Publisher Services